Fiasco

Number Six: DeGolyer Library Series

DeGolyer Library Series

Number One
Troubles in Texas, 1832 (1983)
Edited by David J. Weber

Number Two
Ten Days on the Plains (1985)
Henry E. Davies. Edited by Paul A. Hutton

Number Three
Alamo Images: Changing Perceptions of a Texas Experience (1985)
Susan Prendergast Schoelwer

Number Four
The Defenses of Northern New Spain: Hugo O'Conor's Report to Teodoro de Croix, July 22, 1777 (1994)
Edited by Donald Cutter

Number Five
Trading in Santa Fe: John M. Kingsbury's Correspondence with James Josiah Webb, 1853–1861 (1996)
Edited by Jane Lenz Elder and David J. Weber

FIASCO

George Clinton Gardner's Correspondence from the U.S.-Mexico Boundary Survey 1849–1854

EDITED BY
David J. Weber and Jane Lenz Elder

SOUTHERN METHODIST UNIVERSITY PRESS
WILLIAM P. CLEMENTS CENTER FOR SOUTHWEST STUDIES
DEGOLYER LIBRARY
Dallas

First edition, 2010

Cover image: "Detail of Brownsville, Texas, survey crew" from W.H. Emory, *Report on the United States Boundary Survey* (1857–59). Courtesy of the DeGolyer Library, Southern Methodist University, Dallas, Texas.

Jacket and text design: Tom Dawson

Library of Congress Cataloging-in-Publication Data

Gardner, G. Clinton (George Clinton), 1834-1904.
Fiasco : George Clinton Gardner's correspondence from the U.S.-Mexico boundary survey, 1849-1854 / edited by David J. Weber and Jane Lenz Elder. — 1st ed.
p. cm. — (Degolyer Library series ; no. 6)
Includes bibliographical references and index.
ISBN 978-0-87074-562-1 (hardcover : alk. paper)
1. Gardner, G. Clinton (George Clinton), 1834-1904—Correspondence. 2. Surveyors—United States—Correspondence. 3. Surveyors—Mexican-American Border Region—Correspondence. 4. Mexican-American Border Region—Surveys. 5. United States—Surveys. 6. Mexico—Surveys. 7. United States—Boundaries—Mexico—History—19th century. 8. Mexico—Boundaries—United States—History—19th century. 9. Southwestern States—History—19th century. I. Weber, David J. II. Elder, Jane Lenz. III. Title. IV. Title: George Clinton Gardner's correspondence from the U.S.-Mexico boundary survey, 1849-1854. V. Series: DeGolyer Library series ; no. 6.
F786.C219 2010
526.9092—dc22

2009047160

Printed in the United States of America on acid-free paper

10 9 8 7 6 5 4 3 2 1

Contents

To George Bayoud and Paul Neubach.
They know why.

Illustrations

Maps

Acknowledgments

This project began in the mid-1990s when Jane Lenz Elder worked as my research assistant. Jane and I had finished editing a remarkable collection of unpublished letters from Southern Methodist University's DeGolyer Library that resulted in the 1996 book *Trading in Santa Fe: John M. Kingsbury's Correspondence with James Josiah Webb, 1853–61*. That work behind us, we began to edit another group of letters from the DeGolyer Library's manuscript collection, the personal and business correspondence of George Clinton Gardner from the U.S.-Mexico boundary survey, 1849–54. I had seen the Gardner correspondence soon after coming to SMU in 1976 and noted that scholars had consulted the letters and cited them. They seemed worthy of publication.

Our editing of Gardner's correspondence proceeded at a stately pace as Jane and I turned our attention in 1996 to creating a new research center at SMU, the Clements Center for Southwest Studies, and as both of us put most of our energy into writing other books. In 2001 Jane accepted a position at the Bridwell Library at SMU's Perkins School of Theology. Somehow she carved out time between her new job and her family to write *Alice Faye: A Life beyond the Silver Screen* (2002). As we went to press with this book on Gardner, she was completing *Basil Rathbone: In and Out of Hollywood*. Before she left for the Bridwell and Hollywood, however, she had carefully transcribed Gardner's correspondence, then found collateral collections (including seventeen more letters by Gardner in the private collection of Wade Shipley) and began preparing annotations.

In the summers, between other writing projects, I tried to pick up where Jane left off but two things became clear. First, I would need a large block of time to complete the editing of Gardner's correspondence, and I would need to spend some of that time consulting related papers at the Beinecke Rare Book and Manuscript Library at Yale. Happily, the Beinecke provided both in the form of a generous yearlong fellowship. I spent 2007–2008 as the Frederick W. and Carrie S. Beinecke Senior Fellow with an office in the Lamar Center for the Study of Frontiers and Borders. There, Howard and Shirley Lamar, Frank Turner, George Miles, Johnny Faragher, Jay Gitlin, and the staffs of the Beinecke Library and the Lamar Center provided good company and ideal conditions for bringing this long project to an end. I am deeply grateful.

During the years we worked on Gardner's correspondence, we acquired many debts and we acknowledge these in the notes. We want, however, to single out several people for special mention. The staff of the DeGolyer Library, presided over by David Farmer in the early stages of our work and by Russell L. Martin in the final stages, has been supportive in every way. My former student, long-time friend, and expert on San Diego history, Charles Hughes, expedited our research by providing invaluable advice, guidance to sources, copies of some of the correspondence of Edmund Hardcastle from the Maryland State Archives, and the letters of H. H. Robinson, which Charles is preparing for publication. Mark Herlong, a Byzantinist who is completing a biography of Gardner's boyhood friend John McCalla, generously shared his knowledge of Washington, D.C., with complete strangers and steered us to more of Gardner's correspondence. In 1996 Wade Shipley of Lovington, New Mexico, kindly photocopied seventeen of Gardner's letters from his private collection and gave us permission to publish them. He subsequently put those letters up for auction, and they are now in the Smithsonian Institution. Of those seventeen letters, we have included eleven that pertain to the U.S.-Mexico boundary survey, citing them to their present home at the Smithsonian.

We are also grateful to Joseph Werne, *the* expert on the boundary survey, who scoured the manuscript for errors; the always cheerful Ruth Ann Elmore, who assisted with research and with preparation of the final manuscript and the illustrations; and to Robert Foxworth of the Edwin J. Foscue Map Library at SMU, who rendered the cartography with care and imagination. Polly Kummel's expert copy editing made the manuscript cleaner and clearer.

On a personal note I am indebted to the nurses and doctors who prevented a rare cancer from killing me while I was working on this manuscript, especially the oncologist Thomas Fynan (in New Haven), the nephrologist Christopher Herbert, and the oncologists Houston E. Holmes III and Luis Piñeira (the last three in Dallas). Above all I am grateful to my internist, Paul A. Neubach of Dallas, a doctor's doctor. Many friends and colleagues, too numerous to name here, sustained me through this difficult time and have my enduring gratitude. I am most thankful to the love of my life, my wife, Carol Bryant Weber, who not only cared for me through the illness but improved the manuscript, as she has all the books I have written.

David J. Weber
SMU

(Unless otherwise noted, all the letters reproduced in this volume come from the Gardner Correspondence in the DeGolyer Library at Southern Methodist University, Dallas).

Abbreviations Used in the Notes

DAB:	*Dictionary of American Biography*. 22 vols. New York: C. Scribner's Sons, 1928–58.
EP:	Willam Hemsley Emory Papers, 1823–86, Yale Collection of Western Americana, Beinecke Rare Book and Manuscript Library, WA MSS S-1187.
HCMSA:	Henderson Collection, SC 501, Maryland State Archives, Annapolis.
GCG:	George Clinton Gardner.
GCGJC:	G. Clinton Gardner Journals and Correspondence, 1853–59, Yale Collection of Western Americana, Beinecke Rare Book and Manuscript Library, WA MSS S-113.
GCGP:	G. Clinton Gardner Papers, 1854–68, Yale Collection of Western Americana, Beinecke Rare Book and Manuscript Library, WA MSS S-1667 G173.
Harris Papers:	Joseph Smith Harris Papers, Yale Collection of Western Americana, Beinecke Rare Book and Manuscript Library, WA MSS S-1293.
NHBT:	*The New Handbook of Texas*. Ed. Ron C. Tyler et al. 6 vols. Austin: Texas State Historical Association, 1996.
PP:	Charles Christopher Parry Papers. Correspondence. Iowa State University, Special Collections Department, MS 290.
SED 34:	Thomas Ewing. *Report of the Secretary of the Interior in Answer to a Resolution of the Senate Calling for Information in Relation to the Operations of the Commission Appointed to Run and Mark the Boundary between the United States and Mexico*, 31st Cong., 1st sess., Senate Executive Document 34. Washington, D.C., 1850.
SED 119:	Alexander H. H. Stuart. *Report of the Secretary of the Interior Made in Compliance with a Resolution of the Senate Calling for Information in Relation to the Commission Appointed to Run and Mark the Boundary between the United States and Mexico*, 32nd Cong., 1st sess., Senate Executive Document 119. Washington, D.C., 1852.

Introduction

In February 1849 seventeen-year-old George Clinton Gardner, known to his siblings and friends as Clint and to his parents as Clinton, received a remarkable appointment that would take him from his home in Washington, D.C., to the edge of Mexico.[1] A bitter, bloody war between the United States and Mexico had ended the year before. In the peace treaty that followed, Mexico and the United States agreed to send representatives to survey and mark a new international boundary, from the Pacific to the Gulf of Mexico. Clint would join the U.S. boundary commission, serving as a junior

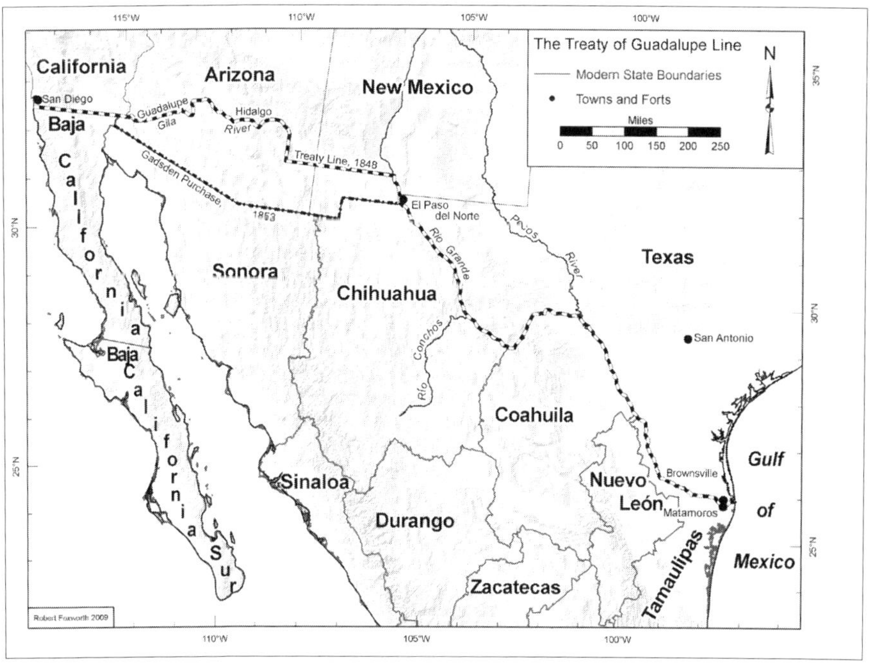

assistant to the commission's chief astronomer and highest-ranking military officer, Maj. William H. Emory. The letter of appointment Clint received that February came from Major Emory himself, with whom his family was acquainted.

For the next five years young Clint traveled with the survey party, working first between San Diego and Yuma and then along the Rio Grande from El Paso to the river's mouth at the Gulf of Mexico. Along the way he wrote letters, mostly to family. Those letters, heretofore unpublished, constitute the body of this book. They provide one insider's view of the remarkable project of surveying, mapping, and marking a line across a dry, remote, and dangerous land.

The task before the surveyors, as Clint soon discovered, was daunting. In the treaty that officially ended the U.S.-Mexican War, signed at Guadalupe Hidalgo on February 2, 1848, Mexican negotiators had been forced to cede the northern half of their country to the United States. Before the Texas Rebellion of 1836, the U.S.-Mexico boundary had run along the western edge of Louisiana, made its jagged way north to the 42nd parallel, and followed that parallel to the Pacific along the northern edges of today's states of Colorado, Utah, and California. The Treaty of Guadalupe Hidalgo dramatically altered the political geography of North America by moving the boundary to the southern edges of today's California, Arizona, New Mexico, and Texas—an expanse of about two thousand miles.

Surveying on such a large scale had to take the curvature of the Earth into account and that required geodesic surveying, which depended heavily on astronomy. Plane surveying, used to measure and mark small portions of Earth's surface, was not sufficient.[2] Long stretches of the new border followed two rivers, the Gila and the Rio Grande. Along those rivers the surveyors had to find the deepest channels, as required by the Treaty of Guadalupe Hidalgo. Between the Pacific Ocean and the Gila River and between the Gila and the Rio Grande, their task was harder. No rivers or other geographical features marked the new border. In the absence of landmarks the surveyors had to mark a line on the ground and erect or place physical markers—although they put up few of these, for they supposed that neither Mexicans nor Americans would ever settle in the arid border region in significant numbers. As one newspaper of the day described it, this was "a country for the most part in a state of nature, and unusually sterile and unproductive."[3]

Today Mexicans and Americans crowd up against the border. Towns and cities straddle it, and the U.S. government believes it necessary to build a wall to maintain the border's integrity. But back then, beyond a handful of small Mexican settlements such as San Diego, El Paso del Norte, and Matamoros, the border region was mainly the domain of independent Indians who had no reason to recognize lines drawn through their territory by distant diplomats. As Major Emory saw it, the U.S.-Mexico boundary region was "infested throughout its whole extent with formidable and hostile bands of Indians."[4] Emory overstated the threat from Indians, but distance from settlements and supplies, blistering summer heat, and difficult terrain made the surveyors' work arduous enough.

However difficult the American surveyors found conditions in the field, the greatest impediment to their work came from Washington. Logic suggested that the work be done by the nation's best-trained and most experienced surveyors, West Point–trained members of the U.S. Army Corps of Topographical Engineers. Domestic politics and regional interests, however, required that a politician lead them. As the historian Robert Hine has explained: "If the proper line were found, a southern railroad might be built and southern economic power rejuvenated. Nothing less than the course of empire might be at stake—too much to trust either to colonels or surveyors."[5]

Article 5 of the Treaty of Guadalupe Hidalgo had specified that each nation appoint a commissioner and a surveyor. In its last few months in office, the Democratic administration of President James K. Polk filled both positions with Democrats with ties to the South. Polk's first choice for commissioner, Emory, was ideal—a Democrat and an experienced surveyor. Emory had graduated from West Point in 1831 and entered the Corps of Topographical Engineers in 1838 when it was formed under the War Department. By 1849 his considerable experience included two years of surveying the Canadian-U.S. boundary. Moreover, he was the only American scientist to have traveled across the Southwest, from Santa Fe to Los Angeles. When America declared war on Mexico in 1846, Emory had been assigned to accompany Gen. Stephen Watts Kearny on a two-thousand-mile trek to New Mexico and California. Along the way Emory had mapped the route, and he produced a scientific report upon his return.[6]

Emory enjoyed the respect of the military community and of the leading scientists of his day, whose meetings he attended. He also had powerful political connections in the Democratic Party. He was the scion of an aristocratic Maryland family, and his boyhood friends included Henry Clay Jr. and Jefferson Davis. A family friend, John C. Calhoun, arranged for Emory's acceptance at West Point. When Emory married in 1838, he chose a spouse from within his social class, Matilda Wilkins Bache, the great-granddaughter of Benjamin Franklin. Her father, Alexander D. Bache, was a director of the Coast and Geodetic Survey and her brother-in-law, Robert J. Walker, a Democrat, was a powerful Washington politician whose offices included two terms as senator from Mississippi (1835–45) and secretary of the treasury in the Polk administration (1845–49).[7] As Emory later recalled, Polk offered him the position as head of the boundary commission on the condition that he resign from the army. Emory refused, but he did agree to serve as the commission's chief astronomer and commander of its military escort.[8]

Unable to persuade Emory to assume the position of commissioner, Polk turned in December 1848 to A. K. Sevier, a former Democratic senator from Arkansas, but Sevier died before the Senate could confirm him. On January 16, 1849, the day that Sevier died, the administration tapped John B. Weller. Weller had no surveying experience. He was a lawyer, three-term Democratic House member from Ohio, and Mexican War veteran who had lost the Ohio gubernatorial election in 1848. This eleventh-hour appointment by a lame-duck president infuriated Whigs, who controlled the House of Representatives. Zachary Taylor, the Whig who had won the White House in November, was to be sworn in on March 5 and eager to dispense patronage.

Unable to stop Weller's appointment, the Whig-controlled House did its best to sabotage him. It amended the appropriations bill that provided funds for the survey, limiting expenditures to members of the Corps of Topographical Engineers. The Senate, controlled by Democrats, refused to accept the amendment, but the commission was left with restricted funds, and the political squabbles were a harbinger of things to come.[9] Meanwhile, fearing Zachary Taylor would replace him with a Whig, Weller sought assurances from the president-elect in a face-to-face meeting. Apparently mollified by Taylor, Weller set out for California on his new assignment, little imagining that the new administration would indeed soon replace him.[10]

The position of surveyor proved easier to fill than the position of commissioner. The Polk administration offered it to a twenty-nine-year-old civilian and experienced surveyor, Andrew B. Gray. The Virginia-born Gray, whose experience included surveying for the Republic of Texas, accepted with alacrity. His civilian status and rudimentary knowledge of astronomy, however, put him at odds with the survey's West Point–trained topographical engineers—Major Emory and his two assistant astronomers, Capt. Edmund L. F. Hardcastle and Lt. Amiel Weeks Whipple.

Conflict between political appointees and topographical engineers began almost immediately, and it plagued the survey to the end. So too did intramural conflict among members of the military and among the political appointees. The commission had hardly reached the shore in Panama when Major Emory began to chafe at Commissioner Weller's lack of organization. In San Diego, Weller and Surveyor Gray got into a fight after some heavy drinking. Gray shot the commissioner through the thigh. The embarrassing incident did not find its way into the official published reports, but private parties like Clint Gardner took note.[11]

Although infighting among members of the commission caused work to stall on occasion, partisan politics in Washington nearly paralyzed it. Even before the Mexican and American commissioners met in San Diego, politicians who put party first and nation second sent ripples westward that caused the U.S. commission to founder. In June 1849 Zachary Taylor appointed John C. Frémont, then in northern California, to replace Weller as commissioner and to notify Weller of his recall. Frémont initially accepted the appointment, but in the end he neither delivered the order to Weller to vacate the office nor filled it himself. Instead, Frémont ran for the U.S. Senate from California and won. From 1850 to 1851 he served as one of California's first two senators.[12]

Weller, unaware of his recall, stayed on as commissioner through the autumn of 1849 but worked under impossible conditions.[13] One problem was soaring inflation. Gold seekers had flocked to California. They drove up prices, wages, and the commission's cost of doing business, diminishing the value of Weller's original appropriation from the 1848 Congress. Another problem was crippling opposition by the Whigs, who took over the 1849 Congress and declined to make further appropriations for the commission until a new commissioner replaced Weller.[14] As it became public that he had lost the government's confidence, a desperate Weller found few who would honor his drafts on the U.S. treasury. To make matters worse, Secretary of State John Clayton, a Whig, refused to pay Weller's drafts on the state department until he brought his accounts up to date. This demand, unreasonable given time and distance, assured the failure of Weller's mission and helped destroy him politically in Ohio, which was precisely the goal of some powerful Whigs.[15]

At the beginning of the new year, an unhappy Weller complained to the secretary of state that he had run out of funds and still owed about $11,000 to his civilian employees. "It is hard, indeed, that gentlemen who have served the government faithfully, at rates less than one-half that paid to other employees in this country, should now be left destitute, six thousand miles from their homes. I can neither pay them their just dues nor transport them home."[16] By then Weller had few men left to transport home. Most had abandoned the survey for jobs that actually paid or had succumbed to the lure of the goldfields.

As one of Weller's Democratic supporters charged, the Whig-controlled administration had treated the boundary commission with "criminal neglect."[17] Or, as Major Emory put it more judiciously, "the government failed to comply with its obligations."[18] Weller made the best of his situation by remaining in California and, like Frémont, winning election to the U.S. Senate (1851–57). Weller went on to become governor of California (1858–a60).

The fiasco in California set the tone for the rest of the survey: gaining political advantage continued to matter more to politicians than fulfilling the terms of the treaty in a timely, cost-effective manner. The entire border might have been surveyed in two years, but it took longer than that to finish the California portion alone.

Clint Gardner, who was among the few civilians who stayed with Emory, did not return to his home in Washington again until the late summer of 1851, after the California phase had ended. His home stay was brief. Less than two weeks later he left to follow Emory to El Paso, where a reconstituted American commission was continuing the survey. In El Paso, as in California, Clint would see first hand the effects of the Whig-Democrat contretemps, the personal rivalries and animosities among members of the surveying team, and the feckless leadership of a new commissioner, John Russell Bartlett, another poorly qualified civilian who had sought federal patronage.

The survey's fieldwork finally ended in the autumn of 1853 with Clint Gardner, now twenty-two, as the senior civilian assistant on the survey and one of the last to leave.[19] He had learned the skills of an astronomer and had won Emory's confidence. The irascible Emory was notoriously difficult to work for and on one occasion in San Diego almost pushed Clint into resigning. Clint had stayed on, though, to become Emory's trusted secretary and, in the final stages of the survey, one of four section leaders.[20] In the summer of 1853 Emory put Clint in charge of building and operating an observatory at the mouth of the Rio Grande. Clint took readings at the observatory through the autumn while Emory returned to Washington. As survey parties completed their work, Clint collected their instruments, shipped them from the mouth of the Rio Grande to New York, and sold the commission's remaining property at auction before heading home himself.

High-ranking officials from both the Mexican and American sides left a substantial written record of their difficulties and achievements in surveying the border, and historians have written superb overviews. Clint Gardner's letters give us a new vantage point, that of a young man who described himself as "but small potatoes on the Boundary."[21] Through his eyes we get a narrow but fine-grained view of the survey party's logistical and financial problems, the personal and political rivalries of leading figures, the quarrels between the civilian and military members of the survey party, and the personal foibles that often turned the work of the U.S. survey team into a fiasco. His letters provide homely details that can come only from small potatoes, as when he explained why the commission's rations did not include butter in San Diego and when he tried to follow Emory's instructions to send home a bow and arrow that Emory had acquired for one of his sons. Clint's letters also make it clear that the small potatoes knew what the big potatoes were doing. He bristled at the disgraceful way in which Weller's brother Charles and Bartlett's brother George misused their positions to appropriate government property as their own and sell it to the commission's employees, and he excoriated the Whig administration of Millard Fillmore for failing to fund the survey adequately.[22] "This Commission is the meanest affair that was ever got up under the Government," Clint charged.[23]

Because Clint wrote personal letters to family and friends, rather than articles for public consumption or spirited defenses of his actions, as his superiors often did, his assessments of people and places seem unvarnished. Speaking of one of the assistant surveyors, who planned to visit the Gardner home in Washington, for example, Clint told his father: "You will find Mr. Chandler to be a bag of gas if you give him a chance to explode."[24] Or, as Clint wrote of Commissioner John Weller, "the faults to find with him is that he loves Women, cards and liquor too much but then it does not interfere with his public duties he is vain and selfish but I think he tries to make those around him comfortable and I have no doubt but what he will fix every thing right before he leaves."[25]

If Clint Gardner's letters were candid, they were also crafted, as with all correspondence, to tell his recipients what he thought would interest them and to present himself in a favorable light. A letter Clint wrote from San Diego to his friend John McCalla suggests that women and liquor held the same attractions for Clint as they did for John Weller. "Every night a dance, drunk or something else," he wrote to McCalla while admittedly under the influence of brandy.[26] Clint's contact with women went well beyond dancing—some exchanged their charms for cash.[27]

In his letters to his parents and sisters Clint neglected to mention the less savory side of his social life. Moreover, he seldom complained about the drudgery and discomfort of his work. He usually skipped over difficulties and unpleasantness or treated them cavalierly, as when he described surveying in the rugged, arid backcountry to the east of San Diego as "our pleasant tramps."[28] He had this to say about a long journey to El Paso: "I arrived here . . . after having a very agreeable march of forty one days on the road . . . the weather was quite agreeable when one got used to the cold the thermometer being below freezing."[29] Emory, in contrast, remembered the same trip as "a dreary march across the prairies and uplands of Texas."[30]

Clint may have minimized complaints so he would not cause his family to worry about him, but, because he did not spare details about malaria, smallpox, or danger from Indians, it seems more likely that he wanted to impress his parents and siblings that he was enjoying a splendid adventure. Perhaps he was. Rather than growing weary of life in military camps, traveling through difficult terrain, and taking astronomical measurements, Clint seemed to thrive on it. After nearly four years on the boundary survey with Emory, Clint was still not ready to go home. Instead, he told his father that he hoped for a position that would allow him to see more of new country: "My only ambition being to get in some permanent place where I will have a chance of travelling over some of our extensive country about which little is known at present."[31]

If Clint's letters to his school friend John McCalla differed from those he wrote to his family, so too did letters directed to individual members of his family. Letters to his father, which constitute nearly a third of the correspondence reproduced in this book, were more about finances and politics than were the letters to Clint's sisters. Letters to his youngest sister, Alida, who was nearly his age, were more likely to describe his social life, or lack thereof, and to inquire about mutual friends, particularly young ladies of Clint's acquaintance.

Clint Gardner's letters also reveal much about his own values and those of his time, place, and social class. He was born to privilege in 1831. His pedigree reached back to the American Revolution, in which both his maternal and paternal grandparents had played a role.[32]

Clint's maternal grandfather, John McLean (1755–1821), was born in Scotland, moved to America, and lived near Little Britain in Orange County, New York, at the time of the American Revolution. He enlisted in the Continental Army in 1775 and served through 1780, by which time he had become a second lieutenant.[33] After the Revolution he made his home in New York City and maintained a country home up the Hudson at Newburgh. He served for a number of years as commissary general of the State of New York.[34] In 1784 he married Ann Burnet (1764–1827); Clint's mother, Ann Eliza, was the fourth of their seven children.[35] Two of Ann Eliza's brothers appear from time to time in Clint's letters from the U.S.-Mexico boundary survey—John McLean and Alexander Clinton McLean, who appear in his letters as Uncle John and Uncle Clint.[36]

The family of Clint's father, Col. Charles Kitchel Gardner (1787–1869), had a longer relationship with America than did his wife's family. The Gardners traced their American forebears to one John Gardner, who moved from London to New Jersey in 1680.[37] Clint's grandfather Thomas, also a veteran of the Revolutionary War, was said to have served as a commander of a wagon brigade in the army of the Count of Rochambeau, the French commander.[38] Clint's father, born in Morris County, New Jersey, grew up in Newburgh, New York, where his family moved when he was four. After a stint at studying medicine, the elder Gardner turned to a career in the military and rose to adjutant general of the Northern Division of the U.S. Army at the end of the War of 1812, a position he still held when he married Ann Eliza McLean, ten years younger, in New York City on January 30, 1817.[39] The next year, in 1818, he resigned from the military and moved to New York City. There, in 1819 and 1820 he saw the publication of two of his books on military exercises and regulations, and he continued to do research on reference works for the military. In New York Clint's father held several offices and tried his hand at journalism, but with the 1828 election of Andrew Jackson, who raised federal patronage to an art form, Colonel Gardner, a Democrat, began to receive a series of political appointments that sustained him for the rest of his career. By then the family had moved to Washington City to be closer to the font of government appointments.[40]

In 1849, when Major Emory appointed Clint to the boundary commission, the family lived on New Jersey Avenue on Capitol Hill, a prestigious neighborhood.[41] The Gardner home was only a few minutes' walk from the offices of senators and representatives in the U.S. Capitol and thus ideal for Clint's office-seeking father, who was then completing a four-year term as the city's postmaster during James K. Polk's administration. By then the senior Gardner had made enough connections in Washington's political and military circles that he not only obtained positions for himself but was able to get his son appointed to the boundary commission.

When the Whigs came to power in the spring of 1849, Charles Gardner lost his position. During the Whig administrations of Zachary Taylor and Millard Fillmore (1849–53), years when Clint was on the border, the senior Gardner was unemployed.[42] During those difficult years Clint provided monetary help to his family, as his sister Sarah acknowledged: "It is but due to your generous, disinterested, spirit, to acknowledge, it is fully appreciated by *all of us,* now at home—we have felt it from the time of Father's removal from Office until the present time, and you, I hope may be rewarded for it."[43] Young Clint was also generous to people beyond his immediate family if his loan of a hundred dollars to a destitute forty-niner from New Orleans was characteristic.[44]

Unemployed but not idle, Colonel Gardner turned his attention to compiling a substantial reference work, published by Putnam in New York in 1853: *A Dictionary of All Officers, Who Have Been Commissioned, or Have Been Appointed and Served, in the Army of the United States.*[45] When the Democrats regained the White House in 1853 with the inauguration of Franklin Pierce, Charles Gardner received a job in the General Land Office and then, later that year, an appointment as surveyor general of Oregon, where two of his sons, Clint and Charles, would join him. The senior Gardner returned to Washington in 1856 to become a clerk in the Treasury Department.[46]

George Clinton Gardner, then, was raised in a Washington family attuned to the political currents of the day. His letters suggest greater political sophistication than might have been expected from a teenager. They also reveal that Clint had paid attention to the way his father had used connections to obtain government positions. In Clint's second year on the boundary commission, he began to appeal to his uncle John McLean to get him a commission in the military. He asked his uncle to approach his "friends in Congress, Members or Senators, to use their influence in getting for me a commission in the new Regiment of dragoons." Or, he suggested, his uncle might talk to the president or the secretary of war on his behalf if his uncle happened to be "a good friend" of either of them.[47] Two years later, in 1852, an emboldened Clint wrote directly to California senator William Gwinn and to Oregon governor Joseph Lane, asking them to help him obtain a military commission.

Clint's upbringing also prepared him to play the role of a "gentleman," one who could eat at the same table with Emory and other officers, and his schooling gave Clint the skills that enabled Emory to use him as his secretary and rely on him to take astronomical measurements. Clint had received a basic education at Rittenhouse Academy in Washington City. In the autumn of 1847 he then entered Columbian College (the predecessor of George Washington University) on the outskirts of the city. There, he later remembered, he studied mathematics, mechanics, and engineering. He apparently left Columbian College in his sophomore year to join the boundary commission.[48] Major Emory probably knew that he was not just doing Clint's father a favor when he appointed Clint to the commission but was adding a polished, educated young man to his staff.

Clint's letters reveal his keen awareness of his privileged upbringing. He had high regard for his social status and little interest in those whom he regarded as beneath him. In politics he was a Democrat like his father, but democracy went only so far with him. "I am a democrat politically but not socially," he told one of his sisters.[49] He preferred a world, as he told his mother, where "every one knows his place and keeps it."[50] He tried to live by that maxim. En route to California he expressed dismay at having to eat at the same table with laborers, an arrangement that he called "a democratic mess" (meaning a communal meal) and for which he blamed the Ohioan John Weller: "I suppose it is after the manner in which they live in Ohio."[51] Once in California, Clint declined to associate with people from his own class who did not maintain what he saw as proper social distance. He avoided some of Weller's appointees from Ohio, he told his father, because "they associate with the labours [laborers] and working hands of the commission."[52] Members of the Ohio group whom he shunned included H. H. Robinson and Robert P. Effinger, both of whom wrote extensively about the San Diego phase of the commission's work, and a Tom Wilson.[53] Clint never mentions them by name in his letters, even though he worked with Effinger on a special detail under Capt. Edmund Hardcastle. The Ohioans, in turn, ignored Clint, who knew that he was "rather unpopular among them."[54]

Although he asserted that "the world owes me a living," Clint had to work at keeping up appearances.[55] He often asked his parents to send clothes, some of which, like a dozen fancy shirts "with ruffle bosoms and cuffs" and a frock coat, had little utility for an assistant on the boundary survey. But they had everything to do with fashion and, in the right place at the right time, with catching the "eye of a pretty girl."[56] He also went to some lengths to maintain his wardrobe and bedding. Comparing himself with "other young men of the commission," he explained to his mother and father: "To be sure my wash bill is larger, because I have more pieces; I have sheets, pillow cases, night shirts, &c. that they have not."[57] Accustomed to having domestic help at his family's home in Washington, Clint found San Diego wanting. "I think this would be a delightful place to live," he wrote dolefully to his brother Charles, "if one could only get servants for as it is I have to take care of my own things and room."[58] Later, in Texas, he did have a servant, a matter that he does not elaborate on in his letters.[59]

In San Diego, Clint enjoyed a familiar level of comfort when he lived for a few months with an officer and his wife. "I took my seat at the table," he told his parents, "with silver forks, napkins, &c." When that arrangement ended, however, Clint had to return to Major Emory's table and lamented that he had "now come down to the iron forks and spoons."[60] Clint was apparently as discriminating about his food as he was about his eating utensils. When he operated his own camp with its own kitchen at the mouth of the Rio Grande, a cook named James quit, and another man, William, refused to take the job because, as one of Clint's companions explained, "he did not wish to cook for you, as James had told him that you were harder to please than all the rest of the mess."[61]

Just as Clint's letters suggest that he was a product of his class, they also reveal that he held values of his time and place that cut across class lines. As a Southerner whose father owned slaves, he apparently had a bias against Northerners, judging from his remark that the commission's problems would always exist "when a contemptible yankee is at the head"—a reference to John Russell Bartlett of Rhode Island.[62] Like many of his young male contemporaries, Clint seems to have been an adventurer. "I long to see more of the world," he told his mother after three years on the survey, and his references to Don Quixote suggest that he playfully thought of himself as a knight errant.[63] He also appears to have belonged to a type of American male that one historian has termed "martial men" as opposed to "domestic men." In Clint's day martial men were more likely to be Democrats, as were Clint and his father, rather than Whigs. Martial men were more likely to drink to excess and to believe "that the masculine qualities of strength, aggression, and even violence, better defined a true man than did the firm and upright manliness of restrained men."[64]

Clint's identification with "martial men" comes through loud and clear in his eagerness to get an officer's commission in the army, see action, and fight Indians. In Texas, Clint spoiled for a fight with Apaches and expressed disappointment when the opportunity failed to present itself. He thought that Apaches should be exterminated, but he did not wish for the indiscriminate extermination of all Indians, as did some of his countrymen. To the contrary—he regarded some Indians as useful and others, like Apaches, as irredeemable. Because he believed that American soldiers did not know how to fight Apaches, he advocated bringing "Indians from the East to exterminate them."[65]

Clint's eagerness to fight extended from Apaches to Cubans and Mexicans. In 1853, when it seemed to him that war would break out with Cuba—"that little hacienda," as he patronizingly put it—he told his sister Sarah: "I begin to feel a little pugnatious [*sic*] now and if the war, which must eventually come, does commence before I leave the country I will be tempted to join any party that turns up."[66] The previous year, when a controversy arose over a stretch of the new boundary that seemed to close American access to an all-weather route to the Pacific, Clint told his father: "The only way I see is for the government to take the road by war."[67] Clint was also ready to fight Mexicans himself. Writing from the mouth of the Rio Grande to one of his sisters, Clint "noticed the Mexicans on this work who are stationed just opposite are rather cool. . . . But who cares, I would as leave fight them as any other of their countrymen at this pugnacious time."[68]

Clint's letters do not reveal what he thought about Cubans, but some of his comments suggest that he shared the widespread belief of Americans of his era that Mexicans were inferior and their Catholic religion contemptible. As Clint wrote from Panama, "there are a great many churches here but no religion."[69] For Clint and his countrymen Mexicans were "greazers."[70] Put in charge of a survey party, Clint recommended to Emory that the skilled jobs go to "Americans," and the menial positions to "Mexicans."[71] In specific circumstances, though, Clint could be more discriminating. He enjoyed the society of the Californios in San Diego. "The Mexicans in El Paso," on the other hand, "have had great doings there during the holy days. Bull fights &c. but nothing to interest a civilized person."[72] He was aware that Mexican culture varied from place to place and that he was in rough frontier communities along much of the border. He knew he should not judge the people of the interior of Mexico by what he saw on the frontier.[73]

Clint also understood that class and gender transcended culture and nationality. Those distinctions are clear in his observations about the Mexican elite and Mexican women. In San Diego he rubbed shoulders with local oligarchs. Even though his remarks about them were condescending, he also found the Mexicans charming and enjoyed their company. In San Diego he had no aversion to dancing with women of any class, even women who were "black" or those who "were the color of a chew tobaker [tobacco] or rather the color of a dark indian squaw."[74]

Although his letters reveal much about him, he also left much unsaid. He worked on the boundary at a time when the American expansionism that led to the U.S. war with Mexico still had vigor and filibusters moved into Mexico and Cuba from American soil. Clint's pugnacity and his identification as a Democrat—the party that favored expansion—suggests that he would have sympathized with the filibusters, but his letters say nothing explicitly about them.[75] Political crises resulting from the acquisition of new territory from Mexico were in the air in 1849, resulting in the Compromise of 1850, but Clint has little to say about this. Closer to the scene of his work, Clint is also silent about the activities of other members of the commission. He admired the naturalist Charles Parry and recommended him to his family, but Clint did not describe Parry's work at collecting botanical specimens. Although he was a low-level employee of the boundary commission, Clint did become Emory's clerk and so knew of the work of the Mexican surveyors who worked in tandem with the Americans. In his correspondence, however, Clint had little to say about the Mexican team, even though, as one historian concluded, it "completed the same essential work on the boundary as the Americans."[76]

When Clint wrote about Mexicans and Indians he tended to describe rather than analyze. His letters offer no explanation of *why* Apaches fought or *why* Indians he encountered in San Diego's backcountry were poor. "We came across a great number of Indians," he wrote to his sister Sarah from San Diego, "but they are a poor miserable set of human animals. I hate to call them to my remembrance again, so I will say no more about them."[77] Perhaps because he was writing letters and not a journal or piece for publication, his discussions of geography and local customs were also spare and became sparer as he became more jaded. He sent richly descriptive letters to his family from

Panama and San Diego, but by the time he reached Texas his letters focused more on personal and family matters than on the world around him.

Clint addressed many of his letters to his parents, but he also sent a substantial number of letters to his seven siblings, and he expected that they would share the news in his letters, if not the letters themselves. In order of birth his brothers and sisters were John, Emma, Jacob, Franklin, Sarah, Charlie, and Alida. Clint, born after Charlie and before Alida, was the youngest of the family's five boys and the seventh child. Because Clint's brothers and sisters and their spouses show up frequently in his letters, as does an aunt who lived with the Gardner family during these years, we provide short sketches of them here.

When young Clint headed to San Diego in 1849, his oldest brother, John McLean Gardner, was a lawyer in his late twenties who still lived at home with his parents. He apparently never married and continued to live with his mother and two siblings after his father's death in 1869, at which time he was employed, as his father had been, as a clerk at the federal treasury.[78] John was apparently a poor correspondent. After addressing two letters to John, Clint seems to have stopped writing to him.

Clint's oldest sister, Emma Kitchel Gardner, had married a U.S. senator from Louisiana, Alexandre Mouton, in January 1842 and no longer lived at home when Clint was on the boundary survey. Mouton, a Jackson Democrat, resigned his position that year to make a successful run for the governorship of Louisiana. After his four-year term as governor expired, he and Emma moved to his sugar plantation, Ile Copal, near present-day Lafayette, Louisiana, where Mouton also maintained a residence (today the house is a museum).[79] He had five children by a previous marriage, and he and Emma had six children of their own.[80] Emma named one son, born in 1853, George Clinton for her brother.[81] Clint and Emma exchanged letters when Clint worked for the boundary commission, but these have not survived. Clint did keep up with her and her family, and he was apparently infatuated with Emma's stepdaughter Odeide. On two occasions when he passed through New Orleans, he hoped to visit the Moutons.

Jacob Brown Gardner, known to his siblings as "Jake," "Doc," or "the doctor," was a physician who practiced medicine in Washington City in 1850, when he was twenty-nine. He married a woman named Rose in late 1851 or 1852, an event that Clint learned about through the mail. By 1853 Dr. J. B. Gardner was a druggist in Washington.[82] No letters survive that were written exclusively from Clint to Jacob or from Jacob to Clint. Jacob usually appears in Clint's letters because he had promised to send Clint copies of a Washington newspaper, the *National Intelligencer*, a promise that Jacob eventually filled.

Franklin Gardner, Frank to his youngest brother, graduated from West Point in 1843, when he was twenty. He fought in the U.S.-Mexican War and won promotion to captain for gallant and meritorious conduct at the battles of Monterrey and Cerro Gordo. In 1850 twenty-six-year-old Frank Gardner married eighteen-year-old Mathilde Mouton, a daughter from Alexandre Mouton's first marriage who had become the stepdaughter of Frank's sister Emma.. Frank remained in the U.S. Army until the outbreak of the Civil War. On May 7, 1861, he entered the Confederate Army as a lieutenant colonel. After his discharge as a major general, he returned to Louisiana and spent the remainder of his life as a planter, near Lafayette.[83] His name appeared in Clint's letters from time to time as Clint encountered his brother's military acquaintances, but like his other brothers Frank was a poor correspondent. "He is without pens or some thing else for he never writes," Clint complained after four years away from home.[84]

Sarah Ann McLean Gardner, who was twenty-two in 1850 and four years older than Clint, kept up a lively correspondence with her younger brother, reporting in particular on the Washington social scene (twenty letters survive from Clint to Sarah but none from Sarah to Clint). Sarah married a naval officer, John J. Almy, on February 2, 1854, apparently delaying the wedding until her younger brother returned from his work on the boundary.[85] The couple settled in Brooklyn and raised a family, and Almy rose through the officers' ranks to become a commodore in the U.S. Navy.[86]

Charles Thomas Gardner, Charlie to Clint, moved out of the family home and took up residence in New York City some months after Clint went west. Attracted to California by the discovery of gold, Charles indicated to Clint, in one of his infrequent letters, that he might head west. Clint discouraged him, and Charles never went to the goldfields. He did, however, join Clint on the Northwest Boundary Survey in 1854.[87] While his brother Frank fought on the Confederate side in the Civil War, Charles Thomas Gardner enlisted in the Union army in 1862. He was honorably discharged in 1865 and returned to Washington, where he went to work as a clerk in the Treasury Department, as did his brother John. He lived in his parents' home with his mother, his brother John, his sister Alida, and his wife, Anna.[88] Charles appears with greater frequency in these letters than do Clint's other brothers. Four letters survive that Clint wrote to Charles while on the U.S.-Mexico boundary survey.

Clint's sister Alida Armstrong Gardner, a year younger than he, no longer lived at home when he left for California. She attended a boarding school, the Patapsco Female Institute, which stood atop a hill in Ellicott City, Maryland.[89] She remained there during Clint's years with the boundary commission, and he clearly appreciated her frequent letters with their news about her school, classmates, and social life. Alida apparently never married. In 1870, the year after the death of her father, she still lived at home with her mother and her brothers John and Charles. Nineteen of the letters in this book are from Clint to Alida.

When he wrote to immediate members of his family, Clint also mentioned two other members of the Gardner household, two "aunts" who lived with the Gardners. "Aunty," whose name has eluded us, was Clint's father's sister and apparently part of the New Jersey branch of the family.[90] In writing collectively to his family, Clint often included "Aunty" among his recipients, and he wrote a least one letter directly to her, on October 16, 1852. That year he urged other members of the family to prompt his aunt to write to him.[91] As he told his father: "I would give something great to receive a letter from Aunty for it would really be a treat never having seen one of her epistles in my life."[92]

The other aunt who lived with the Gardners and appears occasionally in his letters was an honorary aunt, Aunt Sucky or Suckey—Clint spelled it both ways. She was apparently a fifty-three-year-old black slave for whom Clint expressed great affection. In 1850 more than a fourth of Washington City's total population of 40,001 was comprised of free blacks (8,158) and black slaves (2,113).[93] The Gardner household had at least two other slaves in 1850, one a woman aged eighteen and the other a boy of fourteen. The woman may have been either Stephanie or Eliza.[94]

Two of Clint's maternal uncles also appear in these letters, John McLean and Alexander Clinton McLean. Two letters from Clint to John and one to Clinton have survived, and we have included them in this volume, but the two uncles' names appear in several other letters.

Clint believed that John McLean had influence in Washington. Clint supposed his uncle could use his connections to obtain a military commission for his nephew, that Uncle John's opinion would carry weight if he wrote an editorial for a newspaper about the work of the boundary commission, and that he had so much influence that he could break a congressional logjam that blocked an appropriation from flowing to Major Emory.[95] Clint apparently did not overestimate his uncle's influence. Emory told Clint that he "would not be astonished" if his uncle John received an appointment as boundary commissioner.[96]

We have not been able to identify John McLean's occupation, but he appears to have been a businessman who moved around the country as well as in political circles in Washington.[97] When he arrived in New York City in September 1851, Clint told his mother, "I will see Mr. Thompson this evening and find out in what part of the world I can find Uncle John."[98] Clint found his uncle at his home at Newburgh on the Hudson, where some of Clint's cousins also lived.[99] At one point Uncle John was in Charleston long enough to have boots made there for Clint.[100] The next year, in May, Clint rendezvoused with him in New Orleans.[101] That summer, as yellow fever swept into New Orleans, Uncle John remained in Louisiana but moved to the more salubrious climate of Attakapas.[102] Later that year, when Clint's father obtained his position as surveyor general of Oregon, he invited his brother-in-law to join him, "offering him business out there."[103]

If John McLean used influence to get political favors, Clint's uncle Clint—Alexander Clinton McLean—worked to receive them. He began his career in the Marine Corps, appointed a second lieutenant in 1827 "by the president, by and with the consent of the senate." He left the marines in 1833.[104] When young Clint served on the boundary commission, his uncle Clint had another political appointment. After angling to become postmaster at San Francisco, sometime in late 1849 or early 1850 Alexander Clinton McLean obtained the position of U.S. mail agent between New York and San Francisco.[105] In that capacity he accompanied the mail and stopped at the port of San Diego, where Clint hoped to rendezvous with him. Uncle Clint's comings and goings between the two seas were chronicled in the newspapers of the day. For example, the *(San Francisco) Daily Alta California* of July 22, 1850, reported that the ship *Oregon* had arrived on July 20: "She brings a large mail, letter and newspaper, in charge of Major McLean, U.S. Mail Agent."

In early 1853 Alexander Clinton McLean lost the job of mail agent but was reinstated in the summer and served through the end of the year.[106] While he was out of work, he sought another office and hoped that his socially prominent daughter might be of help. Learning of this, Clint wrote to his sister Alida, "Uncle Clint I should think was depending on a broken reed if he waits for his beautiful and accomplished daughter to get his for him."[107] Divorced since 1842, Alexander Clinton Gardner apparently did not remarry.[108] He died at the Gardners' home in Washington on January 12, 1854.[109]

In the chapters that follow, George Clinton Gardner's letters appear in print for the first time. Most original letters are in the manuscript collection of the DeGolyer Library at Southern Methodist University, described as the "business and personal correspondence of George Clinton Gardner while on the United States Boundary Survey, 1849–1853" (A1980.0065). They consist of 117 letters written by him and thirty-one more written to him or about him. No record remains of the provenance of the correspondence, but it appears that Everett DeGolyer Sr., who laid the library's foundation, purchased the letters from the well-known New York book and manuscript dealer Eberstadt and Sons. The correspondence came to the DeGolyer Library in a presentation favored by the Eberstadts—a specially built box that contains the original letters and a neat typescript. Gardner's correspondence did not appear in an Eberstadt catalogue, but our supposition is that Eberstadt offered them privately to DeGolyer, who was one of the firm's most dependable customers for rare, costly items pertaining to the American West and Mexico. Unless otherwise noted, the correspondence reproduced in this volume comes from the DeGolyer Library.

We have augmented the DeGolyer's collection with thirty-seven letters from other collections: thirteen from the Willam Hemsley Emory Papers, 1823–86, Yale Collection of Western Americana, Beinecke Rare Book and Manuscript Library; eleven letters from the G. Clinton Gardner Journals and Correspondence, 1853–59, also at the Beinecke; eleven letters from Wade Shipley of Deming, New Mexico, that have since found their way to the Smithsonian Institution; one letter from the John Moore McCalla [Sr.] Papers, Rare Book, Manuscript, and Special Collections Library, Duke University; and another letter from the John Moore McCalla [Jr.] Papers at the Historical Society of Washington, D.C..

The letters appear in chronological order. Letters written *to* Gardner are in italics. For the convenience of readers we have clustered the letters into chapters, prefacing each chapter with an introduction that provides context. We have also annotated the letters in order to clarify details that might otherwise remain obscure.

We have left Gardner's punctuation intact. Gardner had little patience for the niceties of placing punctuation, as he explained in a letter that he wrote to his sister Sarah on April 23, 1849, from Panama: "I can't stop to put stops nor can I stop to correct-stop for if I were to it would be stop stop all over." It appears that he wrote initial drafts with little punctuation but would go back over a letter if he had time and insert punctuation, mostly dashes and commas. He seems to have exercised more care in his letters to his mother than to any other family member.

We have let stand most of Gardner's misspellings, and we have inserted a clarification in square brackets when his spelling is so eccentric that his meaning is difficult to discern. His meaning usually is clear, as when he spelled words with a single consonant that require a double consonant ("stoping" for *stopping*, and "ribon" for *ribbon*) or words with a double consonant that require a single one ("untill" for *until*). To assist the reader we have capitalized and italicized the names of sailing ships and steamers, although Gardner did not.

Information in Gardner's salutations has been incorporated into the italicized title of each letter. For example, where Gardner has written "St. Charles Hotel, New Orleans," we have deleted it from the body of the letter and inserted it into the letter's header.

---

1. Although a short biography published in 1900 says he was born in 1834, GCG was born on August 18, 1831. Writing from San Diego on August 18, 1849, to his mother, whom he could not have deceived about his age, GCG notes that day as his eighteenth birthday. A memorial of respect from the Sons of the American Revolution, issued at the time of GCG's death, also gives his birth date as August 18, 1831. Wilson (1900, 41) appears to have been the sole source for GCG's obituary in the *New York Times,* which ran on August 15, 1904, and repeated the erroneous 1834 birth year.
2. Members of GCG's generation would have understood their task this way (Davies, 1866, 64).
3. "The Mexican Boundary Commission," *National Intelligencer*, September 27, 1851.
4. Emory, 1857–59, 1:1.
5. Hine, 1968, 3.
6. Timmons (1990) is a convenient facsimile edition.
7. Norris, Milligan, and Faulk, 1998, 1–64.
8. Emory, 1857–59, 1:1.

9. Werne, 1987, 330; John Clayton, Secretary of State, to Weller, Washington, March 15, 1849, in SED 34, 6–7.
10. The creation of the commission and its work has been outlined by many writers, but the most authoritative overviews of its politics appear in Goetzmann (1991, 153–208) and Werne (2007), upon whom we have depended. Weller started quickly for California in order to meet his Mexican counterpart on schedule. Whigs asserted, however, that Weller left hastily in order to avoid a recall by the new Whig administration. The best discussion of the political discord and the freeze in funding is in Werne (1987, 330–31). Weller's career is outlined in the *DAB*, 19:628–29.
11. GCG to his father, San Diego, October 28 and 31, 1849.
12. Chaffin, 2002, 413–14.
13. Weller to Thomas Ewing, San Diego, June 13, 1850, in SED, 119, 76.
14. Emory, 1857–59, 1:5–6; Weller to John Clayton, San Diego, June 16, 1849, in SED 34, 27.
15. Werne (1987, 338) suggests that key Whigs sought to destroy Weller politically because they feared his influence in Ohio.
16. Weller to Clayton, San Diego, January 3, 1850, in SED 34, 38. Weller's brother, C. L. Weller, was in Washington in December to implore the government to appropriate additional funds (C. L. Weller to Clayton, December 15, 1849, in SED 34, 34–35).
17. H.H.R. [Hiram H. Robinson], San Diego, November 1, 1849, in the *Cincinnati Enquirer*, December 13, 1849 (reference courtesy of Charles Hughes).
18. Emory, 1857–59, 1:6.
19. "I now can say I am the oldest Assistant on the work (civil I refer to)" (GCG to his mother, Camp Opposite Presidio del Norte, Mexico, August 13, 1852).
20. For Emory as "not easy to get along with," see Kajencki (1988, 240).
21. GCG to his father, Frontera, Texas, January 11, 1852.
22. GCG complains about graft in a number of letters.
23. GCG to his father, n.p., March 1852.
24. To his father, Frontera, Texas, February 6, 1852.
25. GCG to his father, Camp Riley, October 15, 1852.
26. GCG to John M. McCalla, San Diego, February 3, 1850.
27. See GCG's reference to *mancebas* in GCG to his brother Charles, Frontera, Texas, March 18, 1852, and the discussion of Mrs. White in J. Hamilton Prioleau to GCG, Camp 80 on Rio Grande, September 15, 1853.
28. GCG to his father, San Diego, February 28, 1851.
29. GCG to his sister Sarah, El Paso, November 27 and December 11, 1851.
30. Emory, 1857–59, 1:10.
31. To his father, Ringgold Barracks, Texas, November 28, 1852, speaking of Marine Chandler, a member of the survey party.

32. Later in life he joined a patriotic society, Sons of the American Revolution, which listed him thus: "GEORGE CLINTON GARDNER. Washington. DC (10780). Son of Charles Kitchel and Ann Eliza (McLean) Gardner; grandson of *Thomas Gardner,* Commander of Wagon Brigade Rochambeau's Army; grandson of *John McLean.* Second Lieutenant Seventh New York Regiment" (Clark, 1902, n.p.) The society was formed in 1889. At the time of his death the Empire State Society of the Sons of the American Revolution sent a "memorial of respect" to the family, noting his lineal descent from Thomas Gardner and John McLean, "both in the New Jersey Militia" (October 7, 1904, original in the GCGJC).
33. Clark, 1902.
34. *Register of the Empire State Society of the Sons of the American Revolution* (New York: Empire State Society, 1899), 487. John McLean's home in Newburgh was at the junction of Montgomery and Water streets; he was buried in Newburgh at the Old Town Cemetery of the Presbyterian Church after his death on February 28, 1821 (McLean, John, in "Men Who Served the War of the Revolution," www.oldtowncemetery.org/amerrevbio.html).
35. The names of married women do not appear in the census at this time, but Ann Eliza McLean Gardner is listed in the census of 1870 with her full name, her husband having died the year before (U.S. Census, 1870, Ward 5, Washington, D.C., roll M593_125, p. 76).
36. John McLean married Mary Nicholson Burnet, one of ten children of James and Mary Nicholson Burnet, on January 20, 1784. Ann Eliza's siblings, apparently in order of birth, were John, Mary S. [Ross] (1787–1812), William B., Ann Eliza [Gardner], George Washington McLean, Alexander Clinton, and James (Margaret V. S. Wallace, "'Big' Little Britain: The James Burnets," *Orange County Post,* July 2, 1970, p. 10, http://freepages.genealogy.rootsweb.com/~lizzie/James%20Burnet.htm).
37. *DAB,* 7:141.
38. Clark, 1902.
39. *Albany Advertiser,* February 5, 1817; *Salem Gazette,* February 4, 1817. Neither marriage announcement gives the date of the marriage, but a published poem by B. B. French written to the Gardners on the occasion of their golden wedding anniversary (1867) suggests that the date was January 30 (copy at Brown University).
40. The move took place on or before 1828, for we know that Sarah Gardner was born in Washington that year. Her older siblings were born in New York (U.S. Census, 1870, Ward 5, Washington, D.C., roll M593_125, pp. 75–76). We cannot locate the Gardners in the 1830 or 1840 census. For Jackson's patronage see Howe (2007, 331–34).
41. They lived on the east side of New Jersey Avenue, between B (today, Constitution Avenue) and C streets (Waite, 1850, 33; Hunter, 1853, 39). Their house stood on what today would be the grounds of the Capitol, just to the north of Constitution Avenue, facing the monument to Robert A. Taft. In 1849 the area around New Jersey Avenue, which runs diagonally northwest from North Capitol Street, was considered Capitol Hill. Today Capitol Hill is generally con-

sidered the area east of the Capitol—an area that was in open fields in 1849. We are grateful to Mark Herlong for this description.

42. Millard Fillmore succeeded Taylor, who died on July 9, 1850.
43. Sarah Gardner to GCG, Washington City, August 28, 1853. In an undated letter to his father written in 1849, Clint reports that he had heard that his father had "been turned out" and asks, "Is it so?" It was so. Poore (1886, 1:355–56) notes that during the last year of Taylor's administration 3,046 members of the Post Office Department were turned out of office. Charles Gardner did not give an occupation in the 1850 census.
44. See GCG to his brother Charles, Panama, June 22, 1849. A decade later GCG explained his philosophy about money to his parents, saying that he had not saved as much as his fellow surveyors, who earned less than he did, and that he needed to begin to make some investment in lots in Washington for his future. "I look upon money as only a means of obtaining the comforts and pleasures of life and if it could not purchase those for me I care not for it—yet still in looking forward to the time of our return to the states I should dislike very much to be left without the means of living as well if not better than others" (GCG to his parents, February 1, 1860, Colville Depot, Washington Territory, GCGP).
45. The full title: *A Dictionary of All Officers, Who Have Been Commissioned, or Have Been Appointed and Served, in the Army of the United States, Since the Inauguration of their First President in 1789, to the First January, 1853,—With Every Commission of Each;—Including the Distinguished Officers of the Volunteers and Militia of the States, and of the Navy and Marine Corps, Who Have Served with the Land Forces.* D. Van Nostrand published a second edition in 1860, "with emendations and a Supplement."
46. *DAB*, 7:141; obituary, *New York Times,* November 3, 1869; Heitman, 1965, 1:445. We infer the existence of the job in the General Land Office from Charles K. Gardner to GCG, General Land Office, June 30, 1853. He was appointed surveyor general of Oregon in November 1853 and held that position until January 1856.
47. GCG to his uncle John McLean, San Diego, July 2, 1850.
48. Wilson (1900, 41) provides unreliable dates for GCG's early years or no dates at all. Wilson does, however, report that Clint left Columbian College in his sophomore year. We know he was still at the Rittenhouse Academy as late as July 1847, so it seems reasonable to conclude that he entered Columbian College in the fall of 1847 for his first year (1847–48). "Order of Exercises for Rittenhouse Academy, July 27, 1847," listed GCG as participating in two musical events (John Moore McCalla [Jr.] Papers, Historical Society of Washington, D.C.; Mark Herlong). Wilson, and GCG's obituary in the *New York Times*, which drew from Wilson, reported that GCG attended "Columbia," but it is clear from the location that they mean Columbian College, which was then on a parcel at what would today be 14th and Euclid streets, in the county rather than the city of Washington. Reference courtesy of Mark Herlong.

49. GCG, quoting a Mr. Whittney approvingly in a letter to his sister Sarah, Panama, June 12 and 14, 1849.
50. GCG to his mother, Camp Riley, California, August 18, 1849.
51. GCG to his sister, Sarah, Panama, June 12 and 14, 1849.
52. GCG to his father, San Diego, California, November 30, 1849.
53. They also included Dr. E. K. Chamberlin and Oliver Witherby, whom GCG does mention in passing, and others. Several of these men came from Lancaster, Ohio, and thought of themselves as the "Lancaster boys" (Fireman, 2004, 4, 13).
54. GCG to his father, San Diego, November 30, 1849. GCG did not shun all Ohioans. He had a good relationship with Dr. E. K. Chamberlin, whose behavior apparently met his approval.
55. GCG to his sister Sarah, Camp, October 3, 1852.
56. GCG to his father, n.p., April 20, 1849; GCG to his uncle Alexander Clinton McLean, San Diego, July 23, 1850.
57. GCG to his mother and father, San Diego, June 30 and July 2, 1850.
58. GCG to his brother Charles, San Diego, California, April 20, 1850.
59. He mentions this only in passing (GCG to his sister Alida, Frontera, Texas, May 8, 1852).
60. GCG to his mother and father, San Diego, June 30 and July 2, 1850.
61. J. Hamilton Prioleau to GCG, Camp 80 on Rio Grande, September 15, 1853.
62. GCG to his father, Frontera, Texas, March 27, 1852.
63. GCG to his mother, Camp Opposite Presidio del Norte, Mexico, August 13, 1852. For adventurers in GCG's day see Bryan (2006).
64. Greenberg, 2005, 12.
65. GCG to his father, Frontera, Texas, February 6, 1852.
66. GCG to his sister Sarah, Fort Duncan, Eagle Pass, Texas, October 22, 1852; GCG to his sister Sarah, Ringgold Barracks, Texas, June 18, 1853.
67. GCG to his father, Ringgold Barracks, Texas, November 28, 1852.
68. GCG to his sister Alida, Boca de Rio Grande, July 19, 1853.
69. GCG to his sister Sarah, Panama, April 23, 1849.
70. GCG to his sister Alida, Frontera, Texas, May 8, 1852.
71. See our introduction to chap. 7.
72. GCG to his mother, Frontera, Texas, January 8, 1852.
73. See GCG to his sister Alida, Camp, August 9, 1852.
74. GCG to John M. McCalla, San Diego, February 3, 1850.
75. In the presidential election of 1852, GCG declared himself neutral, but that appears to have been the case in that election alone (GCG to his sister Sarah, Camp on Las Moras Creek, September 21, 1852).
76. Hewitt (1990, 186) pointed out that U.S. historians have been particularly biased toward tell-

ing the American side of the story and were led in part by Emory's report to underestimate the quality and quantity of the Mexican surveyors' work. "The Mexican commission accomplished more than Emory and the historians who followed his lead suggest," wrote Hewitt (1990, 173).

77. To his sister Sarah, Camp Riley, October 30, 1849.
78. The 1850 census gives John's age as thirty, along with his occupation. We have relied on that census for information about the family in 1849 (U.S. Census, 1850, Washington, D.C., roll M432_57, p. 27). The 1860 and 1870 censuses show him still living at home.
79. To see the house, go to www.acadiansingray.com, click on "Photo Gallery" in the lefthand menu, then click on "Structures."
80. The 1850 census listed Emma as thirty years old—the same age as her brother John. In 1860 Alexandre Mouton returned to politics, serving as a member of the Louisiana delegation to the national Democratic convention and was later elected to the Louisiana Secession Convention (Dawson, 1990, 118–22). Mouton and his first wife, Zelia Rousseau Mouton, had five children: Marie Céleste, Jean-Jacques Alexandre Alfred, Mathilde, Henriette Odeide, and Cecilia Arcade. In 1850, according to the U.S. Census, at least three of those children still lived at home: Mathilde, eighteen; Odeide (misspelled in the census as Odilde), sixteen, and Cecilia, thirteen. The 1850 census shows that Emma and Alexandre had three more children together: Anne Liza, then six; Charles A., four, and Paul, two (U.S. Census, 1850, Western District, Lafayette, Louisiana, roll M432_232, p. 265), and three more children in the years after the 1850 census: George Clinton, Marie, and William Rufus King (Conrad, 1988, 1:587). Alexandre Mouton's townhouse in Lafayette is used today as a museum (see note 82).
81. Mathilde Mouton to GCG, Ile Copal, September 21, 1853.
82. By the time of the 1850 census, twenty-nine-old Jacob no longer lived at home with the rest of the family. Ten years later the census lists him as a thirty-nine-year-old physician, married to Rose M., age thirty-five, with three children: Eliza E, seven; Charles C., seven; and Joseph N., six months (U.S. Census, 1860, Ward 4, Washington, D.C., roll M653_103, p. 559). Jacob appears in Hunter (1853, 39), listed as a druggist.
83. Heitman, 1965, 1:445–46; U.S. National Archives, Civil War Compiled Military Service Records (online database). Writing to his parents from San Diego, on October 2, 1850, GCG notes that he learned of Frank's marriage, in the "papers." On September 21, 1852, Clint, writing from Texas to his sister Sarah, noted that he had learned of the death of Frank and Mathilde's daughter Cecilia R. The 1870 census lists Frank as forty-seven, a planter worth $2,500, and still married to Mathilde, then thirty-nine (U.S. Census, 1870, Vermillionville P.O., Lafayette, Louisiana, roll M593_516, p. 358). Frank was born in New York City on January 29, 1823, and died on April 29, 1873 (Warner, 1959, 97).
84. GCG to his sister Alida, Ringgold Barracks, Texas, November 28, 1852.

85. *(Washington, D.C.) Daily National Era*, February 3, 1854, p. 3.
86. The U.S. Census for 1850 shows Sarah living with her parents when she was twenty-two. The 1860 census shows that Sarah, then thirty-three, was married to John J. Almy, a forty-four-year-old lieutenant, with three children: Charles, five; William, three; and Augustus, also three. They lived in Ward 11, District 2, of Brooklyn (Kings County), New York (U.S. Census, 1860, roll M653_770, p. 325). The family also appears in the 1870 census, by which time Lieutenant Almy had become Commodore Almy, with real estate worth $12,000 and personal property worth $45,000. Sarah was forty-three by then, and they had an eight-year-old daughter, Annie. They still lived in Brooklyn (U.S. Census, 1870, Ward 6, Brooklyn, roll M593_948, p. 161).
87. Charles expressed interest in California in a letter of February 12, 1850, written from New York; Clint replied to that letter on April 20, 1850, addressing it to his brother at 11 East 11th Street, New York, New York. We have not been able to locate him in the U.S. Census of 1850, which contains a number of Gardners named Charles.
88. U.S. Census, 1870, Ward 5, Washington, D.C., roll M593_125, p. 75. By 1880 Charles and Anna had three daughters, Helen, Gertrude, and Sarah, and he continued to work as a clerk in the Treasury Department (U.S. Census, 1880, Washington, D.C., roll T9_124, p. 81). Charles died on December 25, 1895 (Heitman, 1965, 1:445).
89. Several letters make it clear that Alida no longer lived at home in 1849 and was at the boarding school. See GCG to his sister Alida, Panama, May 8, 1849, in which he asks about "Miss Ginnie and all the other young ladies"; GCG to his sister Sarah, Panama, June 12, 13, and 14, 1849, with his mother's notation to Alida; and GCG to his father, San Diego, November 30, 1849. The 1850 census locates her at the school and lists her age as seventeen (U.S. Census, 1850, Howard, Anne Arundel, Md., roll M432_278, p. 521). The 1870 census for Washington shows her still single, living with her parents (U.S. Census, 1870, Ward 5, Washington, D.C., roll M593_125, p. 76). The institute existed between 1837 and 1891. Today the impressive ruins of its buildings are a historic park.
90. GCG's mother had only one sister, who died in 1812. The only direct evidence that "Aunty" was Charles Kitchel Gardner's sister is in GCG to his father, Little Britain, September 12, 1851: "I should like very much to visit New Jersey to see Aunty and your relations, write soon, my love to all." We presume that Aunty was a blood relative, rather than an "honorary" aunt, because Clint referred to himself as her nephew (GCG to his father, mother, John, Jacob, Sarah, Charlie, and Aunty, Panama, May 13, 1849, and GCG to his aunt, Camp near Fort Duncan, Texas, October 16, 1852). We have not been able to identify all Charles Kitchel Gardner's siblings. In 1861, writing from Brooklyn to his brother John, Clint noted that "Uncle Wash and Aunt Rebecca were here this morning and have just left." They were apparently two of his father's siblings (GCG to John, Brooklyn, September 22, 1861, in GCGP).
91. In letters of February 16, June 20, July 6, July 23, October 8, and October 22.

92. GCG to his father, Camp near Fort Duncan, Texas, October 8, 1852.
93. Green, 1967, 33.
94. The 1850 U.S. Federal Census—Slave Schedules (available at www.ancestry.com) lists the ages of the three slaves but not their names. The names of the two women may be inferred from GCG to his father, New York, April 22, 1853, in which CGC, who has sent money home, asks his father to give Stephanie and Eliza one dollar each as a gift—the same amount he gave to Aunt Sucky. Aunt Sucky was apparently the fifty-three-year-old slave mentioned by the 1850 census, and either Stephanie or Eliza was the eighteen-year old. In this same letter GCG asked his father to give Aunty five dollars, making clear to us that Aunt Sucky and Aunty were two different people. The fourteen-year-old boy may have been Aunt Sucky's "worthy son Step" (GCG to his sister Sarah, Camp, October 3, 1852). The 1860 slave schedules in the census show that Charles Kitchel Gardner owned two females slaves, one sixty and one twenty years old.
95. GCG to his uncle John McLean, San Diego, July 2, 1850; GCG to his father, Frontera, Texas, January 11, 1852; GCG to his father, Camp at the Canon on the Rio Grande, June 20, 1852; GCG to his father, Frontera, Texas, March 27, 1852. Although GCG refers to him only as "Uncle John" in the body of his letters, it is clear from the address that his uncle's surname is McLean, as in GCG to his uncle John McLean, on Capitol Hill, Camp Riley, California, July 23, 1849.
96. GCG to his father, Frontera, Texas, January 27, 1852.
97. The name John McLean was exceedingly common. The most influential John McLean in politics was a Supreme Court justice from 1830 to 1861, but that John McLean (1785–1861) was not directly related to Clint's mother. Weisenburger (1937) remains the standard biography.
98. GCG to his mother, New York, September 7, 1851.
99. GCG to his mother and father, Little Britain, September 12, 1851.
100. GCG to his father, San Elizario, Texas, June 5, 1852.
101. GCG to his sister Sarah, New Orleans, May 6, 1853.
102. J. Mc. D. Taylor to GCG, New Orleans, August 23, 1853.
103. Sarah Gardner to GCG, Washington City, August 28, 1853.
104. *American Annual Register for the Years 1825/26–1832/33; or the 50th–57th Years of American Independence*, edited by Joseph Blunt (8 vols.; New York: W. Jackson, 1827–35), 3:565; Benjamin Homans, ed., *Military and Naval Magazine of the United States* 1, no. 5 (July 1833): 320.
105. For Uncle Clint's efforts to secure the San Francisco job, see GCG to his father, Camp Riley, California, October 15, 1849.
106. *New York Daily Times*, August 2, 1853.
107. GCG to his sister Alida, Boca de Rio Grande, July 19, 1853. The daughter was apparently Jessie Mary, described as "the only child of Col. Alexander Clinton McLean of Newburgh,

New York. She was a great beauty, and had the most distinguished social success in New York, until she went to Europe." See Arnold (1899, 291), who relates a story about her tragic later years.

108. "An Act to Divorce Catharine E. McLean from her husband Alexander Clinton McLean" passed February 26, 1842. See *Acts of the Legislature of the State of New Jersey* (Trenton: Secretary of State, 1842), p. 77.

109. *(New York) Weekly Herald*, January 21, 1854, p. 24.

FIASCO

Number Six: DeGolyer Library Series

CHAPTER 1

To New Orleans and Panama

February 13–June 22, 1849

Clint Gardner's letter of appointment from Maj. William H. Emory, written from Washington City on February 13, 1846, instructed him to report to New Orleans within twelve days to ship out to Panama on the steamship *Alabama*. Clint knew that such an offer would be forthcoming and had prepared for it.[1] Two days after Emory wrote the letter, Clint set out from his family's home in Washington for Cincinnati, traveling by coach. From there he made his way by riverboat down the Ohio to Louisville, where he boarded a vessel that took him down the Mississippi to New Orleans. At least three other members of the boundary commission traveled with him—Major Emory's brother, Frederick, and George F. Hooper, both "sub-assistants" in the surveyor Andrew Gray's party, and Emory's black slave, Robert, who was in charge of Emory's personal baggage and astronomical books.[2] At New Orleans the men planned to join other members of the commission, including its head, Col. John B. Weller, who was scheduled to sail on February 25 for Panama. Major Emory, sailing from New York, caught up with Commissioner Weller in Panama, and from there the boundary commission proceeded to San Diego, where its real work began.

Article 5 of the Treaty of Guadalupe Hidalgo specified the location of the new boundary from a point south of San Diego Bay to the junction of the Gila and Colorado rivers, eastward along the Gila, then along the southern boundary of New Mexico to the Rio Grande and the Gulf of Mexico. Each nation would appoint a commissioner and a surveyor who would "run and mark" the new line, and "the result agreed upon by them shall be deemed a part of this treaty, and shall have the same force as if it were inserted therein." Thus the work of the commission had to be done with great care, for its results, as Secretary of State James Buchanan told John Weller, "will be final and conclusive."[3] Article 5 further specified that the commissioners and their surveying parties would meet at San Diego within one year "of the exchange of ratifications of this treaty"—an event that had taken place at Querétaro on May 30, 1848. The American contingent, whose leader was not appointed until mid-January 1849, had to scramble to get to San Diego by the deadline of May 30, 1849. It failed.

Delayed in part by the freezing of the Ohio River, Gardner's party reached New Orleans on March 5, ten days late, and missed the boat.[4] Commissioner Weller, also delayed, had left three days before on the newly built Pacific Mail steamship *Alabama*.[5] More than two weeks passed before Clint's group could secure passage to follow Weller to Panama. Treasure seekers, bound for the new gold discoveries in California, had swarmed into the Crescent City and made it difficult to secure booking for the popular Panama crossing. Clint, who had expected to receive expense money from Weller, found himself in New Orleans "without a single cent."[6] His pay, like that of other civilian members of the commission, was to begin on the day he left home. In fact, Clint and the other civilians received no salary because Congress had yet to authorize payment for the civilians working for the commission. Surveyor Gray might have helped, but he had only enough of the commission's money to buy tickets to Panama and had depended on getting more funds when he rendezvoused with Weller in New Orleans.[7] Gray did, however, win Clint's gratitude by lending him "his dress coat and a new pair of White Kid Gloves and . . . it was through him that I was invited to the soirées."[8]

Clint's pay, had he received it, would have been $400 a year, only slightly more than the commission paid a servant. Without help from his father Clint could not have maintained the kind of life to which he was accustomed, a life that included the theater and proper dress. In New Orleans, Andrew Gray's loan of proper attire allowed Clint to attend a performance of *Hamlet*. Fortunately, Clint also found an hotelier who advanced him room and board on credit, and he made new friends who loaned him money.

On March 21 Clint shipped out of New Orleans on the *Falcon*, another new steamship bound for Panama, with at least three members of the boundary commission aboard—Andrew Gray, Frederick Emory, and Edward Ingraham. They landed at Chagres, on the Panama coast, on the twenty-seventh and started across the isthmus, ascending the Chagres River by canoe—a journey through exotic jungle that impressed Clint as it did so many travelers. After lingering at the mountain town of Gorgona, where the river ceased to be navigable, they made their way on muleback to Panama City on April 1. There Clint's party planned to take passage to San Diego on a steamship. Instead, they were stranded. So too were other members of the survey party, including Commissioner Weller, who had arrived two weeks before, and Major Emory, who had sailed to Panama directly from New York and reached Panama City just a day after Weller. Emory arrived in a state of annoyance over what he saw as Weller's muddled preparations, which included neglect of important business and unpaid bills, and Weller's apparent lack of confidence in Emory's abilities.[9] Before the commission had even begun its work, the stage was set for discord between its chief astronomer and the civilian commissioner.

The speedy Panama crossing was the most popular route for Americans who traveled to California by sea in 1849, notwithstanding Panama's reputation for a "pestiferous" climate and "wretched" accommodations, as one guidebook put it.[10] Panama became so crowded with gold-fevered argonauts, however, and bookings on vessels on the Pacific side were so difficult to obtain, that its time-saving advantage was only theoretical. Weller had written to the American consul in Panama City on January 30, asking him to arrange transportation to California, but the consul failed him.[11] For the next few months the members of the commission, who came to number nearly forty, languished in Panama City. The tireless Major Emory passed the time calculating latitude, longitude, and magnetic declinations, among other things, and he put his assistants to work to sharpen their skills or to keep them sharp.[12] Clint, whose duties included recording readings on the thermometer, barometer, and hydrometer, reported that he was "busy all the day at the office and very often at night."[13]

A contagion more deadly than gold fever haunted those who crossed the Isthmus of Panama that spring of 1849. Cholera, previously unknown in America, had first swept across the United States in 1832–34 and reentered the country from Europe in the winter of 1848–49. The disease killed swiftly after the appearance of the first symptoms—diarrhea, vomiting, and cramps.[14] Clint had witnessed processions to the graveyard in New Orleans, and the plague followed him to Panama City, where it arrived in April, and he estimated that it killed three or four people a day. Among the dead were one of his roommates, twenty-one-year-old C. W. Hart, and many passengers who arrived on the ship that eventually took Clint away from Panama.[15] Clint apparently staved off cholera by following Andrew Gray's advice to drink brandy and to limit his intake of water.

In mid-May the survey party began to leave the cholera-infested isthmus for California. The first contingent sailed on the *Panama* on May 18. The small wooden side-wheel steamer, which had left New York in mid-February, had continued down the

Atlantic coast, crossed into the Pacific through the Straits of Magellan, and reached Panama City on May 6, a nearly four-month sail from New York.[16] The crowded vessel, which had on board the surveyors' heaviest astronomical equipment under the charge of Capt. Edmund L. F. Hardcastle, could take on only ten of the nearly forty members of the boundary commission who waited impatiently in Panama City.[17] The ten included Commissioner Weller and Major Emory. The *Panama* sailed on May 18, leaving the remainder of the men behind, restless, unhappy, and for a time uncertain about how they would pay for their room, board, and passage.[18] A second group of surveyors sailed on the *Oregon* a week or two after the *Panama*'s departure. Clint Gardner formed part of a group of thirteen that stayed on until June 23, finally leaving on the *California*, a newly built vessel that, like the *Panama*, was operated by the Pacific Mail Steamship Company between Panama and San Francisco.[19] The boundary commission's hastily organized, chaotic journey to San Diego was a harbinger of things to come.

The eighteen letters in this chapter begin with the appointment of Clint to Emory's staff on February 13, 1849. The remainder, with the exception of one letter from Emory to Colonel Gardner praising his son's "intelligence & deportment," contain Clint's descriptions of his journey to Panama and his time there. The chapter ends with the last missive Clint wrote from Panama, on June 22, 1849, the eve of his departure for San Diego.

Major W. H. Emory to G. C. Gardner, Letter of Appointment, Washington, February 13, 1849[20]

Sir

Under authority given me in the letter of the Hon. John B. Weller, Commissioner for running and marking the Boundary between the U. S., & Mexico, dated Jan. 31st 1849, I have selected you as one of my party.

You will accordingly repair to New Orleans by the 25th of this month to sail in the Steamer ALABAMA.

Your compensation will be that allowed the Junior assistants fixed by the Commissioner and your traveling expenses to New Orleans will be paid by him also.

I am very Respectfully

Your Obt. St.

W. H. Emory
Bvt. Major U.S. Army
Chief Astronomer
Mexican Boundary Survey
True Copy
Edward Ingraham

GCG to his mother, Mrs. C. K. Gardner, Washington, Pennsylvania, February 16, 1849, 5 o'clock

Dear Mother

Here I am safe without loss. I had the pleasure of riding in the coach that was built for Gen. [Zachary] Taylor, that is the way I get time to write. I eat dinner here and then start for Wheeling. I have been traveling day and night since I started. Give my love to all

Your affect. son
in haste
G. Clinton Gardner

GCG to his father, Col. Charles K. Gardner, Evansville, Indiana, February 26, 1849

Dear Father

Here I am safe and sound on the steamer *Glencoe* bound for New Orleans without money to pay my passage but that makes no difference you know when *I am knowen all over the world.* I believe the last letter I sent home was from Marietta well, I arrived at Cincinnati on the 23rd about 12 o'clock & put up at the Broadway Hotel until about 12 o'clock on the 24th. I went to the Theatre on the night of the 23rd and wrote a letter to Louisiana after I came back. Dr [John] Kearney was my Room mate, he is a young Doctor from New York and is going to California with Col. [Henry L.] Webb's company.[21] I expect he did not like such late sitting up as he was in bed when I came home, so I had to wake him up to get in. I left Cincinnati on the 24th about 12 o'clock in the Mail boat *Ben Franklin* for Louisville at which place we expected to get a fine boat to New Orleans but had to take this one as we had not time to wait for another although I believe the *Patona* which was to leave the day after we left which was Sunday will get in New Orleans before us. We left

Dr. Kearney in Cincinnati. He said that he was going to take the *Charles Hammond* which was to start from there on Sunday evening and go right through without stop[p]ing. When I get to New Orleans I will be without a single cent and will have to get 25 cents from Emory to pay my passage, for I have but $11.95 to my name. I wish you would write to me as soon as possible and tell me how to send you your $50.00. If you do not write I will get Mr. [C.] Toledano to send it for me.[22] Give my love *to all* and excuse haste as we are just pushing off from this place. Tell Sarah not to talk about nice boats any more.[23]

Your affectionate

son Clint

[George F.] Hooper says that perhaps we will stay in New Orleans some time if we do I am going to try and go to see Emma.[24]

G.C.G.

GCG to his mother, Steamer* Glencoe, *March 1, 1849

Rec'd 26th March 1849

Dear Mother,

We expect soon to arrive at Memphis at which place I intend leaving these few lines for you. We have had quite a pleasant trip so far with the exception of a few delays. Mr. [George F.] Hooper saved me $2.00 coming over the Mountains, which was charged to me at Cumberland, for extra baggage but I am sorry to say he did not save me or himself anything by selecting this boat to New Orleans, for when we left Cincinnati, the steamer *Charles Hammond* was to start on the 25th for New Orleans, which is a far better boat than this although not quite as large, in which we would have had to pay only $12.00 passage, but instead of going in the *Charles Hammond* he thought it would be better to go to Louisville and take a larger boat; so we started from Cincinnati on the 24th about 12 o'clock, which was just one week after Col. [John B.] Weller's departure, in the steamer *Ben Franklin* and had to pay $2.50 passage to Louisville and then take this boat, which is no better than the *Charles Hammond* and pay $12.00 to New Orleans. The worst of it is the *C. H.* has passed us and will get in New Orleans a day or two before us, but such is our luck, it appears that we were doomed to travel slow just because we want to go fast. We expect to get in New Orleans now about sunday evening or monday morning and my only wish is that when we do get there, that the party will stay there for a week or so, then I will go to see Emma. I see by the papers that Mr. Mouton was in New Orleans on the 5th of this month but I suppose he has returned before this.[25] I have an idea of leaving my watch with Emma if I can get it to her. I am going to buy myself a pair [of] pants and gaiters made of corduroy and perhaps a coat, what do you think of it? Suppose Col. Weller has left New Orleans before we get there, what do you suppose we will do, not one of us will have a first red cent. Mr. [Frederick] Emory says that he is going to put up at the St. Charles and I am going to follow his example, we will get along after all, if we are without money we are good looking.[26] Give my love to all including father, yourself, aunty, uncle John [McLean], uncle Clint [McLean], Jake, Charley. O[h], I forgot to put sarah before Jake, but never mind I will put it in big letters at the bottom. Stephany Susan Colbert and all others that may be within hearing.[27]

Your affectionate
son Clint

I expect to get a letter from home when I get to New Orleans and I hope the postage will be paid, or I am a used up man.

S A R A H

Very good printing for a steamboat

GCG to his father, St. Charles Hotel, New Orleans, March 6, 1849

Rec'd 25 March, 1849

Dear Father

We arrived here yesterday morning before breakfast and found out that Col. Weller had sailed the day before.[28] Mr. [Frederick] Emory received a letter on his arrival from Major Emory saying that he wanted his things &c. to go with Col. Weller and he was going to sail from New York I think, but will not be certain, *so you must not say any thing about it*. Not one of us have a single cent and we are staying at this Hotel, on credit, because Mr. Emory knows one of the Proprietors. The Proprietor lent Mr. Emory some money and he has Telegraphed to his brother William and also to Mr. Grey [Andrew B. Gray], so as soon as we can get an answer we will know what to do but until then we will just have to live here. I took that bundle to Mr. C. Toledano and his clerk told me that it would be attended to. I have not been able to find Mr. Durrive yet.[29] I looked after him yesterday and will look again to day [*sic*]. I intend telling him how I am situated and perhaps he would assist me; if I could get out to see Emma I would go there and stay, for I think it is rather expensive to live here and pay $2.50 a day, when we have no money. Mr. [Alexandre] Mouton was in the City about the 10th of last month. I wish you would write to me as soon as you receive this and tell me what are your ideas of this thing, for I think that Major Emory and Mr. Grey will sail from New York thinking that we have gone with Col. Weller. Mr. Hooper proposes going to Chagres and meeting Col. Weller there, but if we do that we might not overtake him and then we will be worst off. If I do not hear from some part of the world soon, I will go and get Mr. Durrive to lend me some money and go out to Mr. Moutons plantation and go to work for I think I can learn to make sugar as well as the next person.[30] I was introduced to Capt. Howard and Lieut. Tyler yesterday—both of whom know Frank well.[31] Write as soon as you receive this, I would have written yesterday but the mail left at 12 o'clock and I did not know it. Give my love to all including Sarah and

believe me Your affectionate

son Clint

GCG to John McCalla Jr., Care of Gen. McCalla, at Washington, D.C., St. Charles Hotel, New Orleans, March 16, 1849[32]

Dear John,[33]

I believe the last time that I troubled you to read a letter from me was from Cincinnati and as I may not have an other opportunity of writing to you, I thought I would avail myself of this one and give you a little longer letter than usual although I do not feel much in the spirit of it now, for I am quite vexed to think that I have got to stay here until the 20th of the month after staying here as long as I have. I arrived here on the morning of the 5th expecting to meet Col. Weller the commissioner and sail with him imediately for Chagres, but when I arrived here I found that he had gone so you see I have had to wait for Mr Grey the surveyor to sail, and I believe he is going on the 20th. When I look back and think how I have spent the time it makes me mad, for I have done nothing but lounge about the hotel, eat and drink for I could not go visiting as I have no fashionable clothing. I have been invited to several soirées but could not go; I did go to one that was given by the young ladies staying at this hotel, as Mr Grey was kind enough to lend me a dress coat and other necessary articles, but that was the only one. I have been to the Theatre twice to see Mcready play Macbeth and his favourite, Hamlet, but that was all, as it is necessary that a gentleman should go in ball dress with white kids &c. every time he goes.[34] I could not go often, as it would not do for me to borrow all the time. I expect you were a little supprised when you received that scrawl that I wrote to you the night before I left, as you had heard nothing of my intentions, perhaps you thought it was rather a hasty move of mine, but it was not, I had contemplated it for some time and that was the only reason. I visited Washington so much when I was at the Alexandria school. I regret very much that I was not able to see you before I left, but when I went down to see you Thursday evening you had gone out to College, I told Tucker Blake to tell you in the morning that I wanted to see you but I suppose he forgot it as he had his dress to think about, I am afraid that he has forgoten how to study since he has been out to that College, when he was at Nourse's school there was not a more studious boy in school than he was, but now I believe his dress takes all his attention.[35] I think he would do well to live down here for all the people here think more of their dress than any thing else and going to operas theatres, balls &c.

I had the pleasure of meeting Johnson the Sunday before I left with his friend Dr. Morgan, I expect he gave you a discription of my appearance as almost every person noticed my California dress, you cannot imagine how many remarks were made about the short tail coat or rather the coat without a tail, when I was coming down the river, I heard a man say that I had on a coat but a—s did not know it, but for all the people made so much fun of it I expect to have a great deal of use of it.[36]

In looking over my old letters I found several from you and all of them were thanking me for writing such long ones, but I am afraid that you will give up reading this one before getting through it, as I am not done yet but you know perhaps this one will be the last that you will ever receive from me so if you find it rather long you can lay it aside and when ever you would like to hear from a an [*sic*] old friend just read a little more, and perhaps by that means you will be able to get through with it. I would like ~~to have~~ very much to have been in Washington on the 5th of this month, not to have seen [Zachary] Taylor made president, but to have seen the people, I suppose you were at *the ball* that is the one that was at Jackson Hall; but I think that if I can only see my parents again and can get a liven [living?] I will never return to Washington for I can be perfectly happy with *Notre Mère*, w[hom] I expect to have the pleasure of seeing soon and if I do not I most certainly will when I return from California.[37] The sunny south has no charms for me but I think I can live here very well, in case some one would not like to live at the north but at present California is the land for me and I think that it is high time that I was moving out of this place.

I suppose you visit the corner of 3rd and C streets as much [as] usual, the next time you go there tell them that I am in the land of the living, but the Lord only knows how long will be there, and wish to be remembered to them; and then if you should happen to drop in a few doors west of that on the other side of the street just do a little for me, although I am not in need of it yet it will be doing them a favour, I am sorry that I could not visit them before I left, but I suppose you and Johnson can do all for me and yourselves too.[38]

We had quite a thunder storm here the other night and the rain fell in torents, when I went to bed I expected to get up and find myself surrounded with water, since the people say that this part of the city is 6 ft below the river, but as it happened the levee did give way and the St Charles was standing hard and dry.

I expect I am tiring you by writing such a string of stuf, so I will bid you good night perhaps for ever.

Remember me to all your friends and believe me as every your sincere friend.

G. Clinton Gardner

GCG to his brother Charles, St. Charles Hotel, New Orleans, March 18 and 19, 1849

Dear Charlie,

I believe I will try you this time and see if you will answer my letter, I thought when I arrived here I would find some letters, but I went to the post office in vain, it will be just two weeks tomorrow since I arrived here and every day I have been to the post office, without getting any letters until the other day. I received one from my friend Custis Lee,[39] so you see I am not entirely forgotten by my friends, if I am by my relations but you know as I am of a disposition to forgive and forget, I will say nothing more about it.

Since I have been here I think I have seen a good deal of the city, but not as much as I might have seen; I have been to the Theatre but twice and then it was to see Macready play Macbeth and his favorite piece, Hamlet, I have been invited to the soirées, but did go to but one, because I had not the clothes that were suited to wear, the only reason I went to that one was because Mr. Grey lent me a dress coat and white kids, he said that he would lend me any thing, a person can not go to the Theatre or soirée or even a church, with out he is dressed up in a full Ball dress, that is the reason I have not been to church since I have been here. I forgot I did go once to day, that is I walked in one door and out of the other only to see the church because the people said that it was a beautiful church. I took a ride out to the lake the other day with [Captain] Howard, who was formerly of Washington, he is a brother to Clem Howard that was in Stoff's drug store but it was not a very pleasant ride, as I was at his expense and had not a cent of money even to offer to pay the tole [toll], we met 6 funerals a greater part of which were persons that died of the Cholera and there are about 6 or 7 burial grounds so you may see how many persons die here. Mr. Grey told me that I must drink very little water and always put brandy in it, so you see every day at dinner I have my brandy by me. I go to see Mr. Durrive every now and then he has given me a standing invitation to dinner, when ever I wish to come he told me the other day that he had some of Franks things that were left at his house amongst, which is a mexican blanket which I am going to take with me. I have written to Frank twice about it but have received no answer, the remainder of the things I think I will send to him at Jefferson Barracks [Missouri] if there is nothing more that I want.[40]

It is said that when a man is hard up without any cash he is in the nine holes, but I think we are in the ten holes this time, as we are without cash and are in debt, all that I owe is for my board and $14.00 for a corderoy coat and a pr. of corderoy pants to pay which, I intend to borrow of Mr. Durrive, if Mr. Grey [Gray] does not get some money from Washington to morrow, for we must start for Chagres tuesday evening in the steamer *Falcon*,[41] which is to sail from this place and scarce will the ruddy Phoebus spread the golden tresses of his beauteous hair over the face of the wide and spacious earth, or scarce will the painted birds with the sweet and mellifluous harmony of their forked tongues salute the approach of rosy Aurora, who, quitting the soft couch of her jealous hus[band], discloses herself to mortals through the gates and balconies of the bright horizon,[42] when the renowned Mr. Grey and his suite accompanied by myself and Mr. [Edward] Ingraham of Major Emory's division will be afloat on the great gulf of Mexico.[43] I think I will have to stop now and bid you goodnight. I will write a short note to father tomorrow which you will find enclosed or rather have found.

I wish you would tell Alcide[44] to tell Custis Lee to send his letters up to father and then I wish you to ask father to send them to me with any others that may be in the post office and oblige your

affectionate brother

G. Clinton Gardner

March 19th. We are to start to morrow evening. I intend taking my last dinner in the United States with Mrs. Durrive and her brother. I am glad to say that I will not have to borrow any money. When Alcide comes up tell him to bring Custis Lee with him for he is a great friend of mine and is a fine fellow. I should like you to know him. Give my love to all and write soon.

believe me your affectionate

brother Clint

in haste

I am in a perfect state of health thank God. The Cholera is in the City.

G. C. G.

GCG to his father, New Orleans, March 19, 1849

Dear Father,

I will not be able to send you the $50.00, that you lent me as Col. Weller has left us without a single cent and Mr. Grey [Gray] has not any more money, than is necessary to take us to Chagres.

I find that Mr. Grey is a very fine man, he lent me his dress coat and a new pair of White Kid Gloves and said that he would lend me any thing that he had; it was through him that I was invited to the soirées for I did not know any of the young ladies. He tried to make me pass off the time as agreeable as possible, so I conclude that he is a fine man.

Your affectionate son, Clint

in haste.

GCG to his father and mother, Steamer* Falcon, *March 27, 1849

Steamer *Falcon* at sea

Dear Mother

Here I am so far on my journey and to night we expect to arrive at Chagres so I take this opportunity of writing you a few lines so as to send them back to the United States. I suppose you received those letters that I sent from New Orleans and also those I wrote on my way to that place. I started from New Orleans on the morning of the 21st (Wednesday) and after we got out in the gulf I was quite sea sick and remained so until Friday morning and since then I have been very well but rather dull as there is no way of enjoyment until yesterday the gentlemen of our party got up a dinner party to Capt. Tobin at which we enjoyed our selves a great deal but the best of it was that this morning I had to pay $5.00 perhaps you may think that I am rather extravigent [*sic*] but I could not stay out.[45] I think this is about the greatest party that ever started for California every now and then we have a little dance and that is the way we pass off our time. You must excuse the brevity of this but you know it is only to let you know that I am still alive for there is no news to write except that we have very fine weather all the time.

Excuse me for writing a little to father at the end of this but I suppose it will not be of any interest to you so you do not care. Give my love to all and believe [me?]

as ever your affectionate

son Clint.

Dear Father

We started from New Orleans in Company with Major [Justus] McKinstry who came from New York and I believe has some thing to do with the Boundary line at any rate he holds all the funds.[46]

I have not had a single cent from the Government yet. I owe to almost all the party, the amount of which I believe is about $29.00.

So you see all that I have spent is $79.00 and all the Government owes me is $55.80 so you can tell what I have spent for myself by taking the difference. I will write at San Diego if I can. *I wish you would get me the president's Message with all the documents that accompany it and send it to me, you can get a member of*

Congress to Frank it. It will be of great use to me.[47]

Excuse this scrawl and believe me

Your affectionate son

Clint

GCG to his father, n.p., n.d., Panama, April 1849

Private

Dear Father

Permit me to return my sincere thanks for those things you were so kind as to send they are very usefull and also thank mother for her kind intention of sending those shirts. I am sorry I did not receive your letter in New Orleans but I hope you will excuse me for writing those letters of complaint.

I have heard that General Order No 1 has been given out and I hear that I am to have $400 (dollars) for the first year.[48] I made out my bill for travelling expenses and it was paid it was $88.65 then I had to pay Mr. Gray $56.20 which he lent me to pay my bill at St. Charles [Hotel in New Orleans] $8.00 of which was for washing, &c., extras; and a suite of clothes which was $14.50. I had to pay Mr. Whitting $5.00 which he payed for me on the *Falcon* for the dinner. Mr. [Frederick] Emory $8.05 for mosquito bar comb & brush bath, &c. Mr. Hooper $1.10 for powder and omnibus ride. $9.05 for 8 shirts of color with studs. I bought them because it was as cheap as they could be had in New Orleans for and when I was there I had no money without borrowing it. $3.25 for washing which is very high but I have had no clothes washed since I left New Orleans before soon I will have to pay some more. $4.00 I lent to Mr. Ingraham who will pay me in a few days. I paid him 50 cents which I got of him in New Orleans to take a bath. 10 cents for blacking 10 cents for a candle and $1.10 I have in my pocket. When I get to San Diego I will send you a draft for $50 or if you are still post master can't I pay it to the post master at San Diego so as you can draw it at Washington if so please let me know. I heard from a New York [word missing] that you had been turned out is it so?[49]

Give my love to all and believe

Your affectionate son

Clint.

Fred Emory and Mr. Gray send their best respects to you.

G. C. G.

GCG to his sister Sarah, Panama, April 23, 1849

Dear Sarah

I have just heard that the mail for the states close tomorrow so I thought I would write you a few lines.

I suppose you have heard before this of my arrival in this beautiful City, as I wrote to father by the last mail. I have just found out my position in the service which I will tell you before I close. I received a letter from Emma just after arriving here in which she calls me Asst. Surveyor, which I understand she heard from you but you are a little mistaken. I do not belong to the surveyors party. Major Emory is my head and he is the chief astronomer we have our office on San Juan de dios street and Head Long Hall which is the house I live in, is at the other end of the same street but has a different name it is called San Felipe street. I expect father will have received a letter from Major Emory before you get this one the reason I suppose so is because he asked me his first name if he has, I would like to hear what he says. Fred Emory wished me to give his best respects to you when I wrote he is over to the Island [Taboga] now with almost all of Mr. Greys party they are all encamped there they say it is a beautiful spot. I hope I will get an opportunity of seeing it before I leave this place a great many describe it as being a perfect garden of Eaden [Eden] it produces all the fruits of this climate.[50] I suppose you would like to hear how I got here before I say any thing about the City so here it comes open your ears Mother Father and all.

We arrived at Chagres on the night of the 27th and landed the next morning about 12 o'clock. I was quite surprised when I came out on deck to look at the place that the letter writers say is such a hardship to cross. I expected to see low marshy ground but instead of that it was rather high just at the entrance of the river is a large Old Spanish fort on quite a high hill which makes a very pretty scene.[51] Chagres is quite a village built of read [reed] houses with thatch roofs they have a St. Charles Hotel an Astor House &c. but I should not like very much to stay at any of those places the meat that they have to eat is cut up in long strips like ribon but it is not wound around blocks although it is sold by the yard. The natives go about without any clothes that is a greater part of them it is quite amusing to see the women carry their children astride of their hips and when they put them in the house they tie them like a monkey to keep them from running away their dogs they let run lo[o]se and it appears that there are more of them than anything else the pigs, chickens,

&c. all live in the same house with the family but in this place they are more refined they only take the chickens in the house with them these are the common people that keep stores &c the first familys, I believe live very nice and clean. We left Chagres in the evening about 3 or 4 o'clock in little canoes. Mr. Gray, myself and Capt. Tobin, the one to whom the dinner was given on the *Falcon*, took one canoe with two boys which we ought not to to have done for they were not strong enough to pole us over some of the rapids and keep up with the other boats and started up the river which was beautiful. I have never seen any thing more beautiful than some of the scenes on that river the banks are high and covered with trees of almost every color from the darkest to the lightest green, and from the deepest crimson to the brightest yellow. I enjoyed the trip up the river very much at night we encamped on its banks the first night we all encamped together but the second night we got left behind and had to encamp by ourselves and as luck would have it we got on an ant hill and the ants came near eating us up for our intrusion. Mr. Gray was very kind to me he took me in the boat with him and as I had no blankets we both slept together we amused ourselves by shooting at the birds, monkeys, crocodiles, &c. of which there are a great number and talking about Washington, he told me that he met father and yourself when he was there and he also made several flattering remarks about you but I will not repeat them for fear you will get vain and finally we arrived at Gorgona and were invited to stay with the surveyors who are surveying for the rail road across the Isthmus which invitation we accepted so you see I gained something by being with Mr. Grey as the rest of the company had to sleep out as they had been doing coming up the river.[52] I also gained coming up the river for we were invited to take breakfast at one of the surveyors camps which was half way up the river between Chagres and Gorgona and as they were so polite to us at Gorgona we stayed until the morning of the 1st of April the rest of the party went strai[gh]t on to this place and we started about 7 o'clock in the morning of the 1st and came though about 3 o'clock which was quite fast. I am glad I did not overtake Col. Weller at New Orleans for I understand that he gave only one horse for two of his party to come from Gorgona on so you see I would have had to walk half way which would not of been very pleasant as it is 14 miles and was very warm when they came over but Mr. Gray and myself had a little rain and found use for that mexican blanket which I got out of Frank's trunk at New Orleans they are fine to ride in they cover you all over. The road is very rough but not dangerous in dry weather the only unpleasant part of it is the dead horses along the side of the road which is up and down hill all the time in some places the road is cut through the top of a hill making a gutter about 2 ft. wide and 1 ft. deep which is rather dangerous to go up in wet weather for it is so slipery that the horses almost always fall as I was going up one of them the horse sliped and fell back but as my legs excuse me my broomsticks were long I stood up with one foot on each side of the gutter and let him slip from under me but took good care to hold on to the end of the rope which is very long tell Charlie he had better tie bricks to his feet and swing himself around the waist for long broomsticks are usefull.

Well after arriving in this place I eat my dinner well I did and went to look at the place and met a procession which was to represent Christ entering Jerusalem they marched through the gates of the city and had an image on a very pretty mule which Mr. Gray said I wanted to send to my sweetheart in Washington but I told him she did not live there was that not write [right]; yes, I would say.[53]

There are a great many churches here but no religion that is a pity ain't it but I will have to stop now the bells have commenced ringing and a person might as well try to jump up to heaven as do anything when they are going[.] there are about 1,000,000,000 to eternity and all are going at once thank he who made them they are cracked for if it were not for that we would be deaf. Adieu, old Sale, good night and when you write direct your letters G. Clinton Gardner United States Boundary Service San Diego, Cal.

Please hand this to father before you turn it over I can't stop to put stops nor can I stop to correct-stop for if I were to it would be stop stop all over.

Major Emory to Col. C. K. Gardner, Panama, April 27, 1849

My dear Sir,

I wrote you some weeks since that your son had arrived & that I had assigned him to my own party. Tonight I learn that the person to whom I confided this and other letters, lost them between here & Gorgona.

Thinking it will be a gratification to learn that your son is well & that his intelligence & deportment give good promise for the future, I take this occasion to do so. Vy. truly yrs, W. H. Emory

GCG to his sister Alida, Panama, May 8, 1849

My last letter dear Alida was to Sarah but I supposed it would be sent to you so I thought it would not be necessary to write to you by the same mail and as I have not too much time to spare I thought I would request that you would send this to mother after reading it for I may not have time to write a letter home again until I get to San Diego and no one can tell when that will be.[54]

I was quite delighted on the 7th when I had the pleasure of receiving so many letters from about [*sic*] home [GCG means "from home" and has inadvertently added the word *about*] my mail was quite large 7 letters and 3 news papers but not too large. I shall never complain of having too many & my only wish is that it may continue as large during my whole absence from home. There were 2 from sarah 1 from Mother with a note on it from father 1 from Emma 1 from yourself and two from the boys at school and I suppose the papers came from John. I am much obliged to him for them and hope he will continue to send them the only thing that I am afraid of is that a good beginning makes a bad ending and my mail will commence to decrease after a while but this is not telling you any thing about the place I live in so I will commence my quarters are on San Felipe st. just opposite to a church so if you ever come to Panama you will not see me for I am to start for San Diego next Saturday I believe or at least some of our party will. The city was in quite an excitement last Saturday night by the arrival of the steamer *Oregon*. I had no idea that the natives were so excitable before the street that the American Hotel is on was some thing like what I have heard broadway, N.Y. to be a perfect mass of humanity all the Americans were out to see the steamer and then the night after the Steamer *Panama* came in from around the Horn,[55] so you see there are now two in and the *California* is expected every day she has been delayed for want of coal and from the attempt of the engineer trying to run away, the steamer tickets which have been very low for some time are now selling for 5 and 600 dollars the steerage are from 2 to 300 dollars so you may imagine how much money is made by speculating.[56] The price of the tickets in the states are not more than 250 and 300 dollars for Cabin passage, but this is not given you a description of the city yet so I will commence again.

The city is quite large and is strongly fortified or rather was intended to be it is surrounded with a wall and ditch the houses are all built of brick with tile roofs the streets are very narrow they are not much wider than the pavement of Penn. Avenue and some of them are not so wide all the houses are built with porches all around them which makes a perfect shed all along the streets and when it rains you can walk around the square with out getting wet but if you try to cross the street you will get wet through, for it never rains here without it pours you would think that the whole heavens were opened on you. I think the rainy season has commenced for we have had a great many showers of that kind some times it will be quite clear and in less than 5 minutes, it will commence to rain in torrents.

There are 10 churches I believe here the largest is the Cathedral which faces the Grand Plaza [Plaza de Independencia] so also does the State house of this place. The San Francisco I believe is the next in size it faces on another plaza just north of our house then comes the San Domingo which is going to ruin or has gone to ruin for there are nothing but the walls left. The others I believe are all used every day and rather too much for me or rather the bells of them are (I wish as I heard a man say the other day I could only hear a bell again for I have not heard one for so long I have forgot how they sound) in a home.[57] There are not many real Spanish people here the most are the half Spanish and negro the real Spanish girls are quite beautiful just opposite our office are several very pretty ones you would laugh to see my note Book of the Barometric and Thermometrical observations of which Major Emory has given me charge under the head of Remarks which is for Remarks about the Weather I have made remarks about the young ladies as they come out to see me take them every morning. The handsomest sight is to go to church and see them all set[t]ling down on the floor the churches have no pews in them. The ladies go to church with a small negro or servant carrying a rug behind them for them to set [sit] on they have beautiful music at the churches and the inside of them are beautiful they must have been magnificent when first built. There is a convent here also but I can not find many nuns in it. I have tried every way to go in but have not succeeded as yet. There is also an old ruin or not exactly a ruin for it was never finished, of a Jesuit College all the Jesuits were driven out of the country before they had time to finish it.[58] I think it makes a beautiful sketch judging from a sketch made by one of the commission. I have sketched the city from the bastion which makes a very pretty scene. I would like to make some more but I don't think I will have time if we start on Saturday for now I have to work every day at the office from breakfast to dinner and often after dinner so you see my time is limited. I made a visit to the Island the other day to see the beautiful spot that is described as a perfect paradise, it is called Taboga a greater part of Mr. Grey's party have been stationed there, it is as it is described a very beautiful spot. I was sorry I could not take a sketch of it but I was sent over on business and had not time it is about 12 miles and we went over there in 2½ hours but was just 8 hours coming back we got out in 3 storms all of which we came through safe and it was raining in torrents about 20 yds in front of us several times but we did not get a drop of it. At the Island are several baths which are natural the water comes down from the mountains and passes over the rock in such a manner as to make baths about 7 ft deep and if you want to take a shower bath you can stand under the spray which falls rocks around.

I have not told you how I live yet so don't get tired. A young fellow by the name of [Edward] Ingraham [and I] have one room to our selves with cots in them and I have a mosquito bar over mine and it looks quite nice the room is not plastered all around like our houses they only plaster the walls and leave the ceiling with nothing but whitewash[.] windows are scarce our room has four but then you know two great gentlemen live it as for window glass such a thing is unknown.

The greatest curiosity here is a wagon brought from the states by an American and an American horse about 16 hands high all the natives come out to see it pass even the Gov. himself stopped to look at it. The water we drink is brought from a spring out side of the city on the back of mules and every other thing of the kind is carried on mules a wagon was unknown untill the Americans commenced coming here. There are a great many thiefs here now since we have been here, the *Crescent City* mail agent was killed and the mail taken not long ago and so goes the people the natives are very honest.[59] Well I must stop now so adieu to Panama.

How is Miss Ginnie and all the other young ladies? is that candy man living yet or have any of you young ladies made a mistake and eaten him for a s[t]ick of candy he is not much larger than one so be carefull. Give love to Miss Ginnie, and my best respects Miss Effie and all others who ask after me and believe

me your affectionate

brother Clint

GCG to his father, mother, John, Jacob, Sarah, Charlie, and Aunty, Panama, May 13, 1849

Private

Dear Father,

I thought as this was sunday I would take a sheet of calculation paper and write every one at home a short letter and send it by Mr. C[harles] L. Weller, as he was so kind as to offer to take any letters home for me; he will leave here in the next steamer bearer of dispatches to the government, and *also to my family and friends.*[60] I wrote several letters by the Steamer *Falcon* but I have heard that the mail did not get to Chagres in time. If so of course you have not received any of them. I was quite delighted on the 7th to receive such a large mail and hope it will continue to be as large nothing could have given me more pleasure; although I have not yet learnt to appreciate so many letters, but still they were very acceptable, I can hardly realize that I am so far from my native home, when I take my evening walk on the N. E. Bastion I imagine it to be Penn. Avenue and as my imagination is very good my time passes very pleasant. I am busy all the day at the office and very often at night so I do not have much time to run out in the sun.[61] My health is very good and I have not been sick but once and that was for a very short time. I had the fever one evening and that is all I have seen of it. I had the tooth ache a little while also on account of one of the plugs having come out, with which I expect to have a great deal of trouble, but I can asure you if it makes too much fuss it will have to walk also. When Mr. [Charles] Weller returns I think that it would be better not to ask him to bring any thing out for me, but if he should offer to do it, it would be another thing; the reason of my telling you this is because he is rather a close man from what I have seen and will have to pay so much a pound for his baggage across the Isthmus and of course the more you give him for me, the heavier will be his baggage and as mother said something about Frank giving her some advice about clothes &c. I thought she might try to send some by him. This I think will be the last time you will hear from me while I stay in Panama there are two steamers now in the bay and others are expected every day a great many sailing vessels have started for San Francisco and soon we will be off.

Please hand this over to Mother now and let her read my short note to her, but I beg of you do not let any of the others grab for from what Sarah says I expect it will be torn up before it is read, so I need not take any trouble to be very minute. Believe me, your affectionate son, Clint.

Dear Mother—I feel myself highly honoured to receive such a long letter from you as you so seldom write. I can imagine how you were seated by the centre table in the parlour on the first day of April writing to him, who was travelling from Gorgona to this, the first city built in the new world,[62] in company with Mr. Grey and a young man from Penn., who is going out to San Francisco to live I believe, his name is Mr. [C. W.] Hart, he is now rooming with me and has had the fever but is almost well a young man by the name of [Edward] Ingraham, who belongs to Major Emory's party in the same position of myself,[63] is also in the same room with me he is a very fine fellow and a nephew of Major [Lawrence P.] Graham he is to go with Lt. [Amiel Weeks] Whipple and I with Capt. [Edmund L. F.] Hardcastle as near as I can understand. I think I would have been better pleased with Mr. Grey, although every thing is for the best so I have nothing to say. I think it will be hard running to get up to San Diego in the time the treaty mentions.[64] I have just heard that 10 of the commission are to go in the *Panama*, which I believe will start first 10 in the *Oregon* and the rest in the *California* which has not arrived here yet. I believe the *Senator* will be here soon which I think is a government steamer,[65] if so we will more than probable go in her, it is said that the *Panama* starts on Tuesday and the *Oregon* will not go until the *California* comes for there must be some steamer here to take the next mail from the states which will get here about the middle of next month. (There are I believe 40 in the Commission, that is there are 40 now in Panama but I expect there will have to be some more added to it as they do not think there are sufficient labour[er]s. I suppose Col. Weller will try to get men on [the] *California* but he will miss it, I think.)[66] The rainy season I think has commenced for it rains a little now every day and some times it pours, you would think the streets were young rivers in a hard rain if you should see them the water has been 3 or 4 inches deep while running down the streets, I do not mean in the gutter for they never have such things[.] [H]ere the whole street is the gutter. Now I will request of you to hand this to John so as he may read the short note which I am going to write to him and believe your affectionate Son Clint.

Dear John,—In my mail which I received on the 7th of this month, I received 2 newspaper in one bundle and 1 in an other for which I suppose I am indebted to you for so you will please except [accept] my sincere thanks for them. I was quite delighted to receive them for they were the first papers I had seen from Washington, although, I had seen some of later dates from other citys yet they did not tell me sufficient of what was going on in my native place. I was quite delighted to hear that William and Jim Barke[r][67] had gone to the gold region but I am afraid that they will wish themselves back again before long, for from what I hear of the over land route it is not a very pleasant journey to take but then you must not believe all you hear, for from the paper a person would judge that the journey across the Isthmus was a very dangerous and unpleasant one but from what I have seen of it I think it is very pleasant and I am sure you will not often find more beautiful scenery than there is on the Chagres River. The only objection I had to it was that there was not variety enough the scenes were too much alike and as for the road from Gorgona here I think it is not at all dangerous in dry weather but in wet weather it is I should judge a little slippery in some places but with care there is no danger the only objection that can be raised to it is the expence, although I think the natives earn what they get, for they work hard[.] [T]he price from Chagres to Gorgona is only 8 dollars and the natives have to pole you up in canoes over rapids in some places, which are quite strong and from Gorgona to this place you have to pay 10 dollars for a horse, or something they call one for yourself to ride upon then you have to pay a good deal for your baggage take it all together and I think it is a very pleasant trip but rather expensive. I will have to ask you to hand this over to your most affectionate brother the Doctor, by doing which you will much oblige your younger brother Clint.

Dear Doc[68]—You being the next in order I will try and write you a few lines, so as you will not have an excuse for not writing to me, but don't get angry if it should be very few indeed. As you take the most interest in the drug business I will try and give you a little idea of the drug stores in this great city they are 4 in number I believe and are very small the room in which they are is not more than 8 x 10 feet. The shelves are all around the room with a little counter in the middle about 4 feet in length which has a small railing around it. I believe the names on the bottles are the same as they are in the States the shelves are supported by pillars which are painted all colours and take them all together they are quite pretty. And now as that finishes the drug business you will please hand this to your affectionate sister Sarah and oblige your little bub Clint.

Well Dear Sarah how do you do. I don't think it has been long since I wrote a long letter to you but since I am to thank you for two long and interesting letters which I received on the 7th of this month, I will continue and try to give you a little description of Panama, which I have not done in any of the preceding letters so here goes it but don't be in a hurry. I have got to go and compare cronometers as Major Emory has sent for me and will go it as soon as I come back. Here I am again all ready to tell you about this great city which is as near as I can learn, about 150 years old. Old Panama which is about 9 miles N W of this is the oldest city of the new world, it was built but a short time after the discovery of this continent, and is now in ruins, the ground as near as I can find out is some distance above the streets the City was destroyed by an earth quake and was afterwards built here as this was a better sea port.[69] This place is surrounded by an immense wall and ditch which is principally built of stone the houses are very close together and when it rains the streets are like a perfect pond of water we amuse ourselves here in the evening by walking around and looking at the ruins of which there are a great many an[d] also listening to the bells which are continually going; "O that I could hear as I heard a man say the other day, a bell ring once more, so as to remind me there was such a thing for I have forgotten how they sound,"[70] on the N E part of the wall is quite a long prominade about 20 ft wide and every evening it is filled with the Americans who go out to enjoy the sea breeze, which I can assure you is very refreshing after a hard days work. There are a great many large buildings such as the State house[,] the Jesuit College, which by the by never was finished all the Jesuits having been driven out of the country before they had time to do it, & several churches of which there are 8 or 10, the largest is the Cathedral which faces on the grand plaza then I think the San Domingo comes next but it has gone all to ruin, the walls are still standing and within them are some arches which are most beautiful, the vines have grown all over them and one of them is about 40 ft in length with not more than 3 ft rise. The San Francisco is the next in size it is also a beautiful church, but I think not so well finished inside as some of the others; but it is strange to find that this race of people have built such a large city as this is, (although it does not cover much ground) without knowing the use of a wagon or cart, I expect all the paving stones were brought here on the backs of men and mules for I know they bring water, bricks, hides, &c. now on the backs of these animals; some of the Yankees have made and brought one or two wagons from the states and you cant imagine the excitement which they created at first, every body was running to see the wagon, it was followed by a string of boys just the same as the boys would follow soldiers in the states and after all it was nothing more than a water cart with large solid mahogany wheels, bound with iron and contained 18 kegs of water which they sell for a dime a piece, drawn by two oxen, a great thing to create an excitement was it not?[71] but now they have two or three spoke wheeled wagons in town, which some American has with him. I expect the mules will thank the Americans for coming across, since they have introduced that which will save them a great deal of labor after awhile. Well I think now it is time for you to hand this over to Charlie so as he may read that which is for him and wait until another time and I will give you a little more, so as to keep you always in mind of your little bub Clint.

Now for it Charlie hello how are you old fellow, I admire your spunk in trying to get in the Coast Survey.[72] I hope you have succeeded before this I should not mind taking a place there myself when I return to the States, that is if I ever do, for I am thinking about filling my pockets full of rocks and going across the Pacific, what do you think of it? But I am afraid that if I ever get at the other end of the line after starting from San Diego, I will be tempted to go and see my sweetheart dont you think so? the attraction will be great I can assure you, but that is going too far ahead dont say any thing more about it what I am thinking about now is going to San Diego. I have almost become a citizen of this place already and I am afraid that if I stay here much longer I will become one sure enough for all the Spanish Girls have fallen in love with Ingraham and myself and they say they wont go to California so how am I to leave them? one of them sent me up an immense stone basin for me to wash in in the mornings, what do you think of that? and when I go up to the office to record the Thermometers, barometer, Hydrometer &c. they all come out to see me on the opposite porch, and if boards were not so costly I would buy one and make a bridge from one porch to the other, but the carpenter of the Commission says that they charge a dollar to look at one and the lord only knows how much to take it away, so I had to give up the idea so also will I have to give up the idea of writing you any more, so pass it over to your dear Aunty and you will oblige your hermano Clink as Pussy says.

Now dear Aunty[73] I know you are laughing, so you will have to stop for you cant read & laugh too, for if you make such faces as you used to do, you will tear the paper, which is strong I know, but it cannot stand such faces. What do you think? I went to church the other day to see the young ladies and they were all seated on the floor (for they have no pews) saying their prayers in Spanish which I could not understand but as they had some delightfull music I could hear so I stayed there until one of them came out and she caught a glimpse of my cross so she stoped. Ingraham who was with me said that I was a priest, so she looked at me right hard and then took off my hat and pulled my head down (so you see I am taller) to look if I had a bald spot on my head (so I found out that all the priests have part of their heads shaved) and said that I was no priest was that not a shame they would not own me as one of their congregation, which I wished them to do very much, for I wanted to get permission to go in the convent to see the beautiful nuns, but since then I have seen a funeral of a little child which gave me enough of their religion, they carried it out to be buried, with a Guitar 2 fiddles 1 Tanabour [tambour?] and 1 flute playing all the dances they could find the coffin was beautifully made of bright tin with brass tacks in it, they are not of the same shape as ours. They are of this shape some

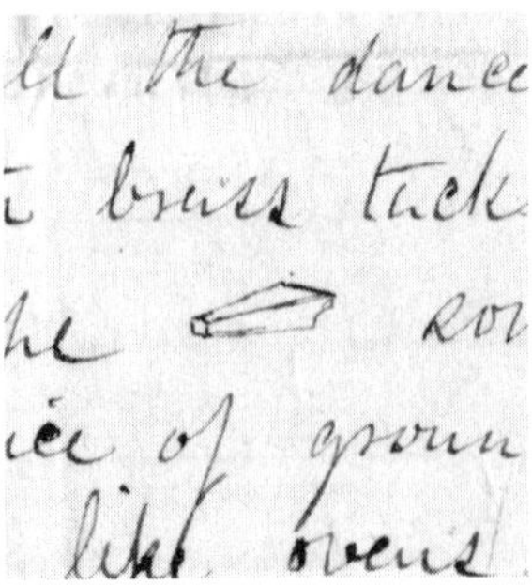

times with a rounding top, their grave yard is a small piece of ground surrounded with a high wall and all along in the wall are holes, like ovens, in which they stick the rich persons and plaster them up for a year or two, then they take them out and put the bones in one corner and the hole is ready for another, but the poor people are stuck in the ground or rather those, that they do not consider christians and when they saw the Americans bury their dead in the ground they said "poor Americans they are as good Christians as we[,] they [*sic*] ought to be put in the wall". Such are the people of this world and as you are the last but not the least I will have to ask of you a favour which is to Give my best respects to all my friends Mr. Nonsse, Mrs. Whitwell &c. then to Old Susan tell her that she had better come out here to live and be a washwoman, she can get 15 cents or 18¾ (which it passes for here) for washing one shirt and 10 cents or 12½ for socks &c. which require nothing but washing. I am afraid that if I stay here much longer, I will be flat down, but I can't get away, the Girls will hold fast to me. Tell Alcide when he comes up to tell Custis [Lee],

Castleman[74] and himself that I have received their letters and will answer them when I get to San Diego, or sooner if I get a chance. Tell all I am in haste [as] one of my sweethearts want to kiss me sweet child. Believe me as ever your affectionate nephew Clint. Good by old woman.

[P.S.]

(There has been the greatest quantity of gold and silver carried over the Isthmus which has been brought here by the ships from California and Peru the land which I intend going to if my sweetheart don't write me to come home.[75] The gold that was carried over the Isthmus the other day was stolen the men were killed &c. Some persons say that it belonged to the United States and that it was from California but I have not heard positively. believe as much as you chose [choose] and let the other alone.)[76]

GCG to his father, Panama, ca. May 20, 1849[77]

I expected, dear father, that the letter dated the 13th would be the last that you would receive from me while at this place, but since then I thought that this would be a good chance to send you that fifty dollars, which I hoped to do when I was in New Orleans, for which you will find enclosed a draft if such it may be called perhaps it would have been one if it had have been made out by a business man. I would have sent you more but as yet I can not spare any more, for my expenses here, such as Washing, &c., takes a great deal my wash bill comes to one dollar a week and some times more. I have often been tempted to wash my own clothes, but the water and soap will cost almost as much as the washing, so I thought there would not be much saved. I have now $25. which I have lent to Lt. Whipple and taken his note for it payable on demand, which I expect to keep until I get to San Diego. I have also 4 or 5 dollars in pocket. I have sold my [gun] kit without the case for $17, which was the cost of it with the case, the reason of my selling it was, because almost every person said that it was a very unsafe one and the greater number of those belonging to the commission have Colts, which are considered the best. I asked $20 for it but came as near selling it for that as the man did his pig, of which I suppose you have heard tell, so I thought it best to make something by it, so I kept the case which I intend converting into a paint box and the box which is now my paint box back to its original purpose which was a dressing case, what do you think of it?

I will try and send you some money by every safe opportunity. We are now living well since Major Emory has left he started in the Steamer *Panama* last thursday night[78] with the Commissioner, Mr. Gray and several of the commission.[79] Lt. Whipple has been left in charge of the remainder of Major Emory's party and has brought [us?—small part of page torn] to his Quarters to live and mess[80] with him, which I can assure you is much better than eating at the American Hotel which we had to do, but now soon we will leave in the steamer *Oregon*, which is to start next Wednesday evening as I heard.[81]

I told Mother that only ten were to go in each of the steamers that were in port and the others were to wait for the *California*, but I believe more than ten have gone in the *Panama* and I have heard that the remainder will leave in *Oregon* & will not wait for the mail from the States. I would have no objections to staying here now for some time, since we have such a fine mess, if it were not for the loss of my clothes. I have lost several articles of clothing but not as many as of the others. Mr. Gray lost 5 or 6 shirts and I think if [I] get off without losing any more, I will do well. I mentioned in my last letter that I was to be under Capt. [Hardcastle—small part of page torn], but I was mistaken. I believe both Ingraham and my[self] will be with Lt. Whipple, whom I like very much.[82]

I think if you could get me a Spanish and English Dictionary and send it to me as a pub. Doc. or printed matter, if it does not cost too much, it would be of great use to me for I am without one and they are selling for from 5 to 10 dollars my Ollendorff has [been] of great use to me.[83] I do not know what I could have done without it.

I have just been told that only 10 went in the *Panama* and that ten are to go in the *Oregon*, but no one knows they are all suppositions; but I think that by tomorrow it will be decided who are to go and who are to wait for the *California*. I will keep this [letter] open until I hear.

I forgot to tell you in what great demand I have been, about two weeks ago there was a party of 15 that started for San Francisco in a large metallic life boat and they had no coast Chart, they came to me to copy one for them and as luck would have it, before I could get the paper they found a printed one, I intended charging $25 which I could have got very easily.[84] Also the other day Lt. Whipple asked me to send a letter to Major Emory, who had gone on board of the Steamer, so I started down the street and meet Lt. [George Horatio] Derby Top. Eng. who was just starting out for the steamer.[85] I did not know him, but as I thought he knew Lt. Whipple I asked him to take it, which he did very kindly and as we parted he very politely bid me good bye and said he hoped he would see me in San Diego and if so he would be very glad to have me join his party. I told him that I was with Major Emory he said he knew that but in case I should leave him, which was as much as to say if I got turned out, he would take me.

There has been a great talk here, about Col. Weller of which I suppose you have heard, as I heard the other day that it had been mentioned in some of the papers in the States. I suppose it must have been before this as I heard that all about it before I left New Orleans. I will tell you about [it] some other time if you wish. I will now bid you adieu for a while.[86] your affectionate son

Clint Gardner

P.S. I have heard that I am to go in *Oregon* but then I heard again that I was not to go so I do not know any thing about it. You can ask Mr. [Charles] Weller if you desire to know.

G. C. G.

GCG to his mother, Panama, May 21, 1849

I send, dear Mother, enclosed in this short note a box which contains a specimen of Panama jewels, which I had made to order expressly to send to Washington they do not keep such things for sale here, but they will make any thing of the kind for you, that you may order; so Lt. Whipple and myself drew the plan on paper and I managed with the assistants [assistance] of my Ollendorff, to make them understand me. I believe the pearls and Gold are native but where the emerald came from I do not know. I would like you to get the Doctor to tell me if it is a true emerald, for I got it for one.

I intended making these a present to you, but knowing that you have such a fine set of pearls, I don't believe you would value them. Then I thought of sending them to Sarah, but they look so old fashion that I don't think she would value them and the same way with Alida. I think they are too old fashioned for her. Then I thought they would exactly suit Aunty but then she does not go out enough to show them, so I came to the conclusion after a long consideration on the subject that I would send them to you and let you arrange straws for you all to draw for them, or draw lots in any other way that you all think best, but I beg of you do not scrabble for them. Who ever wins will please accept them as a gift from little Clint.

P.S. I mean no others to draw for them than those mentioned.

G. C. G.

GCG to his sister Sarah, Panama, June 12, 13, and 14, 1849

As I have nothing to do at present, dear Sarah, I thought it would be well for me to commence a letter to you and finish it some time before I leave this beautiful spot there are now five of us here waiting for the Steamer *California* to sail, she is at the island of Taboga which is about nine miles south of this place, taking in water and getting some of her berths &c. replaced that were taken out to burn for coal, of which they had run out.[87] Mrs. Gen. [Persifor F.] Smith and Mrs. Major [Edmund Augustus] Ogden came down from San Francisco in her and were here for several days; and as Mr. [Richard] Rust who is Major Emory's private secretary was acquainted with Mrs. Smith he went to call on her and I sent my *card* by him, to which she sent me an answer, saying that she would be glad to see me at such a time, so I did myself the honour of calling, and as soon as I entered the room she got up from her seat and said she would know me in a moment by my resemblance to my sister, so when you write to Emma you can tell her what a great compliment was paid her, for I can asure you any one that I resemble, or that resembles me *must* be very handsome.[88] Perhaps you would like to know who Mr. Rust is, he is from New Orleans if you ask Father I think he can tell you, he was the editor of the democratic organ during the time Mr. Mouton was running for Governor.

I said that the steamer was at the island of Taboga, but do not remember of telling you what kind of place that was, but perhaps you have seen before this a letter in the New Orleans delta which was written by Capt. Tobin, who describes Taboga as a perfect garden of Eden, it furnishes all the fruits grain &c. for this market, and has a splendid spring of water on the top of its mountains, which furnishes all the ships that go out from this port, and not only that but runs down the sides of the mountain over rocks, which form basins and are delightful to bathe in, the one that I took a bath in is about six feet deep and six feet square. I did not go on top because I had not time, some of the party that lived over there told me that near the top of the mountain was another basin just as large, which the beauties of Taboga used for themselves. Just at the bottom of the mountain is a little village which is built of read, like the huts in Chagres & Gorgona it was burnt down some years ago and it is strange to me that it has not been burnt since for they have no chimney in any of the houses but build their fires in the middle of the house, and let the smoke get out the best way it can, some cook in the streets and even in this large city you will see the people building their fires in the streets to cook with, there is not a house in Panama with a fire place in and you very seldom see a stove here if you do it was brought here by some American.

The steamer *Crescent City* arrived at Chagres some days ago and I got a few Boston papers to read, by which I find the Cholera is raging all over the United States. I did not see any accounts of it being in Washington, so I suppose you are free from it as yet, it has been raging here for some days past and is still as bad as ever, but more among the natives than the Americans, the native doctors do not try to treat it, as soon as a man is taken with it every body runs from him, so I have been told, and after he dies you can hardly get the priests to go through their form of religion over him, they are so much afraid of taking it themselves. There is one American doctor here now who I believe has been very successful in his treatment, having cured several, and thinks if it is taken in time any case of it may be cured. The other day the wife of one of the workmen on the *California* was taken with the Cholera and the husband was sent for, but she died before he had been here long, and not long after that before he could return to his work again, he was taken down with it, and died in a very short time; they left 4 children, and one of them is so ill that it is not expected it will live, some of the Americans are raising a subscription to take the remaining ones back to their friends in the states, what a loss to them it must be, little did they think when they left home their children were to be left helpless in a distant land, but so it is and as I have been told, "every thing is for the best". The Cholera I believe was brought here from New Orleans by a sailing vessel, first at Chagres, then Gorgona and last but not least the city of Panama; some of the passengers that came down in the *California* after making a fortune were cut down by this disease they returning home with all the gold they wished, to enjoy it with their friends died when about half way home. Mrs. Gen. Smith was afraid of the Cholera when she started from here, so she went by the way of Crucis [Cruces].

I have not heard whether she got off safe or not, but I suppose she did as I have not heard of the Cholera being at Crucis.[89]

We are expecting every day the arrival of the steamer *Falcon* at Chagres by which I expect a great many letters, but I have heard that the San Diego mail will not be opened here, as there is no one here to take the responsibility. Col. Weller was the one that had it opened before; it will be rather bad if we have to travel up on the same steamer with our letters without being able to get them until we get to San Diego. We expect to sail from here about the 20th but if the steamer *Falcon* don't make a little more haste, I think we will have to wait until the 1st of next month, as the steamer must wait for the mail.

There appears to be more religion here than any thing else, a person to come here for a little while would take this to be a holy city, but after knowing the habits of the priests a little and looking in at the gambling houses, you would find that their religion does not keep them from going to such places, a great many of the priests take off their robes and shovel hats, as a friend of mine calls them, and put on citizens dress to go to such places, and in order to hide the ball [bald] spot on their heads they tie their hair up in a knot over it. In such places you will see all the high officers of the government amusing themselves.

I forgot to mention that the natives have a great idea of the Cholera, they think that since the Americans brought it here that it is dangerous to have an American in their houses. I think if the *Falcon* brings many passengers they will find a little trouble in getting rooms, they appear to be much afraid of them, not long ago the American doctor was telling me what they thought of him, he says he was called in to see one of them, & the priests had rubed ointment all over him which they called holy and he ordered them to rub it all off, for they had stoped up all the pours in the skin, they called him a devil for abusing their holy ointment, but he cured the man and ever since then, his name has been going around among them as some great person some are afraid of him, and others are willing to pay him any price if he will take care of them, the usual fees for a doctor are 4 ten cent pieces or 50 cents, but he charges them $5 a visit and $3 for his medicine, tell the Doctor [Jacob Gardner] that he can make money out here, but I think he would be about the first to be sick, for I have been neither well nor sick ever since I have been here, and I tell you what I shall be glad to get away for I am sick of this place. I have nothing to do now except to take the tide. I ought to go out twice a day but since the Cholera has been here Dr. [Charles Christopher] Parry, who has been left in command of our party, told me he thought it would be best to keep out of the sun as much as possible. Dr. Parry is one of Major Emory's computers,[90] he has studied medicine and took his diploma, but does not practice as he does not like it, he being next in rank when Lt. Whipple left has the command.[91] I think I will stop for today and finish some other time so, adios para [el] dia [goodbye for today].

June 13th. In looking over my little Journal that I have been trying to keep I find several little things that might be of some interest. June 7th was ascension day as it is called here and about 9 o'clock the church bells commenced ringing, guns were fired off, the soldiers were out in full dress, and so of course I was out too to see what was going on. The priests came out of the Cathedral with candles in their hands four of whom held a canopy over the head of the bishop who marched in the procession with a large cross made of gold. They marched around the city stoping about every square to burn incense sing and pray at the different alter that were erected for the occasion, every body had their best bed quilts, shawls, petticoats, &c. they had I suppose in honour of the bishop, who does not honour them with his presence very often as I have never seen him before, the soldiers were in the procession also. I tell you what they did look proud, just imagine about 100 negros drest up in fine style and every now and then looking at their dress but whenever they looked low enough to see their feet their pride fell, as they were without shoes, some of them had shoes and all of those it appeared had white gaters. I thought at first this was to drive away the Cholera and it was afterwards that I found it was ascension day;[92] the reason of my thinking so was because they had planted 2 guns on each side of one of the church doors to fire off for the purpose of driving away the Cholera and they had also been burning brush &c. so I thought the procession would do as well, the reason of this proceeding is because it was directed to be done by a meeting of some of the first people of the city they also ordered the weeds to be cut down that were growing about the city.

I believe I will tell you now what kind of a mess we have since Lt. Whipple left. I think I told some of you how nice we lived when we were with him but since then instead of rising in the world we had to come down. There were 13 of the commission left here after the *Oregon* left and we were all put in the same house to mess at the same table according to Col. Weller's orders which makes quite a democratic mess. I suppose it is after the manner in which they live in Ohio. Mr. [Joseph] *Kaufman,* the private secretary of Mr. Gray was made quartermaster for the occasion. Dr. *Parry* who was left in command of Major Emory's party: Mr. [Richard] *Rust* the private secretary of Major Emory[,] Mr. *Ingraham*, who is in the same position as myself G. Clinton Gardner son of Col. Gardner late post Master at Washington[,] Mr. [C. W.] *Hart* a young man from Tenn. who came out with us in the steamer *Falcon* and after living here until the *Oregon* left he lost his servant boy (by the bad treatment of a native doctor) whom he had brought out with him he then came to live with us he intends going to San Francisco to live[,] *Moses* [Conner] the carpenter[,] *Frank Stone, George Barnad,* [William D.] *Miller* all employee[s] of Mr. Gray[,] & *Frank [Francis] Holly* [Holley] employee of Major Emory compose the mess.[93] I think Col. Weller might have arranged it a little different so as not to put us all together. I am in hopes soon to get out on the line so as every person will know their places for I am tired of living in such a way as Mr. Whittney says "I am a democrat politically but not socially." I forgot to mention that as I walked out to take a bath in sea this morning I found the city gates locked and upon inquiring the reason I was told it was to keep the cholera out such is their idea of this disease they have stoped firing guns now as they have killed that part of the monster what they will get at next I can not tell all the "upper ten" that can go away from the city have gone some to the island of Taboga and some in country I think those that go to the island of Taboga will fare the best as they say there is not a single case of sickness there the Americans are flocking there in great numbers and have told terrible tales about this place and the cholera some that 20 or 30 die here in a day and others more but I do not know where they find so many cases and I know not more than ¾ die out of the whole number that are taken with it and the other day some went over to Taboga and said that one of us were dead but I have not as yet missed any one from the table.

It is reported now that the steamer will not leave here until the 30th of this month so it is a hard matter to tell when it will go but the day before [it] leaves I will finish this for the mail. I am afraid I will tire you out if I write much more so adios, tango que marcharme [*adios, tengo que marcharme*—goodbye, I have to go].

June 14th. Again I will commence I find that this is the only way I can pass my time agreeable we play cards, chess, &c. to pass off time but then one soon get tired of them and so I return to writing. I expect you will not take the trouble to read this as it is so long but it makes no difference to me so you dont tell me you did not read it. The Cholera is still here but I think it is on the decrease for I have not seen any funerals pass since I last wrote and have not heard as much talk about it. I forgot to mention that at that meeting they ordered the gates of the city to be shut so as to keep the cholera from coming in. The way I found it out was yesterday I started out to take a bath in the sea and had to get through a break in the wall but as I was coming back I saw a little boy creep under the gate but I dont believe the Cholera will think of coming in in that way he was one of the smart fellows and I expect the Cholera is a big fellow if so he cant get under the gate.

I have just found an opportunity to send this in haste to go in the *Falcon* with the last letter. So give my love to all Mother, Father, &c.

your affectionate brother

Clint

I have not time to read it over so you must overlook all mistakes.

At the end of this letter Clint Gardner's mother appended the following note to Alida, who was then away at boarding school, at the Patapsco Female Institute.

> *Dear Alida we have just received this from Clinton and as Sarah is on her way to New Port with Mrs. Butterworth*[94] *I send it for you to* read first *you must send it back to me that I may enclose it to her. Mrs. Margaret French has a Son two days old. James McLean*[95] *was married. Julia Hobbie*[96] *went on Wednesday to school she sent a great deal of love to you we are all well. No more news.*
>
> *Your affectionate mother,*
>
> *A. E. Gardner*

GCG to his brother Charles, Panama, June 22, 1849

Dear Charlie

I did not intend writing again until after my arrival at San Diego, but as I have a few moments yet, I thought I would say a few words by the way of leaving Panama. I am quite delighted at the idea of starting from this place, and before long we will be on the wide ocean tomorrow is the day appointed for the departure we are all engaged in firing our truck[97] we have been here so long that it is as much trouble to collect our things as it was when I left home. I have increased my stock a little as well as lost some things.

The cholera it is hard to say whether it is on the increase or decrease but I have not heard much about it of late, so I suppose it is going down. I suppose you remember in some of my letters I mentioned the name of Mr. [C. W.] Hart, a young man who had gained the friendship of every one he became acquainted with, he was but 21 years of age, well educated and a large and fine looking fellow, but alas he is no more; he died on the island of Taboga while on a visit to that place, of the cholera. He left here the evening before in good health and remarked that he was the happiest man in the world and arrived at Taboga some time in the night, the next morning he took his breakfast in good health but before an hour had past he was taken sick and in about 12 hours died. I never was more surprised in my life than when I heard of his death, he has no relations but sad to say he was engaged to be married, and has often told me that when he was 30 years old he was going to be married. He had the young lady's likeness with him. We have written to all of those that we could find directions to amongst his papers and fixed up his business as well as possible.

There has been given to me a lower cabin ticket in the steamer with the others of the commission, all the labourers have a steerage passage and I believe that is the only distinction between us. There is a young man here from New Orleans with whom I became acquainted soon after arriving here he came here under good circumstances but lost from his companion, until he had just enough to take him to San Francisco and the other night he had that stolen from him so he is without any money so I concluded to go in the steerage and let him have one hundred dollars of it which he is to give me his note I suppose father will say I am foolish, but I have not time to explain any thing as yet but when I get to San Diego I will tell him all about it then he will think different.

I will write from San Diego next give my love to all in great haste

your affectionate brother Clint

Tomorrow I expect to be off. Glory [?]

This young man from New Orleans has not determined whether to take my offer yet. G. C. G.

1. GCG to John McCalla Jr., New Orleans, March 16, 1849.
2. Frederick Emory later acted as secretary of state in the "cabinet" of the notorious filibuster William Walker during his expedition to seize the Mexican states of Baja California and Sonora in 1853. Walker was arrested, entered a guilty plea, and was fined, but he was excused from payment on the ground of poverty (May, 2002, 41, 140, 162). Emory (1857–59, 1:6) identified Robert as his servant and a slave.
3. James Buchanan, instructions to J. B. Weller, Washington, February 13, 1849, in SED 34, 4.
4. Andrew Gray notes the problem of the iced-over Ohio when writing to Secretary of State John Clayton on March 7, 1849 (SED 34, 24).
5. Goetzmann (1991, 159) mistakenly says that Weller left for Panama by February 28 on the *Panama* and that William Emory was with him on the same vessel. William Emory shipped out of New York and met Weller in Panama (C. L. Weller, "Col. Weller's Delay Explained," *New York Daily Tribune,* July 12, 1849).
6. GCG to his father, New Orleans, March 19, 1849.
7. Gray to John Clayton, Sec. of State, March 7, 1849, in SED 34, 24, and 6–7.
8. GCG to his father, New Orleans, March 19, 1849.
9. Weller reached Panama City on March 17, and Emory arrived on March 18 (Emory to Sec. of State James Buchanan, March 25, 1849, copy in Emory's letterbook, box 1, f. 2, EP). Emory sailed from New York to Chagres on the *Northerner* (Norris, Milligan, and Faulk, 1998, 68–71).
10. Ware, 1972, 49; Kemble, 1972, 147–48.
11. Weller, explaining his predicament, to the secretary of state, John Clayton, Panama, March 20, 1849, in SED. 34, 24. Several writers (including Goetzmann, 1991, 159) have accused Weller of making no provisions in advance, but being stranded in Panama does not appear to have resulted from Weller's poor planning.
12. Emory to J. J. Abert, Panama, May 10, 1849, copy in Emory's letterbook, box 1, f. 2, EP.
13. GCG to his father, mother, John, Jacob, Sarah, Charlie, and Aunty, Panama, May 13, 1849.
14. Rosenberg, 1962, 114–15; GCG to his sister Sarah, Panama, June 12, 1849.
15. Emory (1857, 2) notes the arrival of cholera toward the end of April with the coming of the rainy season.
16. In a letter of May 8, 1849, to his sister Alida, GCG notes that the *Panama* arrived at Panama City on the previous Sunday, which would have been May 6. For the *Panama* see Kemble (1972, 242) and Wiltsee (1938, 17–18).
17. See Emory (1857–59, 1:3) regarding the equipment.
18. See Whipple to Weller and others, Panama, May 14–23, 1849, in SED 34, 40–43.
19. GCG identifies the group of thirteen in his letter to his sister Sarah, Panama, June 12, 1849.
20. In addition to the copy in GCG's papers, a copy of the letter also exists in Emory's letterbook, box 1, f. 2, EP.

21. Webb, a veteran of the U.S.-Mexican War, commanded a party of gold seekers bound for California that included John W. Audubon, who published an account of their eventful journey from New York through Texas, Mexico, and Arizona to the goldfields, including a description of the circumstances that led Webb to abandon the party (Audubon, 1852, 14).
22. C. Toledano was a commercial merchant at 97 Magazine Street, according to the 1849 New Orleans city directory. Reference courtesy of The Historic New Orleans Collection, Williams Research Center.
23. The reference is to his sister Sarah.
24. Emma Mouton, the oldest of GCG's three sisters, was married to a Louisiana planter and politician. See the introduction.
25. GCG's brother-in-law, Alexandre Mouton, who had married Emma Gardner. See the introduction.
26. GCG was not stinting himself in his choice of lodging. The St. Charles Hotel, located in the 200 block of St. Charles Avenue just across Canal Street from the French Quarter, was described by Lady Emmeline Stuart-Wortley in 1851 as looking "a little like St. Peter's at a distance as it is surmounted by an immense dome" (122). Gen. Benjamin Franklin Butler used it as the headquarters of his occupation force in 1862, at which time the correspondent for the *New York Times* called it "unquestionably the finest model and most imposing hotel structure on the continent" (June 20, 1862).
27. Stephany appears to have been a household servant, and perhaps Susan and Colbert were as well. See the introduction.
28. Two contemporaries who sailed with Weller reported that they left New Orleans on March 2. See H.H.R. [Hiram H. Robinson], Panama, March 18, 1849, in the *Cincinnati Enquirer*, April 12, 1849 (reference courtesy of Charles Hughes), and Effinger, who provides a vivid account (Fireman, 2004, 13). Goetzmann (1991, 159) mistakenly says Weller left by February 28.
29. *Cohen's New Orleans and Lafayette Directory . . . for 1850* (New Orleans: Office of the Delta, 1849), 52, lists three different men named Durrive: Edward, Frederick, and J., an officer at the customhouse (reference courtesy of The Historic New Orleans Collection, Williams Research Center). J. or Jules Durrive was the only adult male Durrive in the 1850 U.S. Census (New Orleans Municipality 3, Ward 1, Orleans, Louisiana, roll M432_238, p. 72), but GCG was in search of Frederick, with whom his family had a connection. In a letter to GCG, dated Washington, July 5, 1853, Colonel Gardner says that he can send Clint a letter in New Orleans, care of Frederick Durrive, Esq. GCG mentions Mrs. Durrive in a letter to his brother Charles from the St. Charles Hotel, March 18 and 19, 1849.
30. The reference is to his brother-in-law's plantation, Ile Copal, near Lafayette. See the introduction.
31. We have been unable to identify Captain Howard, who appears only one more time in these letters; Lieutenant Tyler, who makes no additional appearance, could be Charles

Humphrey or William Henry (Heitman, 1965, 1:976, 977). "Frank" was GCG's brother, the army captain.

32. John Moore McCalla [Sr.] Papers, Rare Book, Manuscript, and Special Collections Library, Duke University. Reprinted here with permission, and with appreciation to Mark Herlong, of Washington, D.C., who called this letter to our attention and transcribed it.
33. John McCalla Jr., born in 1835, was GCG's junior by four years, but their correspondence suggests that they were close friends. They attended the Rittenhouse school and Columbian College together. McCalla went on to become a medical doctor. Mark Herlong, of Washington, D.C., is writing McCalla's biography, drawing from a trove of letters and diaries.
34. One of nineteenth-century England's foremost Shakespearean actors, William Charles Macready began his third and final American tour in 1848. A few weeks after GCG saw him in New Orleans, Macready appeared in New York, where a long-standing rivalry with the renowned American actor Edwin Forrest led to a disturbance. On the evening of May 10 a mob attempted to hiss Macready off the stage. When the police tried to arrest several individuals for "riotous behavior," tension escalated to the point that the military was called in and fired on the crowd. Thirty-one people died and more than one hundred were injured. This startling event was reported as far away as San Francisco, whose paper ran the story in July (Moody, 1958, 12; *Alta California*, July 19, 1849).
35. Mark Herlong notes that Tucker Blake, whose first name was Ebenezer, went on to become "one of the more flamboyant characters in 19th century DC. His jokes were repeated in the society columns and he was once described as a tourist attraction equal to the Washington Monument. He was trained as a doctor but never practiced; he was a 'man about town' apparently living off his father's successful investments" (Herlong, e-mail to editors, June 23, 2004). Blake continued to be a fastidious dresser. For more information about him, see Zevely (1903, 117). The Rittenhouse Academy was a one-story school that opened in 1840 and was run by Rev. Joseph E. Nourse and his brother Charles (Zevely, 1903, 110). GCG attended the Rittenhouse Academy.
36. What GCG meant by "a—s" is not clear to us.
37. Several inaugural balls were held in Washington on March 5. Jackson Hall was a gathering place for Democrats. GCG's reference to *Notre Mère*, our mother, is obscure.
38. In 1849 the cornerstone was laid at the northeast corner of Third and C streets NW for the soaring gothic revival Trinity Episcopal Church, which had previously been on Fifth Street, across from the old city hall. GCG's cryptic references to the corner of Third and C, and to the place "a few doors west," however, suggest something other than a shared religious experience (Goode, 1979, 199–201).
39. George Washington Custis Lee, born in 1832 and raised in and around Virginia, was the oldest son of Robert E. Lee and a contemporary of GCG's. Custis Lee entered West Point in 1850,

graduated in 1854, and served in the Army Corps of Engineers before resigning his commission at the outbreak of the Civil War (*DAB*, 9:105–6).

40. Jefferson Barracks was about ten miles below St. Louis. Frank was GCG's brother and a captain in the army.
41. The *Falcon*, operated by the U.S. Mail Steamship Company, ran the route between New York and Panama, with stops in New Orleans and Havana (Wiltsee, 1938, 10).
42. GCG is closely approximating a passage from the English translation of Cervantes' *Don Quixote* that would have been available to him in various editions, including *The Life and Exploits of Don Quixote de la Mencha*, trans. Charles Jarvis (Exeter, England: J. & B. Williams, 1827), pt. 1, chap. 2, p. 8. GCG, then, is joking with his brother, who must have studied *Don Quixote* with him, that Clint and his companions are, like Quixote, knights errant off on a romantic adventure. Translator Jarvis, in an editorial note, describes this passage as Cervantes' "ridicule on the like affected descriptions, so common in romances." GCG, who either had access to a copy of *Don Quixote* or a splendid memory, nearly quotes the passage verbatim, moving it into the present tense and changing the horizon of La Mancha for a bright horizon. GCG makes another reference to Quixote in GCG to his sister Alida, n.p., September 21, 1852.
43. Edward Ingraham and GCG became close friends. Both owed their appointments to family connections. For more information about Ingraham, see chapters 3, 5, and 9.
44. Alcide, mentioned on several occasions in these letters but with no surname given, was Joseph Antoine Alcide Mouton, the stepnephew of GCG's sister Emma. Alcide, as he was known to family and friends, was born on January 1, 1831, the son of Antoine Emile Mouton (the brother of Emma's husband, Alexandre) and Marie Magdaleine Gadrate Rousseau (Alexandre's first wife was Marie Magdaleine's sister, Celeste Zelia Rousseau). The sisters' father was, perhaps inevitably, Jean-Jacques Rousseau, a descendant of a French Acadian who settled in Louisiana. In the U.S. Census of 1850 Joseph Antoine Alcide Mouton is listed as Joseph A., nineteen (U.S. Census, 1850, Western District, Lafayette, Louisiana, roll M432_232, p. 249). In the U.S. Census of 1860 he is identified as Alcide, a planter in Vermillionville, or what is today Lafayette, Louisiana (U.S. Census, 1860, Lafayette, Louisiana, roll M653_412, p. 785).
45. Tobin was not on the payroll of the survey party. He does not appear in Weller's abstract of disbursements (SED 34, 50). He may have been Capt. John William Tobin, the young, well-educated son of a wealthy Alabama cotton planter who enlisted in the Gaines Rifle Regiment during the U.S.-Mexican War and settled in New Orleans after the war. His papers are in the Historic New Orleans Collection, Williams Research Center. For a biographical sketch, see Alcée Fortier, ed., *Louisiana: Comprising Sketches of Parishes, Towns, Events, Institutions, and Persons . . .* (Madison, Wisc.: Century Historical Association, 1914), 3:431–33.
46. Quartermaster Justus McKinstry.
47. We believe GCG is referring to the *Message of the President of the United States to the Two Houses*

of Congress, at the Commencement of the Second Session of the Thirtieth Congress . . . , December 5, 1848. It contained documents, reports, and maps pertaining to the prosecution of the Mexican War, descriptions of the geography, and resources of the newly acquired territories of California and New Mexico, as well as official confirmation of the gold discoveries in California, with related maps and reports.

48. Weller's General Order no. 1, Panama, April 11, 1849, copy in Emory's letterbook, box 1, f. 1, EP, specified annual salaries for all the key civilian positions on the commission and allotted all other employees $400 a year. Servants, paid at the monthly rate of $30, could make $360 a year, so GCG's salary was toward the bottom of the scale. An interpreter and a draftsman made $1,000 a year, and the principal assistant surveyor made $1,800.
49. Charles Gardner had lost his job as postmaster at Washington as Whigs replaced Democrats.
50. Taboga, with freshwater springs and cooler temperatures than on the mainland, was regarded as more salubrious and agreeable than Panama (Kemble, 1972, 133). GCG offers more details in his letters of May 8 and June 12, 1849, to his sisters Alida and Sarah, respectively.
51. Fort San Lorenzo, whose construction began in 1597 on the right bank of the Chagres on a cliff above the river's mouth.
52. Gorgona was a town in the hill country, about forty miles up the Chagres. In the dry season (December through April), when the river was low, travelers disembarked there and continued overland to Panama City, a distance GCG estimated at fourteen miles. In the rainy season they continued two miles farther up the river to Cruces.
53. The reference to Jerusalem tells us that he was writing on Palm Sunday. In 1849 Easter fell on April 8 so Palm Sunday occurred on April 1.

    Clint's sweetheart at this time may have been a Miss Ginnie who attended the Patapsco Female Institute along with his sister Alida, in what today is Ellicott City, Maryland (GCG to his sister Alida, Panama, May 8, 1849). GCG apparently attended a ball at Patapsco in May 1848 at which he might have met Ginnie. While he was working for Emory, his "sweetheart" was engaged to be married to someone else (GCG to his sister Alida, Camp near Initial Point of U.S. & Mexican Boundary, June 1, 1851).
54. Alida was away from home at the Patapsco Female Institute.
55. The historian John H. Kemble estimated that as many as two thousand Americans in Panama City required passage to San Francisco at this time. The arrival of two steamers at once must have been a welcome event. Three steamers, the *Panama, Oregon,* and *California*, had been built in 1848 for the Pacific Steamship Company to provide regular transportation along the newly acquired west coast of the United States and had arrived in Pacific waters only that year. A single trip between Panama and San Francisco once a month proved wholly inadequate after news of the gold discoveries became widespread. The *Panama*, which had been launched in New York in December, had engine trouble that caused it to return to port for repairs. It had

set sail again on February 17, 1849, carrying the commission's instruments (Kemble, 1972, 31, 36, 218, 239, 242; Emory, 1857–59, 1:3).

56. The *California*'s delay had occurred on its maiden voyage up the Pacific coast. It had set out from New York on October 5, 1848, for service in the Pacific and arrived in San Francisco on February 28, 1849. Before reaching San Francisco, the *California* exhausted its supply of coal off Point Conception but made it as far as Monterey by burning the furniture on board. It spent a few days off Monterey while the crew chopped enough wood to take the steamer the rest of the way—after the wood had been cut, crew members found additional coal stored under the floor of the cabins. At San Francisco Bay the goldfields proved irresistible to the crew, which jumped ship en masse and left the captain with a schedule to meet and no one to run the ship. It was May 1 before he gathered enough men and a supply of coal to take the *California* back to Panama, where it arrived on May 23 (Kemble, 1972, 28, 120; Berthold, 1937, 60–61, 65).
57. Either an obscure literary allusion or a private joke. GCG uses this same expression in his letter to his family on May 13, 1849, when he put it in quotation marks. He added "in a home" after writing the line, inserting it outside the parentheses. Perhaps he meant to put it inside.
58. In 1767 Carlos III expelled members of the Jesuit Order from Spain and its overseas possessions.
59. The *Crescent City*, owned by the J. H. Howard & Sons Line, was the first large steamer to leave New York for Panama after news of the California gold discoveries reached the east coast (Wiltsee, 1938, 10–11).
60. GCG to his father, October 15, 1849, suggests that Charles Weller was sent back to Washington to obtain funds for the commission.
61. During their protracted wait for transportation to California, Emory had his assistants unpack astronomical instruments and make observations. He hoped to help his assistants to maintain their skills, as well as contribute to the existing body of knowledge about Panama's geographical position and climate (Emory, 1857–59, 1:2).
62. Old Panama, founded in 1519, was the oldest European city on the American mainland. In terming it the oldest city of the New World, GCG forgets earlier indigenous cities and Spanish municipalities on Caribbean islands.
63. Their position was recorder. As Emory put it: "Mr Gardner and Mr Ingrahm [*sic*] are at present employed as recorders, they as well as Mr Ludlum are undergoing instructions and I hope in time they will become sufficiently proficient to take the place of assistant computers" (Emory to John B. Weller, Panama, April 9, 1849, copy in Emory's letterbook, box 1, f. 2, EP).
64. Before May 30, 1849.
65. The *Senator* did not reach Panama until September 14, 1849 (Kemble, 1972, 247).
66. These two sentences, which we have placed in parentheses, appear in the margin of GCG's letter. The context suggests they belong here.

67. W. N. and J. H. Barker, both from Washington, D.C., traveled overland to California, setting out on April 2, 1849 (Haskins, 1975, 22).
68. GCG's brother Jacob.
69. Difficult to defend, Old Panama was destroyed by the English pirate Henry Morgan in 1671, rather than as the result of an earthquake, and was reestablished at its present site.
70. Apparently a literary allusion, which GCG also used in his letter to Alida of May 8.
71. Bayard Taylor reported that this cart was made in Panama by "a Yankee mechanic detained for want of money to get further" (1949, 22).
72. GCG refers to the Pacific coast survey, one with which he himself would later hope to obtain a position.
73. Aunty lived in the Gardner household throughout the time that GCG was on the boundary commission. See the introduction.
74. We have not been able to identify Castleman.
75. Gardner eventually did make it to Peru but much later in his life.
76. These four sentences, which we have placed in parentheses, were written vertically in the margin in pencil. They appear to represent a postscript.
77. The original manuscript has been damaged and repaired at this point. The month of May is legible, but the day of the month and year are not. The content makes clear that the year was 1849. The date is likely May 19, 20, or 21. He refers to the sailing of the *California* last Thursday (the seventeenth). Had he been writing on the eighteenth, he would have said "yesterday." He also refers to the imminent departure of the *Oregon*, "which is to start next Wednesday evening." The *Oregon* did leave Panama on Wednesday, May 23. If GCG had been writing on Tuesday, the twenty-second, he would have said "tomorrow." From this internal evidence we infer that he wrote to his father on May 19, 20, or 21.
78. Kemble (1972, 36) and Wiltsee (1938, 18) reported that the *Panama* sailed on May 18, which was a Friday. This does not contradict GCG's claim for a Thursday night departure, if the *Panama* finally got underway after midnight. Haskins (1975, 509) put the *Panama*'s departure date as May 11, 1849, but is a week off. GCG wrote on May 13, Sunday, that the *Panama* had not yet sailed, and Weller, who sailed on the *Panama*, was still in Panama City on May 14.
79. Weller initially intended to take twelve men, including himself and two unnamed individuals described as "servants (colored)"; the others were Dr. E. K. Chamberlin, James Noon[e]y Jr., William A. Taylor, D. Gahagan, A. B. Gray, William R. Kinder, H. H. Robinson, Charles J. Whiting, and Gilbert Murdoch (Weller to Capt. Stout, May 14, 1849, in SED 34, 8). Adding Emory to his party must have been a last-minute decision.
80. Throughout these letters GCG uses the word *mess* to refer to communal meals, a common military usage then and yet today.
81. Weller had GCG's name on a list of thirteen members of the commission to go on the *Oregon*,

but if fewer than thirteen tickets could be purchased, GCG was to be dropped from the list (Weller to Emory, San Diego, June 19, 1849, copy in Emory's letterbook, box 1, f. 1, EP).

82. GCG did not sail on the *Oregon*. When Lieutenant Whipple arrived in Panama, he decided that he and six civilian assistant astronomers and "computers" (including GCG) were essential for beginning the survey in San Diego since the survey could not get underway until they had fixed the initial point. Whipple petitioned William Nelson, the U.S. consul in Panama, to find passage for them on the already crowded steamer *Oregon*. Although Nelson was able to do so, based upon Whipple's assertion that the men were indispensable, Whipple's list conflicted with that of Quartermaster O. S. Withersby, whose list included Moses Conner, the carpenter. Withersby, who held the purse strings, detained the whole group "rather than any should go other than those selected by himself" (William Nelson, U.S. consul, Panama, to Whipple, May 23, 1849, in SED 34, 43, which also contains the related correspondence).

83. The "Ollendorff" to which GCG is referring probably is *Ollendorf's New Method of Learning to Read, Write, and Speak the Spanish Language* (1848) by Mariano Velázquez de la Cadena.

84. In his classic description of the gold rush, *Eldorado*, Bayard Taylor referred to some argonauts who had tried this at about this time: "A few months previous [May], when 3,000 persons were waiting on the Isthmus, several small companies started in the log canoes of the natives thinking to reach San Francisco in them." Needless to say, they did not succeed (Taylor, 1949, 24).

85. Lt. George Horatio Derby, an 1846 graduate of West Point, topographical engineer, and veteran of the U.S.-Mexican War, was assigned to California on December 15, 1848. There he served on the staff of Gen. Bennet Riley and surveyed the goldfields and the lower Colorado River (Stewart 1937, 57). GCG does not mention Derby again in these letters.

86. Perhaps the talk to which GCG refers was Whig criticism of Weller's last-minute appointment by Polk, as well as the haste with which he departed for California, from whence it would be more difficult for the new Whig administration to recall him (Goetzmann, 1991, 158–59).

87. Heavy tides in the Bay of Panama also made it possible to beach a ship on the island for cleaning and repairs without building a dry dock (Kemble, 1972, 133).

88. Mrs. Smith, Frances Jeanette Bureau, whose husband led the army's Pacific Division, was the daughter of François Bureau of New Orleans and most likely became acquainted with Emma Gardner Mouton there (*DAB*, 17:33).

89. Cruces was two miles closer to Panama City than Gorgona, and travelers continued up the Chagres to Cruces in the rainy season, from May to November, when the river was higher and the road from Gorgona turned into a quagmire.

90. Computers were responsible for calculating geographical positions based on the astronomer's and surveyor's observations.

91. Charles Christopher Parry was born in England in 1823 and moved with his family to the United States in 1832. He received his M.D. from Columbia College in 1846 and began to

practice medicine in Davenport, Iowa, that year, but Parry had come under the influence of the botanist John Torrey at Columbia, and his interest in botany surpassed his interest in medicine. He went on to become one of the nation's leading botanists. In 1848 Parry had served on a geological survey of Wisconsin, Iowa, and Minnesota. In 1849 he was appointed to the boundary survey and spent the next few years doing geological and botanical fieldwork along the border. He wrote several sections of the survey's final report (Emory, 1857–59, 1, pt. 2, 1–24 [authorship not attributed], 1, pt. 2, 49–61, and 2, and pt. 1, 1–26; *DAB*, 14:262–62; Weber, 1997, 173). Emory stated that Parry's responsibilities also included those of "assistant," apparently a position similar to GCG's and Ingraham's, and Parry assisted Lieutenant Whipple and his party in determining the "other extremity of the straight line forming the boundary at the junction of the Gila and Colorado" (Emory, 1857–59, 1:4).

92. One of the major holy days of the liturgical year, Ascension commemorates the fortieth day after Easter, when Christians believe that Jesus ascended into heaven.
93. Frank Stone and Frank Holley were civilian employees "in subordinate positions" whom Emory singled out, along with his slave, Robert, as particularly loyal, since they had not deserted him for the goldfields in California in the autumn of 1849 (Emory, 1857–59, 6). GCG, writing to his father from San Diego on November 30, 1850, mentions Holley as a laborer.
94. Mary Butterworth, a relative. She and her husband, Sam, appear from time to time in GCG's correspondence.
95. The 1850 census reports a James McLean, thirty-two, who was living in Washington in 1850. He was a draftsman, married to Ann, who was twenty-nine (U.S. Census, 1850, Ward 2, Washington, D.C., roll M432_56, p. 122). However, it is more likely that Clint is writing about a McLean from the Little Britain branch of the family. Margaret French may have been the wife of B. B. French.
96. Julianna Hobbie, age sixteen in the 1850 census. She was the daughter of S. R. Hobbie, the first assistant postmaster general, and his wife, also named Julianna (U.S. Census, 1850, Ward 2, Washington, D.C., roll M432_56, p. 81).
97. Today he would say he was packing his things.

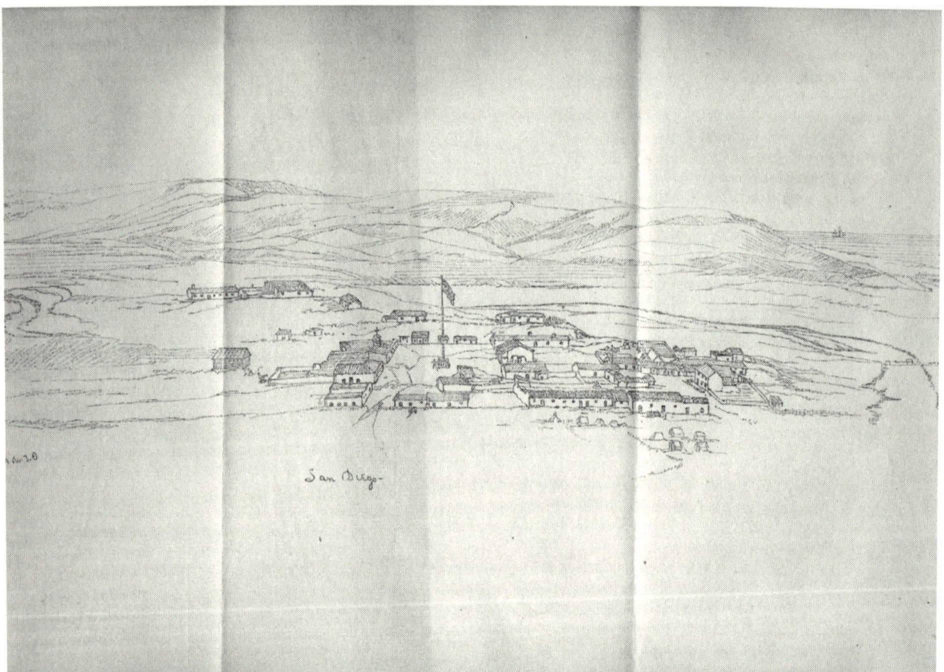

San Diego, December 1849. From H. M. T. Powell, *The Santa Fe Trail to California, 1849–1852.* Courtesy of the DeGolyer Library, Southern Methodist University, Dallas, Texas.

CHAPTER 2

Initial Points—Camp Riley to the Gila

July 23–October 31, 1849

The Treaty of Guadalupe Hidalgo required the American and Mexican commissioners to meet in San Diego, the Pacific starting point of the survey, on May 30. The first contingent from Panama City, headed by John Weller, reached San Diego on June 1, two days late. Weller's Mexican counterpart, Gen. Pedro García Conde, arrived a month later, on July 3. A veteran military engineer who had surveyed the boundaries of the state of Chihuahua, General García Conde had served as secretary of war and in the Mexican congress. He brought administrative and practical experience that his American counterpart lacked.[1] By the time the steamer *California* brought Clint to San Diego on July 13, the commissioners of the two nations had begun to talk.[2] They agreed to work independently of one another but to meet from time to time to compare notes and seek consensus.

A few days before Clint reached San Diego, his boss, William H. Emory, had left Commissioner Weller's headquarters at the site of present-day downtown San Diego, just south of Old Town, and established a base camp along the *camino real* at the south end of San Diego Bay. He selected a spot on ranch land belonging to Santiago Emigdio Argüello and named it Camp Riley for the military governor of California, Gen. Bennet C. Riley.[3] There, amidst "flees . . . rattle snakes, wolves, rabits, &c.," Clint joined Emory.[4]

In establishing himself at Camp Riley, Emory had responded to Weller's order to "proceed without delay to establish a temporary observatory at some convenient point within one marine league south of the southernmost point of the port of San Diego with a view to ascertain the precise latitude and longitude of the initial point upon the Pacific Ocean referred to in the Treaty." Weller's written instructions also required Emory to send some of his men to the second point in California fixed by the Treaty of Guadalupe Hidalgo, the junction of the Gila with the Colorado, and to set up an observatory there.

Then, stating the obvious, Weller told Emory: "You will see that the object I have in view is to connect these two points by an astronomical line and you will take the necessary steps to this effect."[5] Weller's instructions to Emory, however, were simply a formality. Emory, responding to Weller's request for a plan, had effectively written his own instructions six weeks earlier, and Weller merely reiterated them.[6]

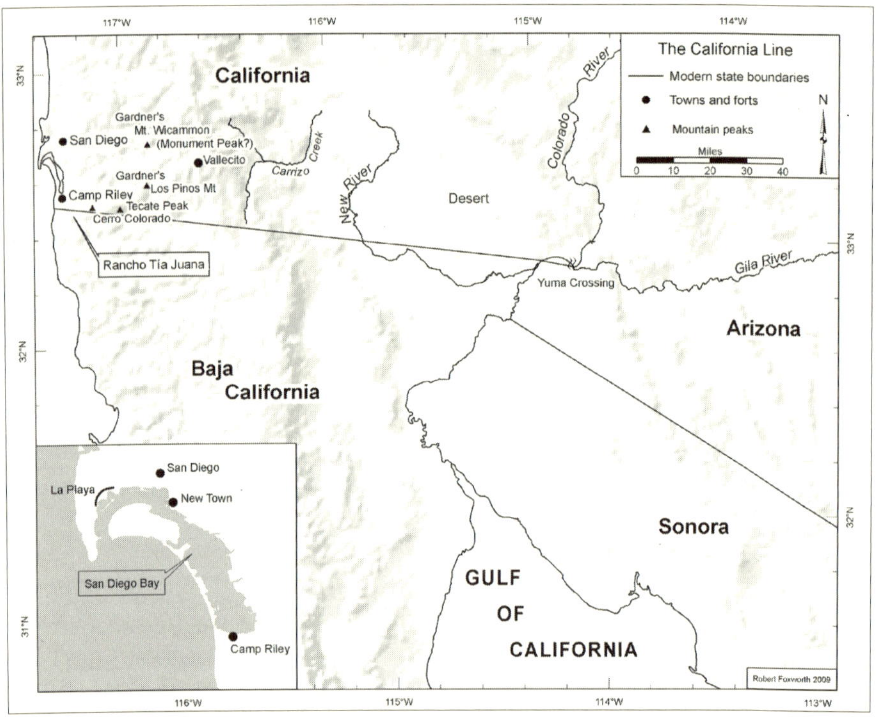

Major Emory had three large astronomical tasks before him. All turned out to be more complicated than anyone had imagined. First, he had to locate the beginning point of the survey. The Treaty of Guadalupe Hidalgo specified that the starting point should be "one marine league due south of the southernmost point of the port of San Diego, according to the plan of said port made in the year 1782 by Don Juan Pantoja." That point proved more difficult to locate on land than it had been to commit to writing. The shoreline had changed in the intervening years, and there was no international standard for a marine league, so the task of establishing the starting point required diplomacy as well as astronomy.[7]

Second, the astronomers and surveyors had to locate the exact point where the middle of the Gila River entered the Colorado River, about 150 miles to the east. Emory entrusted that task to one of his two assistant astronomers, Lt. Amiel Whipple, who set out for the Gila junction on September 11 over a familiar but difficult trail that took him across the coastal range, by way of Santa Isabel and San Felipe, and into a desert still scorching in the late summer. By October 2 he had reached the mouth of the Gila and began to build his observatory. Once again, however, local conditions muddied what had seemed a clear task on paper. The shifting shoreline made it difficult to tell precisely where the Gila entered the Colorado.[8]

Third, Emory had to draw a straight line between the Gila and the Pacific, a task that he later described as the most arduous part of the work in California. The astronomical observations fell to Emory's other assistant astronomer, Capt. Edmund Hardcastle, whom Emory dispatched on a preliminary reconnaissance on July 20.[9] Hardcastle reconnoitered the country to the east of Camp Riley hoping to find a single peak that could be seen from both Emory's observatory on the Pacific and from the mouth of the Gila, where Whipple would establish an observatory. Hardcastle found no peak that filled the bill.[10]

Initially, Clint worked on the first of these three tasks, assisting Emory in locating the starting point for the survey at Camp Riley on the Pacific. As Emory put it, "Aided by James Nooney and George C. Gardner, I took the personal charge of the determination of the latitude and longitude of Camp Riley, and the triangulation by which that determination was to be carried to the initial point on the Pacific."[11] Emory had chosen Camp Riley as his headquarters because it was the closest place to the starting point where he could find wood, water, and protection from winds and drifting sand. Although official reports do not mention it, Clint's letters indicate that Camp Riley consisted of three camps, one headed by Surveyor Gray, one by Capt. Julius Hayden, and the third by Emory. The three camps stood a mile apart, forming a triangle.[12]

Clint's letters say little about his day-to-day routine, but a letter to his mother dated August 18, 1849, is an exception. He explained that they took meteorological and magnetic readings early in the day and rested in the afternoon. In the evening he hoped for cloudy skies because clear skies meant he would have to assist Emory as he took readings with a Zenith telescope into the small hours of the morning. These readings of the sky were essential for finding the precise latitude and longitude of the base camp from which they would, by triangulation and negotiation with the Mexican team, establish the starting point of the survey.[13] The meteorological observations that Clint recorded helped Emory fulfill his orders to contribute to scientific knowledge of the border region in general.

While Emory oversaw the work of the astronomers and surveyors south of San Diego Bay, Weller journeyed north to San Francisco to try to raise money for the cash-starved commission. He left in August and did not return until early October. By then the Mexican and American teams had determined the starting point on the Pacific. At a formal meeting on October 8 Weller and García Conde agreed on the starting point, then traveled to the spot for a ceremony on October 10. They marked the starting point with a post until a permanent monument could be erected, buried an explanatory document in a hermetically sealed bottle, and ended the day with a party.[14] By then Whipple had established his observatory near the mouth of the Gila, so two of Emory's three tasks were done.

Clint Gardner missed the party on October 10. That day he was traveling back to San Diego where he had been working on Emory's third task. When Hardcastle's reconnaissance in July had failed to turn up a single peak where flashes could be seen from the two observatories, Emory settled on a more complicated plan, which he summed up this way: "Capt. E. L. F. Hardcastle, corps Topographical Engineers, assisted by Mr. George C. Gardner, and escorted by Lieut. [A. W.] Slaughter . . . was placed in charge of the party to reconnoiter the country, gain a knowledge of the topography, and select elevated points by which the extremities of the line, or the two observatories in charge of Lieut. Whipple and myself, could be connected in longitude by flashes of gunpowder."[15] Hardcastle selected three high points that he called Cerro Colorado, Los Pinos, and Wicammon.[16] From those points, as one scholar has explained it, "Flashes of gunpowder would be used as signals that could be observed simultaneously between stations, the local time of the signals being recorded at each station and the differences between the times giving the longitudes."[17] In this way the astronomers could calculate longitude at Whipple's observatory by measuring it from the longitude at Emory's observatory.

One of Hardcastle's contingents, which included R. P. Effinger, was stationed at Cerro Colorado, closest to the Pacific; Hardcastle and Gardner went to Los Pinos, and Lieutenant Slaughter to Mount Wicammon. For five nights, between October 5 and October 9, the men climbed their respective heights, did their pyrotechnics, then descended in the darkness—with Clint losing his way and tearing his pants one night. The line of sight worked between the Pacific and Mount Wicammon, but it broke down between there and the Gila. Fog rolling in from the Gulf of California blocked Whipple's view. "No signs of rockets or flashes have been seen," Whipple told Emory.[18]

Captain Hardcastle's group returned to Camp Riley in stages, with Clint arriving soon after the joint commission ceremonially marked the starting point on the Pacific. Whipple, though, stayed at the Gila, planning to take more astronomical measurements during a lunar month, or lunation, which extends from one new moon to the next.[19] "Even without the flashes," Whipple assured Emory, he would be able to establish the latitude and longitude "with respectable accuracy."[20]

The Yuma crossing of the Colorado River just south of the Gila had become the site of a traffic jam. It filled with forty-niners courageous or foolish enough to follow the Gila westward through Apache country. Freshwater and food were in short supply, and the military escort assigned to Whipple found itself aiding forty-niners as well as surveyors. Demands on the survey party reached a crescendo when Surveyor Andrew Gray headed toward the Gila in mid-October to work with Whipple and to meet the Mexican surveyor. After crossing the mountains and striking the main migrant trail near Carrizo Creek, Gray's party encountered a bedraggled group headed by James Collier, en route from Fort Leavenworth, Kansas, to assume his position as tax collector for the port of San Francisco. Rather than continue on to the Gila, Gray decided to turn back and guide Collier's party into San Diego over a route he believed he had just discovered.[21] He entrusted the work on the Gila to his assistant, John Harris Forster. Gray's decision annoyed Commissioner Weller. On October 29, Clint tells us, Weller and Gray, after drinking too much, got into an altercation that left Weller with a gunshot wound to his thigh. Whipple, rather than Gray, finished the survey and arrived at an agreement about the eastern point with the Mexican surveyor, José Salazar Ilarregui, on November 30.[22]

That tempers ran high is not surprising. Work proceeded under exceedingly trying conditions. The same northern California gold that lured forty-niners across the continent beckoned the commission's employees, who began to drift away to the goldfields or to jobs that offered highly inflated wages in California's superheated economy. Weller had foreseen this problem in Panama. He had announced that, given the government's investment in transporting men to California, each employee "will be considered as pledging his honor to remain in the public service at least one year if required."[23] Pledges paled next to profits. By mid-August, Clint noted, Surveyor Gray's "labouring men" were gone, and he could not continue his work.[24] So many men continued to desert that summer that only "four or five of the civil employés remained," Emory reported. They included Clint and, Emory remembered with gratitude, his black servant, Robert, "a slave."[25] Emory's trained assistants had also deserted him. By autumn, Emory lamented, "I have now lost nearly all my assistants, and the computation of the azimuth of the line and the tracing of the line on the face of the earth were done with no other aid than that of Captain Hardcastle and assistant Gardner and the infantry soldiers of the escort. Lieutenant Whipple, with the cavalry escort, was still at the junction of the Gila and Colorado, faithfully aided by the escort and assistants Parry and Ingraham."[26]

Workers also left because the American commission was out of money. The new Whig administration had fired Weller, and Congress refused to make further appropriations to the survey until a new commissioner replaced him. Reports of Weller's dismissal soon reached California, and Weller's credit was frozen. Work slowed, and civilian members of the commission, like Clint, who had yet to be paid, had no prospect of receiving a paycheck until Weller's replacement, rumored to be John C. Frémont, took charge.[27]

Clint's nine letters in this chapter, all written from Camp Riley, start with a long letter to his sister Sarah, written on July 23, 1849, and end with a short note to his mother, dated October 31. In the first of these letters Gardner tells his sister of his harrowing voyage from Panama on a cholera-infested ship. As Sarah shared the contents of the letter with her family, the story must have sent a chill up their collective spine, even though they knew that Clint survived to write the letter and tell them of his safe arrival.

GCG to his sister Sarah, Camp Riley, California, July 23, 1849[28]

Dear Sarah,

The last letter I wrote home was to Charlie, and as Major [Justus] McKinstry has had the kindness to offer to take a letter to San Diego for the next steamer, I thought I would write you a few lines and inform you in what part of the world I am.

June the 23 just one month ago I was getting ready to leave that ancient city, every thing was in confusion—we sent our baggage down to the market house gate—and from there we were taken by a row boat to a small craft. We started from our quarters about 2 o'clock and it was 5 before we got on board of this sailing craft, owing to the number of passengers. Oh! how glad I was when we left that city. After raising sail we soon left Panama in the distance—and were on board of the steamer [*California*] about 7 o'clock—although I was a strange passenger. I slept in the saloon that night. The next morning about ½ past 6 o'clock we started for Taboga—at which place we arrived about 7 o'clock; and after taking some water and a few provisions, we got under way again about 2 p.m. stearing [*sic*] our course nearly south. Our cook that night was complaining a little of being sick—the next morning he got no better, and about 8 o'clock that night died of the cholera notwithstanding all the medical aid: he was buried at sea and 12 o'clock mid night. He left one little son—who did not appear to miss his father much, after a week or so. In the night of the 27th a Mr. Paine died: he was brought on board at Taboga very ill—he is a man that lived in N. Jersey, and was very well off indeed, but still wished more: he had one son about 15 years old with him: he was buried about 9 o'clock the next morning. This was the first burial at sea that I had ever seen—and I was in hopes it was the last, but my hopes were in vain, for on the 29th a Mr. Lamb died about 3 p.m. and was buried soon after. The next day a small craft came in sight—the Capt. said he thought she was in distress—but soon found it to be one of those small "dry outs," as they call them, that did not care for any one.[29] We spoke to her, and found she had been out 35 days from Panama—all well on board—had had but 2 deaths—bound for San Francisco, with 34 men and were in want of nothing but would take a barrel of bread and meat—which was given them. We were then off Acapulco at which place we arrived about 11 o'clock a.m. on the 2nd of July after passing it and going back to it: that night another man died and was buried at Acapulco. The next day about ½ past 12 we got under way again—after putting two of the crew on shore, for some of their conduct. The next day was the 4th the 73 anniversary of our independence—but alas what was there to remind me of it? The only thing that happened was one poor fellow jumped over board—no one could tell for what reason: and on the 5th two more men died. About dark that evening we came in at San Blass—the next morning about half of the passengers went on shore, but as the sun was very hot I staid behind untill evening. San Blass is situated on a high hill which is perpe[n]dicular on the side towards the sea—it commands a view of the harbour: there is a small village at the foot of this rocky hill which is also called San Blass—the people live down there part of the time and move up on the hill during the sickly season, which was just commencing.[30] After looking around with my friend Ingraham I found it grew rather dark and we couldn't get on board again that night without paying 1 or 2 dollars,[31] so we went to a fandango—then slept on the table in one of the grog shops—and went on board again about 6 o'clock the next morning.

7th that night we had quite a storm—I never have seen such lightning before: there was a little brig out in the harbour, which had just taken all their passengers on board and were about to sail when the wind turned, which the lightning struck and took away the main mast. Well we started from there about 3 P.m. on the 8th. Ever since we left Panama Ingraham and myself had been living in the saloon, but as there were so many passengers come on board at San Blass, we had to go in our places in the steerage.[32] Capt. [Thomas A.] Budd was very kind to us, and it was from his acquaintance with Lt. Whipple that he permitted us to live as we did—assisted by Capt. [Richard] Meade's influence, who is going to San Francisco to take command of one of the government vessels: he is an uncle of Ingraham's.[33] After we came off Cape St. Lucas [Cabo San Lucas] it became very cold, and the sea was very rough, and on the evening of the 13th we arrived at San Diego, about 5 o'clock. we met Col. Weller, Mr. Gray, and others—all there ready to meet us.[34] Major Emory had sent a wagon down for those of his party: and in a few moments a young officer road [rode] up and asked for Mr. Gardner—the Wagon was sent down to report to Mr. Gardner: Oh! how big I did feel. Well we all got in and up we went to San Diego about 4 miles.[35] I took dinner with Major Emory, Capt. [Edmund] Hardcastle, & others that evening: the next morning I took breakfast with the same—after which we started for this camp, in the same wagon, and arrived here about 2 o'clock after riding about 15 miles. The Major has called it camp Riley after Gen. [Bennet] Riley—the flees are in abundance, rattle snakes, wolfs, rabits, &c., &c. also Mr. Gray's party is camping about 1 mile from us. Capt. [Julius] Hayden is also about 1 mile from us, but in a different direction the three camps form a triangle. Capt. Hayden is an old school mate of Frank's and also I believe of Jake's—his first name is Julius: he has command of one company of the escort. The Mexican escort are deserting [for the goldfields] every day, 14 and 15 at a time. I believe all the Americans that are going have gone, & there are still enough left.[36] Capt. Hardcastle has gone out to explore.[37] Lt. Whipple will go over to the Gila soon: but Mr. Gardner will remain here, with the Major [Emory].[38] Charlie speaks of going to sea before the mast in order to get out of Washington—but tell him he had better take a flat boat, and bring wood up the river; for those boatmen live better than sailors—the steerage passengers fare better than the sailors—and I can assure you I would not go in the steerage again if any man would pay me twice the passage money.[39] I will send my ticket to him in order to show him what nice tickets they have. If it had not have been for that poor devil that had his money stolen in Panama, I never would have gone in the steerage: he gave me his note for $100 which he is to send me as soon as he arrives at San Francisco:—he could have died in Panama, if we had not assisted him. Oh what fools, to start from home! I have seen several that have come across the country, almost starved to death, with out clothes or any thing. I pity Jim & Bill Barker.[40] Some of the fellows that have come across the country have told us that there has risen in the desert a *river*—the Mexicans about here say it is true—and that it is owing to the rise of the Colorado.[41] It is good news for us, for when we start from here, we will have to cross there—but they think it will disappear when the water falls. So good by Sall. In haste, Your Brother Clint

I wrote the family at Newburgh yesterday—All well.[42]

Our pay is increased to $50 per month commencing July and I am going to buy a horse on the strength of it.[43] I would like you to send me the instructions for making bread with cream Tartar & soda for we have nothing but milk to put in our bread.

GCG to his uncle John McLean[44] on Capitol Hill, Camp Riley, California, July 23, 1849

Dear Uncle

I hasten to inform you that nothing has given me more pleasure than to read your most interesting letter of the 24th of May in which you speak so highly of your most humble servt. I can assure you that the ancient city of Panama did not fail to be fruitfull in objects of interest to such a youthfull mind as that of my self, often during my stay in that ancient city did I walk around the old ruins of the churches &c. with their moss covered walls and think of the great nation that must have once existed there, but I soon became tired of it; but will never become tired of receiving letters from you, so I beg of you write often, excuse haste, I hope soon to answer your letter a little better, after I get a little more time—believe me

your affectionate nephew

Clint Gardner

Tell Charlie to tell Johnny Ingle, Alcide, John McCalla & himself that I will write soon to them.[45]

Then tell him to tell Alcide to tell Custis Lee I will write him also give my love to all I suppose Mother is riding in her new carriage all the time.

Let me know all that is said about the commissioner in the states. I have received the treaty, Fathers[,] John McCallas[,] Charlies[,] yours and sarah's letters after coming all the way from Panama with them.[46]

Clint

I also received two papers for which I suppose I am indebted to John[47] so you will please give him my sincere thanks for them and also for that small book which is very interesting and I hope he will send the other numbers when they come out.

GCG to his mother, Camp Riley, California, August 18, 1849

MY BIRTH DAY

My Dear Mother,

This being my birth day I thought I would make a short reply to your most acceptable and interesting letter of the 25th and 26th of June, which I received in the evening of the 16th accompanied by one from Sarah and Alcide. On the morning of the 16th about day light several persons in Camp heard the guns fired off, which are a signal of the arrival of a steamer; so during the day we were conjecturing as to what steamer it was—some thought it to be the *Senator* others the *Panama*,[48] &c., but as for myself, the only thing I thought about was the letters from home, which I received just before dark through the kindness of our Quartermaster that was at San Diego at the time—and was sorry to hear that Charles Weller had not arrived [in Washington], as I was anxious to hear from those little fixtures which I sent by him, although I have since been almost ashamed of having sent them, but the one that receives them must consider them as only a curiosity. You say you were quite disappointed at not receiving a letter from me by that steamer; but I am afraid that you were a little fatigued to read the foolishness I wrote when the next one came, I intended sending them in the same steamer with Charles Weller but they did not get over the Isthmus in time. You mention also that you sent word to Emma and Frank to write to me, but it seems that I am never to hear from them:—I did get a letter once in a while from Odeide, but since Emma writes that it would be better to direct all my letters to her instead of Odeide I have done so, and the consequence is that I have not heard from there since; she says that "I will perceive *upon reflection* that it is most advisable and proper", but as yet I have not come to that conclusion, although I have been reflecting upon it for some time—perhaps the reason is that her youngest brother has not as yet arrived at the age at which he is capable of reflecting upon so important a subject, although he is 18 years old this day of our Lord, 1849.[49] I am quite glad to hear that the Cholera has not as yet arrived in the City and that all the neighbours are well. You must remember me to Mrs. Oelirch [name unclear] congratulating her on her marriage. I suppose she will take up her residence in Baltimore, then who will take her house? I am sorry to hear that you are going to sell Briar: but I hope "Old Lack" will give his permit and father will buy a very gay horse, for when I return I should like to cut a dash with a Gold mounted saddle, Bridle, and spurs—he must be very gay or else I will not enjoy myself—for after riding these wild horses of California, it would be too much like mounting a cart horse to get on one of the Washington stock. I intended to buy a horse for myself after they increased my pay, but It seems that no money can be had; by the by, in my last letter home I said my pay of $50 a month commenced the first of July, but it was a mistake.[50] It commences the first day of August. I was trying for 3 weeks to get some money to buy a few little things such as a bucket, tub, &c., which it is the place of the Commissioner to furnish, and finally succeeded in obtaining $10. When I left home I also understood that bed[d]ing was to be furnished us, but now it seems, when we get in the field we have to buy our own blankets, and as there are no blankets to be had here, we have to purchase them from Col. Weller's store which he has opened in San Diego and pay $15 for an inferior article, by the by I believe those blankets &c. were brought out at the expense of the Government as commissary stores, (but it would not do for me to say any thing about it).[51]

He has now gone up to San Francisco in the steamer that your letters came in, to see if he cannot get some money and provisions—for it seems we have run short of every thing—he has put us on rations which are equal to one and a half of soldiers rations, for which there are a great many things said about him.[52] I judge he is not liked very much and I must confess I have not taken a great fancy to him since he mixed us all up together in Panama, workmen &c all at one and the same table, but I have heard that it is the same thing up to San Diego: when he was there he had his carpenter &c. seated next to him at table, and no one can move them, if one of the Sub. Asst. Surveyors happen to come in he has to wait for the second table; so go things, but thank my stars I am, as you may say, all most seperated from them. Major Emory has a little more order in his camp, and every one knows his place and keeps it—the only thing I have to complain of is the dirt, which we find continually in our food, the ground about camp is very hard and has this fine dust on it, which blows even through the Canvas and our cook is none of the cleanest, but we are quite glad to keep him, for in Mr. Grays camp they have no workmen at all, they have to cook their own dinner and supper.

I should like to hear what is said of the Col. [Weller] in the states I have heard that the New York Herald has some thing in about him and his party, and a great party it is—the principal Quartermaster came to the conclusion, as he did not like butter, that he would have none in his stores; so as he came up in the Steamer *Oregon* he disposed of all his butter, and we have to do without it; and the other day a vessel arrived at San Diego from Boston, they had some very nice mackerel on board, of which Mr. Gray proposed to buy—as he had nothing but some corn beef at his Camp and very little of that, so he asked the Col. if he would not get some for his camp but the Col. said in reply "Why Mr. Gray I never eat mackrel there is plenty of salt beef in San Diego I will buy that"—so you see because the Quartermaster don't eat one thing, the Commissioner another, we have to do without, but thank heavens we have a little Quartermaster at our camp who will always have some[thing] to eat for us if it is to be had about "these diggins," he said the other day when he went back to the states he would be ashamed to say he was Quartermaster in the Commission, for fear the people would take him to be Mr. W—[i.e., Oliver S. Witherby], the principal Quartermaster, I could tell you a great many other things, but I think I had better stop here as I belong to the Commission and Col. is my superior, I will have a *little* respect for him—so *say nothing* of what I have said, although you must know (as the "Gold diggins" are so near) that one feels a little Independent.[53] I expect there will be a stampede here soon, any how, as the news from above gets better and better at every arrival, some have gone already—but I think they will be glad to get back for from what I can hear very little can be made there without a great deal of labour. There was a soldier from this post that went up there on a leave of absence with $900 in his pocket, and returned with $700, just $200 out of pocket—although he says he was quite fortunate, but had to pay as much as he dug to live there—and so it is. I think a great many will find it so, but it will be just too late. I know I have seen a great many that wish they had never started from home:—to be sure there are some that have made fortunes, but how did they do it, why by trading and gambling, there are some lucky ones that go up in one steamer and return in the next with immense fortunes made by playing monte, &c. In Father's last letter he askes me to give a little description of the people, their manners, customs, &c., but as yet I have not seen sufficient of them, for I have not been able to spend a single day in San Diego, being employed all the time down here. In the mornings when I get up, which is about 8 o'clock, I dress myself, eat my breakfast and then go and record the Metereological Observations for nine o'clock, after that I go out in one of the Observing tents and observe with the Magnetic Instrument until 12 o'clock, then record the mete[o]rological observations again for 12 then I take a nap of about 2 hours then prepare for dinner, after dinner I take a walk for my health or take a bath stop[p]ing every three hours to record those observations, and then about 6 o'clock the Major commences observing with the Zenith Telescope and I mark time for him until about 3, 3½, 4, 4½, 5, 5½ or six o'clock the next morning, then I go to bed.[54] I have always been glad to see clear nights until I got out here, but now I must confess a cloudy night is always preferable to me, for then I take my rest. Mr. Whipple will leave here the 6th of next month for the Gila with a few of our party,[55] then I think we will have fine times, for it will commence getting cloudy & I expect I will be sent up in the Mountains, with Capt. Hardcastle or perhaps by myself, to make signals with rockets.[56] Mr. Gray's party I don't think will do much until another appropriation is made, as they have no assistance, all their labouring men have gone to the "gold diggins"[57]—the more I see of Mr. Gray the more I like him—I think he has been very kind to me, he has lent me an India rubber blanket which could not be got out in this country, then I got one pair of blankets from Col. Weller and that Mexican blanket of Frank's make up my bed[d]ing, but I expect I will have to get an other pair of blankets as soon as it gets a little colder. The day is very warm the thermometer stands 85 today, and at night it gets quite cold again. We have had no rain of any importance yet but I expect when it does commence, we will have it to some purpose. Give my love to *all*—your affectionate son *Clint*

This has been the warmest day yet. I will write Sarah a letter if I have time before the steamer gets back.

GCG to his brother John, Camp Riley, California, September 15, 1849

I was quite delighted, dear John, at receiving a letter from you, although not very well pleased with the news it contained, yet "*it was just as I expected*," as the old woman said when her husband told her that the pigs had eaten up the grind-stone, for I have heard that old [Zachary] Taylor intends to make a clean sweep of it, which I suppose he has done before this.[58] In your letter you speak of Mr. William Thompson having sailed some time ago for the "gold diggings" in the barque [or bark] "*Brothers*", which I was sorry to hear, for I have no doubt he will never reach his destination; but I will try and find out by the next Steamer if such a vessel has arrived; there has been a great number of vessels wrecked on this coast and I would advise any one who travels to San Francisco, never to get in a sailing vessel, for they cannot coast it up even from Panama.[59] I have seen since we encamped here more Americans in distress than I have ever seen before, if I had have [*sic*] read in a book the distress which is experienced on this coast by the shipwrecked emigrants, or rather gold seekers, I would not believe it, for it is not such as a person would suppose *could* exist in any country that is to travel two and three days without water and nothing but birds, rabbits and such things as they could kill on the route. There has been some of the larger vessels that started from Panama that have been lucky in getting to San Francisco in 100 days but the only reason is because they have been able to carry provisions and water for that length of time, and have put out some thousands of miles at sea, got in the trade winds which will take them straight up there. Who do you think I met the other evening here at camp; as a great number of these ship-wrecked travellers came along, my attention was drawn to one, whom I thought I had seen often before and who reminded me of the pleasant evenings I have spent in eating strawberries, raspberries, &c. at home and now can't you guess who it was—why it was no one else but Mr. Truck's son, of whom I have often bought them, he came up like all the rest that have passed here wearied and hungry, but in good health & and spirits, which at first a little astonished me, for I should have thought that any person under going such distress would look miserable, but instead of that it seems to strengthen them far more; there have been but few that have passed here who have been sickly, it seems the climate has entirely restored, even those who were quite sick at Panama, but then see what men must under go to get here, and then when they get to San Francisco they are no better off than they were at home, some perhaps may make a fortune, but I think I would sooner live at home with enough to live on than to move out in this country; although, since I have come out, I do not intend to return until I have seen the great attraction, that is the "gold mines" and when I return I will tell all about it; perhaps you think that will be a long time to look forward to, but you do not know I may be home quite soon. If I were to advise any one coming out I would tell him to provide himself with good substantial clothing and start in the steamer for Chagres, then take a canoe and travel up the river, I would not advise any one to take the steamboat for it goes past the most beautiful part of the river without ones being able to see it, but in the first place I neglected to say that one or two blankets wrapped in an India rubber blanket would be found indispensible on this river, and one should take care never to use the India rubber blanket as a cover for it keeps in the perspiration and makes one very cold while the other blankets absorb it. After arriving at Gorgona a horse, or rather a four legged animal, which I do not think can be called a horse, must be hired, which will cost $10 and ones baggage can be sent over by Leach & Co at the rate of $6 pr. hundred, perhaps the charges are less than they were when I crossed; and after arriving at Panama ones blankets become useful again, for if one goes to the Hotel I believe he will have to use them, but I think the best way is to hire a room and eat at the hotel, for it is much cheaper. Then take passage in a steamer for San Francisco, or if one can get a ticket in New York to carry him through to San Francisco it would be better, for then he need not leave the states until he is certain that he will meet the steamer on the Pacific side. I think no one ought to leave the states with less than $500. for I think it will take every cent of that and more too. I neglected to say that when one leaves the Steamer, it would be well for him to have a small basket with some ham, sardines, crackers &c. such as he likes, to eat on the way up the river, for he will not be able to get any thing more than coffee at the small ranches—without it is a great deal improved since I came along. If you wish to give any one advise on this subject I would advise you to tell him *to stay at home* if he is making any sort of a good living, or without he is a hard working man and can stand hard labour, then encourage him—but don't advise him to bring any thing to speculate on, without it is a number of tents for which I think he will find ready sale, and also find them very usefull to himself, for when I last heard from there, it was stated that the city of San Francisco was increasing very fast and that people who landed that had no tents would have to put their trunks in a pen which is called a corral, and sleep in the streets, so it is said, but as soon as I return from there I will tell you all about it.

So I will have to bring this to a close but will write on the back a few lines to father with your permission believe me as ever your affectionate brother

G. Clinton Gardner

GCG to his father, n.p., September 15 and 20, 1849

Private

Dear Father

I [am] much obliged to you for the Spanish Dictionaries, which I received by the last mail. We have I think ascertained positively that Frémont is the next Commissioner, but he as yet has not arrived.[60] What Major Emory will do I cannot say but I think he does not intend to remain here if he can help it,[61] having overheard some conversation, I am led to believe that the command of the Pacific coast survey has been offered to him: if so I should think he would accept it, since Frémont has been made Commissioner, although I as yet know nothing about it—but if he does I think I can be employed on it also; at any rate I intend asking him to appoint me Junior Ast. which if he does will be a place that I can rise from, according to my merits and will also be perminent for I think the coast Survey will always be in operation and then I will be following the profession which I have selected.[62] An other advantage may arise from it, for I think if he is appointed to that place he will, before commencing operations I think, return home to see his family, if so I can get him to appoint me from Washington and the commission is bound to pay my expenses home, so I will be able to make a visit home and come back at the expense of the government; but perhaps this is all fancy *so I would request that nothing be said about it* and as there is no steamer to send this by for some time I will keep it open, and perhaps be able to give you some positive information, the reason I write now is because I think I will be sent to the mountains, if so I will not have time to write again before I go—so adieu for the present.

Sept. 20th. I am off to the mountains tomorrow morning early, so I will bid you adieu. I have heard of nothing new. I expect the new commissioner will be down in the next steamer but some say Frémont will not accept. I know nothing about it as yet. Give my love to all—remember me to Mrs. Whittwell, Meehan, &c. and believe me as ever

your affectionate
son G. Clinton Gardner

GCG to his father, Camp Riley, California, October 15, 1849

Dear Father

I was quite delighted this evening on receiving a box from you which was left at San Diego by Mon. Meehan.[63] I should have received it immediately on my arrival from the mountains as it has been some time since the last steamer went up but it was kept at San Diego until the Qr. Master [quartermaster] of our party went up there however every thing was safe and better late than never, the papers were of the latest date. I also received from San Francisco a letter from Thelisman[64] which I suppose was taken up there in the San Francisco mail it was accompanied by a small book (Spanish without a master) which I supposed from the hand writing at the top to be from John so you will please return my sincere thanks to him.

I have had quite a pleasant trip up in the mountains around the Indians but as yet we have not heard sufficient from the Gila to know if the flashing succeeded.[65] The initial point of the Boundary has been established and they have put under the post "that marks the spot" a bottle with a paper in it signed by the commissioners and surveyors of the line after which they had quite a nice little dinner given by Mr. Gray at his camp all of which happened before we returned from the mountains and since Colonel Weller returned from San Francisco.[66] Colonel Frémont is on his way here by land I suppose electioneering as he comes along both he and Col.

Weller are going to run for the U.S. Senate.[67] It is supposed here that when Colonel Frémont [arrives] he will have the money that C[harles]. L. Weller was sent back [to Washington] for and will pay us all off[.] some say brake[break] up the commission others say that he will give those that he intends to have remain furloughs until the next appropriation. Some of the members of the commission have left already. Fred Emory has gone to San Francisco among them but I think if Colonel Frémont is going to set us adrift it is better to wait a while and get our transportation home although I do not intend to return home. I think I wrote you in my last letter some thing about the Major [Emory] going on the coast survey but I have heard nothing said about it since and as to myself if Colonel Frémont sends us off I shall stay out here until I make a profession or some thing else that will take me through this world. I think if nothing turns up I will go to San Francisco.

You will see from the enclosed which is a true copy of a bill that was presented to me on my return from the mountains that some of the members of the commission are determined to make money. I paid this as you will see from the receipt without saying any thing about it because the other members of the commission who were without beds did the same, but I consider it as they do a great imposition. It seems the young man that receipted it was brought out as one of the Commission and receives a pay as I have been told from the commission to attend to the *private property* of C. L. Weller at San Diego which was brought out here as public stores. I see marked on the tents in use the number of blankets that were put up in them I suppose by the disbursing officer of this commission (C. L. Weller) who I am sure would have had his back side kicked if he had have come on to San Diego as I have heard several threaten [to] do it.[68] I only thank God, I have had nothing more to do with him, and that I did not start with him from Washington. Colonel Weller I think will pass[,] the faults to find with him is that he loves Women, cards and liquor too much but then it does not interfere with his public duties he is vain and selfish but I think he tries to make those around him comfortable and I have no doubt but what he will fix every thing right before he leaves. I understand that General Condé [Pedro García Conde], the Mexican commissioner has, since the recall of Colonel Weller, written home to his government requesting to be recalled also, as he does not wish to serve with any one but Col. Weller; so it is said.[69]

About the political news of the country I can say but little although supposing it would be of the greatest interest to you I will write you word for word that which Dr. [E. Kirby] Chamberlin the surgeon of the commission who I believe has figured a little in that line about Cincinnati has told me.

"Judge Burnet [Peter H. Burnett] and Mr. [W. Scott] Sherwood (formerly from New York) are the prominent candidates for Governor. Colonel Weller, Dr. Gwynn [William M. Gwinn] of New Orleans, Colonel Frémont and Hon. T. Butler King are the candidates for the U.S. Senate and their chances stand in the order of their names. The election is to be held on the 7th of November.[70] The Legislature to meet in December. In the new Constitution Slavery is prohibited as well as the introduction of Free negroes into the state. This clause passed almost unanimously. So much for the Great Wilmot proviso bugbear, which has disorganized both political parties in the states. It has proved a mere shadow, at which politicians have frightened themselves to distraction."[71]

This is what Doctor told me and I believe he got it from good authority.[72] I think the gentleman that has been made public printer or rather the state printer [J. Winchester] told it to him when he passed here on his way back for a press he was in the convention that was held at San Francisco all the time he says that ten to one in the convention were in favour of Colonel Weller.

Your affectionate son Clinton.

I remember Uncle Clint wanted to make postmaster at San Francisco. Tell him there is no post master at San Diego he had better try for that. California I think is to be a democratic state.

GCG to his father, San Diego, Sunday, October 28 and 31, 1849[73]

This morning we see several of the party of Col. [James] Collier the col[l]ector of customs for the ports of California straggling into San Diego like a great many of these starving Emigrants it seems they have run short of provisions.[74] I have heard that he started from home with one hundred thousand dollars and has spent it in bringing his party out to this country.[75] He was escorted by a company of dragoons under the command of Capt. Thorn who got drowned by being capsized in crossing the Gila with two Mexicans and one dragoon it seems that Capt. Thorn was quite a good swimmer but the two Mexicans who were drowning also got hold of him—one by the leg the other by his arm and carried him under.[76] From what I have heard from the Gila I should think the Emigration into this country was now at its highest stage. Lt. [Cave Johnson] Couts, who is in command of the escort to Lt. Whipple at the Gila & Colorado it is said is continually applied to for provisions and as it was thought when he left that he would not be gone more than three months they did not provide themselves with more than enough rations so he has been compelled to send back for more as he says he could not resist the supplications of starving Women and Children as long as there was any thing eatible in his camp you have no idea of the great deal of suffering which these poor Emigrants have to go through to this country which were it not for the Gold would be worthless.[77] What seems strange to me is that these Emigrants think when they come to a Government Officer or in to one of their Camps they are bound to support them and never do they think that others can be put in the same position as themselves. I should not be surprised to see the whole party from the Gila coming back saying that they had been eaten out and had to return for provisions cannot be taken across there in a day. I just begin to see the edge of my paper so I will have to bring this to [a] close and finish by writing to Sarah. Love to all.

Your affectionate Son Clint

[PS]

I open this again [on October 31] to tell you of a melancholy occurrence which took place day before yesterday, the 29th. Mr. Gray started to go across to the Gila by a short route in the direction of the Boundary line and got some where on the desert when he met Col. [James] Collier, he was so much pleased with his route that he turned back and guided Col. Collier and some of his party over it, saying it was a good deal shorter[78] but the remainder of his party (Col. Collier's) that did not come on it got in San Diego some days before, so the members of the commission commenced to run him about his new route, and sad to say a great many of them were a little drunk, among them Mr. Gray and Col. Weller both of whom got in a dispute and Col. Weller attempted to choke him and in the scuffle, Mr. Gray shot Col. Weller through the thigh he is now getting better but will be confined to his bed for some months. Pretty thing to be talked of in Washing[ton].[79]

G. C. G.

GCG to his sister Sarah, Camp Riley, October 30, 1849

Dear Sarah,

I suppose before this reaches you[,] you will have finished your tour at the north, and as there is no news from this part of the world to interest you, perhaps a short description of my journey in the East, would be sufficient to fill a few pages so as you can say I still write to you but rather dry letters.

We started from here on the morning of the 22nd of last month about 7 o'clock and after travelling S by SE, SE E by SE and E we arrived at Old Arquillo's ranch about 12 o'clock,[80] which at that time to our great satisfaction was visited by all his daughters. I believe I have told you the number of his family (21 children, 26 grand children and 3 or 4 great grand children); our escort was there in readiness to move with us at any moment, so after spending a very pleasant evening we returned to our camp, which we had taken care not to put very far off and which I had taken the liberty to call Theresa, after, in my opinion, one of the handsomest young ladies in California and one with whom I had the pleasure of winking, blinking and walking all the evening, which was the only way I could entertain her, as I have not mustered enough courage to speak Spanish yet, her tongue was going like a mill race. Well after a pleasant dream we were all up and ready to start (long before you had any idea of such a thing. I wonder if you remembered me when you went to church that sunday morning while I was mounted on a beautiful gray charger? I don't think I thought of you once) about day break but as the mules had never been packed before, it was about 9 o'clock before the whole train was on the march and then we had to move very slow, as some of the packs were continually coming off, for the reason that the soldiers had never had much experience in packing and some of the packs were rather unhandy to pack, but notwithstanding all our troubles we made quite a good march and arrived at a place called Tecate about 9 o'clock p.m., rather late marching, but there was no water to be had for our animals any nearer, after pitching our tent and taking a good nights rest we got up on the 24th rather tired, so we thought we had better spend the day there which we did and amused ourselves by hunting, so the next day we started again on our journey and after travelling until about 15 minutes after 3 o'clock p.m. we found ourselves in a very handsome valley just at the foot of Mt. Los Piños one of the mountains selected for the flashing, and as Capt. Hardcastle and myself were to remain there we left half of the escort, and about ½ past 7 o'clock on the 26th we got under way again, and after travelling until about 12 o'clock we came to very pretty valley and quite a large spring, the spring seemed to be at the junction of two vallies and the Indians called it the Junta, we stopped there about an hour to water all the animals, then proceeded to a place called Iacasei, at which place we arrived about 5 o'clock p.m.: but before coming into camp the Interpreter told Capt. that the guide did not know any farther on the road, so we had to send the guide off after an Indian, who was about 10 miles off and that detained us at that place all the next day, which I spent in looking after curiosities and, by the by found as I thought two precious stones but as luck would have it, lost them before I could get in camp to examine them we found several minerals and came to the conclusion that it was the proper formation, in which gold is found but I don't think we gave it the proper test, we should have washed in the mountain streams and the reason we did not, is a decided case of "quien sabe" [*¿quién sabe?*—who knows?]. Well about 10 minutes before seven every one was ready for a march on the 28th, so we moved on and arrived at our destination, Mt. Wicammon about 2 o'clock p.m. Mt. Wicammon, which is quite a tall mountain being over 5000 ft above the leavel of the sea, was selected so as the flashes could be seen from the junction of the Gila and Colorado Rivers. Los Piños is also quite as high and is about half way between Wicammon and Cerro Colorado which is near the ranch we came to the first day out,[81] the flashes made at Wicammon were [to be] observed by Lt. Whipple at the Gila and Capt. H[ardcastle] & myself at Mt. Los Piños, and those made at Cerro Colorado were observed by the Major at Camp Riley and us at Los Pinos &c. it was in this way we were to connect the points by an astronomical line.[82] On the day after arriving at Wicammon Capt. went on top of the mountain but on account of the fogs could see but little, the next day I went up but it was almost the same, so on the next day about 2 o'clock we both went up to signal Lt. Whipple, but got no answer so on the 2nd of Oct. about 7 o'clock Capt. and myself with three dragoons, started back to Los Piños, so as to be ready for the flashing which commenced on the 5th, and arrived at that place about 8 o'clock p.m.. so we had two days to rate the Chronometers and on the 5th went up on Los Piños and could see the flashes very good, we went up 4 nights in succession and got lost coming down the last time, the consequence was I tore all my pants off.[83] On the 8th of Oct. Dr. Chamberlin came in from this place, to see how we were and brought me a very interesting letter from Thelisman, and on the 9th the party from Wicammon came into our camp so on the 10th, about 7 a.m.. we started for home, that is this place, we came across a great number of Indians, but they are a poor miserable set of human animals. I hate to call them to my remembrance again, so I will say no more about them[.][84] we came back here in three days and I am glad to say that it agreed with me in every respect, except my pants for I now look much better than before I went.

I am afraid I am taking this subject too far, so I will close and bid you good night.

Your affectionate brother

Clint

P.S. I neglected to thank you for the purse but it is better late than never.

G. C. G.

I must thank Alida for her note and will write to her next steamer if Mon. Meehan had not have brought that box it would have been four months since a mail at San Diego. Thelisman's letter I should have received four months ago but it went up to San Francisco. G.C.G.

Diegueños. Gardner had nothing good to say about Indians in the San Diego area, but the artist Arthur Schott, who later worked for Emory, portrayed the Kumeyaay, or Diegueños, more sympathetically. From W. H. Emory, *Report on the United States and Mexican Boundary Survey* (1857–59). Courtesy of the DeGolyer Library, Southern Methodist University, Dallas, Texas

GCG to his mother, n.p., October 31, 1849—on same page as end of previous letter

Dear Mother

Before I close this scrawl, I must return my sincere thanks to you for that box which contained almost the whole family when I received it, it was like making a visit at home and it calls to my remembrance the by gone days when we were all together which seems but yesterday, how true as Sarah's favourite author says, "To be pained for a minute, to fear for an hour, to hope for a week—how long and weary!

But to remember four score years, is to look back upon a day."[85]

Just before going to the mountains I opened those socks and took three pr. out as my old ones are almost gone, and one pair of the gloves I have entirely worn out the trip in the mountains have finished them. I am trying to increase my wardrobe by some flannel which I think I can get from the soldiers so goes my clothes and my [word missing] now tells me to stop so I must obey.

your affectionate son Clint.

---

1. Hewitt (1990, 177) sketches a biography of García Conde from several sources.
2. The commissioners held their first official meeting on July 6. The *California* made the trip from Panama to San Diego, with a stop in San Blas, in twenty days. It continued to San Francisco, arriving on July 15 (*[San Francisco] Alta California*, July 19, 1849; Hewett, 1990, 181).
3. The property, known as the Rancho Melijó, was acquired by Santiago Emigdio Argüello in 1833. Emigdio's father, Santiago Argüello Moraga, had owned a rancho directly to the south, the Rancho de Tia Juana, since 1829 (Padilla, 2004, 30, 33; Pourade, 1965, 65, 151). Pourade identifies Santiago Emigdio as Santiago Emilio.
4. Emory, 1857–59, 1:4. The phrase quoted is from GCG to his sister Sarah, July 23, 1849.
5. Weller to Emory, San Diego, July 9, 1849, copy in Emory's letterbook, box 1, f. 1, EP.
6. Emory to Weller, Panama, May 7, 1849, copy in Emory's letterbook, box 1, f. 2, EP.
7. The best account of the controversy and its resolution appears in Hughes (2007, 126–47). Weller became aware of the problem at the end of July, by which time he was already planning to go to San Francisco to get funds for the commission (Weller to Emory, San Diego, August 1, 1849, copy in Emory's letterbook, box 1, f. 1, EP).
8. Several sources explain the problem, including Lesley (1930, 11). Whipple's journey is unusually well documented: Whipple (1961), Couts (1932), and Starr (1983, 165–84). Scharf (1973, 18–31) provides a good overview.
9. "The no less scientific, but more arduous service of conducting the intermediate operations" (Emory to Alexander H. H. Stuart, Secretary of the Interior, New Orleans, September 25, 1851, in SED, 119, 77).
10. Emory to Hardcastle, Camp Riley, July 20 and July 26, 1849, in HCMSA; Rebert, 2001, 102.
11. Emory, 1857–59, 1:4.
12. GCG to his sister Sarah, Camp Riley, California, July 23, 1849.

13. A guide to surveying of the era described the method of establishing stations and triangulating between them, then went on to say, "Observations are also made at the principal stations upon the pole-star, and other stars near the pole, for the purpose of determining the angle, made by the sides of the triangle with the meridian" (Davies, 1866, 179). For a cogent description of astronomical and surveying methods, see Rebert (2001, 28–40).
14. Hughes, 2007, 135–36.
15. Emory, 1857–59, 1:4.
16. Cerro Colorado stood on the eastern edge of the Rancho Tijuana, about four miles from the ranch house, which was on the Mexican side of the border. The name is still in use for this hill, or *cerro* (Fireman, 2004, 51; Padilla, 2004, 33–34). Los Pinos and Wicammon are hard to identify. GCG, writing to Sarah from Camp Riley on October 30, 1849, reported that both were more than five thousand feet high and that it took just a day to travel from Tecate to Pinos and from Pinos to Wicammon. To find taller peaks the party would have had to move some miles beyond the boundary line itself. Los Pinos, which GCG wrote incorrectly as Piños, may have been present-day Los Pinos, whose elevation is 4,805 feet; it stands about ten miles due north of the border near Buckman Springs. On the other hand, Rebert puts Los Pinos in northern Baja California (2001, 61). The name Mount Wicammon is lost to memory. The historian Charles Hughes, who is deeply knowledgeable about San Diego County, suggests that it may have been what is today Monument Peak (alt. 6,321 feet) in the Laguna Mountains, about thirteen miles (as the crow flies) to the northeast of Mount Pinos. From that peak one can see the mountain ranges flanking the Colorado River. Clint's party traveled between each of these mountains in a day—arduous travel, given the terrain.
17. Rebert, 2001, 30.
18. Whipple to Emory, Capital Hill, Right bank of Rio Colorado, opposite the mouth of Rio Gila, October 10, 1849, copy in Emory's letterbook, box 1, f. 1, EP. For Effinger on Cerro Colorado, see Fireman (2004, 44) and Emory (1851, 9).
19. Whipple to Emory, Junction of Rio Gila with Rio Colorado, October 17 and 23, 1849, copy in Emory's letterbook, box 1, f. 1, EP.
20. Whipple to Emory, Capital Hill, Right bank of Rio Colorado, opposite the mouth of Rio Gila, October 10, 1849, copy in Emory's letterbook, box 1, f. 1, EP.
21. Foreman, 1937; Etter, 1998, 56–57. Hughes points out that locals already knew the route (2001, 137).
22. Whipple started back to San Diego the next day.
23. Weller's General Order no. 1, Panama, April 11, 1849, copy in Emory's letterbook, box 1, f. 1, EP.
24. GCG to his mother, August 18, 1849.
25. Emory, 1857–59, 1:6.
26. Ibid., 1:8.

27. The reports appeared first in the San Francisco press and then traveled south (Couts to Emory, San Diego, September 5, 1849, copy in Emory's letterbook, box 1, f. 1, EP).
28. This letter is in the Smithsonian Institution.
29. *Dry out* does not appear to be a nautical term. Perhaps GCG referred to a becalming, but this can only be inferred.
30. San Blas had been the major shipyard for supplying Alta California in the Spanish period.
31. The ship was apparently in the harbor, not tied up at a dock; they would have to pay to be rowed out to it.
32. GCG was traveling steerage because he had given $100 to a fellow American stranded in Panama, as he explained in a letter to his brother Charles, Panama, June 22, 1849 (see chap. 1), and later in the present letter.
33. Richard Worsam Meade II was one of Ingraham's mother's three brothers. Another of her brothers was Gen. George Meade (Cleaves, 1960, 355). Haskins identifies a Captain R. Meade, who sailed for California on the *Crescent City*, which ran between New York and Chagres (1975, 304).
34. The *Alta California* of July 19, 1849, noted that Weller and his party had arrived in San Diego on the first of June; the Mexican commissioner and his party had arrived on July 3.
35. Capt. Cleveland Forbes of the steamer *California* described San Diego as "the meanest place of all, not a morsel of fresh provision and the town is 4 miles from the anchorage" (Pomfret, 1954, 234).
36. Major Emory echoed GCG's impression in his personal account of the survey: "The news of the discovery of gold in the northern part of California, produced less commotion in that quiet town [San Diego] than in New York or Panama. Fortunately for us, it did not feel the effect until the reaction came from the Atlantic side, some months after our arrival. Had it been otherwise, all attempts to keep together the enlisted men and laborers of the survey would have been idle, and the commission would have been disorganized before doing anything (1857–59, 1:3).
37. On July 20 Emory ordered Hardcastle to explore to the east of San Diego, sending him with a military escort. See the introduction to this chapter.
38. Here GCG is referring to himself in the third person.
39. "Charlie" is GCG's brother Charles. For a vivid description of the crowded conditions, see Kemble (1972, 153).
40. Two argonauts from Washington, D.C., whom GCG mentioned in an earlier letter to his family, May 13, 1849.
41. Taylor (1949, 38) also mentioned the appearance of this slough, which became known in 1849 as New River, a name that GCG would later learn (see his letter to Alida, April 28, 1851). Water in New River was particularly high in the spring of 1849 (DeBuys, 1999, 39; Forbes, 1965, 299). See too Sweeny (1956, 115–16), and Rebert (2001, 107).

42. Newburgh on the Hudson. Both of Clint's parents had grown up there and relatives remained in the area.
43. According to the abstract of disbursements made by Weller and filed with the fifth auditor of the treasury on February 9, 1850, GCG's salary as assistant to Major Emory remained at $400 per annum (SED 34, 51). But that appears to be an error. Weller had written to Emory while still in Panama to say that "Messers Gardner and Ingraham I presume you intend to employ as computers. If such be the fact let them be so reported and their compensation will be fixed at once," that is, raised from $400 to $600, which was the rate for computers (John Weller to Emory, Panama, April 8, 1849, copy in Emory's letterbook, box 1, f. 1, EP). Emory not only failed to take the initiative but prevented GCG from being promoted (GCG to his mother, San Diego, April 6, 9, and 20, 1850). Weller raised GCG's salary to $600 a year nonetheless, along with all other salaries, "in consequence of the extraordinary prices paid for labor in this country" (Weller's General Order no. 2, July 11, 1849, copy in Emory's letterbook, box 1, f. 1, EP). On August 1 Weller increased Clint's pay to $700 a month, where it stood for the next year (Weller to Emory, list of employees, San Francisco, April 1, 1850, box 4, f. 25, EP).
44. His mother's brother, John McLean. See the introduction. GCG's salutation on this letter is "Dear Uncle," but he addressed the letter to "John McClean, Esq., Capitol Hill."
45. In 1850 Johnny Ingle was the fifteen-year-old son of the Gardners' next door neighbor John P. Ingle, a clerk. in Washington, D.C. (U.S. Census, 1850, Ward 5, Washington, D.C., roll M432_57, p. 27; Hunter, 1853, 53). Together with Joseph Antoine Alcide Mouton, John McCalla, and Custis Lee, these young men apparently formed GCG's circle of friends in Washington. Ingle, GCG complained three years later, never responded to his letters (GCG to his sister Sarah, Camp, October 3, 1852).
46. GCG means to say that the *California*, which brought him to San Diego, also carried the mail, including a copy of the Treaty of Guadalupe Hidalgo, apparently sent to him by his uncle John.
47. GCG's oldest brother.
48. It was the *Panama* that arrived in San Francisco on August 18, 1849 (Kemble, 1972, 242, 246–47; Wiltsee, 1938, 10, 19, 21).
49. Odeide was in her midteens. The 1850 census lists her as age sixteen (U.S. Census, 1850, Western District, Lafayette, Louisiana, roll M432_232, p. 265). She was Emma's stepdaughter, and Emma apparently did not think it proper for GCG to be corresponding with her directly. See the introduction.
50. Because of the rumors that Weller would be replaced, he could not cash a draft on the government to pay his men (Goetzmann, 1991, 164).
51. Commissioner Weller appears to have misappropriated government funds. His "abstract of disbursements [purchases]" includes all the necessities of camp life: soap, candles, "camp equipage," cooking utensils, groceries and supplies, blankets, hardware and tools, hospital stores,

etc. (SED 34, 50). Perhaps Commissioner Weller bought these items to disperse to his personnel or to reimburse them for their outlay. For example, there was a disbursement of $42 to his brother Charles (the commission's disbursing officer) for blankets. GCG's statement, however, suggests that Commissioner Weller used government funds to buy supplies (from his brother, among others) and then sold items to his employees that he should have supplied to them. In a letter to his father dated October 15, 1849, GCG accused Charles Weller of similar behavior. George Bartlett, the brother of the next boundary commissioner, John Russell Bartlett, was accused of the same corrupt behavior.

52. Weller traveled on the *Panama.* The journalist Bayard Taylor was also aboard and chronicled the voyage in his best-selling book, *Eldorado.* See Taylor (1949). Weller had gone to San Francisco to negotiate, with Frémont's help, a $10,000 bank draft using his private property as collateral. When the draft was protested because of Weller's removal from office, the San Francisco banks seized Weller's property, although Emory states that "at that very time, as the settlement of his accounts afterwards showed, he was in advance to the government." In the meantime Major Emory managed to borrow from local army posts to maintain his troops (Goetzmann, 1991, 164; Emory, 1857–59, 1:5–6).
53. Goetzmann succinctly described the temptation put before survey personnel: "The men had to work without pay in a country where everyone else had a chance for easy wealth" (1991, 165).
54. The Zenith telescope, designed to observe stars as they pass directly overhead, was used to determine latitude.
55. Whipple left on September 11.
56. Emory had assigned Whipple, Parry, and Ingraham to "determine the other extremity of the straight line forming the boundary at the junction of the Gila and Colorado," in addition to which Parry was to carry out his geological and botanical observations (Emory, 1857–59, 1:4).
57. Emory's account of the survey reports that word of the gold discovery "reached our hitherto quiet and secluded camp" in September, but GCG's letters make it clear that it occurred much earlier (Emory, 1857–59, 1:5).
58. A reference to Taylor's sweeping out Democrats?
59. The *Brothers* never reached its destination. Although the exact location of its wreck is not known, it is presumed to have gone down in the vicinity of the Farallons to the west of San Francisco, where three more ships were lost the following year (Marshall, 1978, 90–91). Prevailing winds and currents came from the north, making it laborious to tack up the coast.
60. GCG was right. Frémont had received the appointment, along with instructions to deliver Weller's recall, "when you are about to enter upon the duties of the office" (John M. Clayton, Dept. of State, Washington, to Frémont, June 28, 1849, in SED 34, 10). For Frémont's acceptance, see Frémont to Clayton, San Jose, August 1849, in SED 34, 28.
61. Emory, who had testified against Frémont in his famous court-martial, regarded Frémont's appointment as an insult. On September 15, the same day GCG wrote this letter to his father,

Emory sent a letter of resignation as astronomer to Secretary of State John Clayton and another to Col. John J. Abert, Emory's immediate superior, asking to be relieved of command of the military escort. Clayton would have acceded to his request, but Secretary of War George W. Crawford denied it. Emory was forced to stay on in both positions (Goetzmann, 1991, 163–67; Norris, Milligan, and Faulk, 1998, 87–88, 93–94). For Emory's version, and his resignation letter, see Emory (1857–59, 1:6–7).

62. The acquisition of California following the Mexican War required a survey of its twelve-hundred-mile coastline; the discovery of gold and the concomitant traffic along the coast gave the project unprecedented urgency. Emory's father-in-law, Alexander Dallas Bache, was superintendent of the Pacific coast survey, but Emory was not put in charge of field operations. The position went to George Davidson, who, like Emory, quickly discovered the difficulty of maintaining a party of low-paid government employees in such proximity to the placers (Smith, 1987, 18–19).
63. The 1850 census lists a C. H. W. Meehan, age twenty-eight, living in the fifth ward of Washington, D.C., and working as a clerk (U.S. Census, 1850, Ward 5, Washington, D.C., roll M432_57, p. 11). Haskins lists a W. Mehan as a member of the Pittsburgh Enterprise Company, which left for California on April 13, 1849 (1975, 305). Mon. Meehan was apparently a relative of the Meehan whom GCG mentioned in the previous letter.
64. The 1850 census does not identify anyone in the United States with the surname of Thelisman and only one individual with this given name: twenty-one year old Thelisman Bienvenu of St. Martinsville, St. Martin Parish, in Louisiana (U.S. Census, 1850, St. Martin Parish, Louisiana, roll M432_240, p. 198). He may have come into GCG's acquaintance through Clint's sister Emma, who had married into a Louisiana family in 1842.
65. GCG and Hardcastle had just returned from Los Pinos, where they had set off rockets and flashes of gunpowder. Clint described the journey in detail a letter to his sister Sarah, Camp Riley, October 30, 1849. See too the introduction to this chapter.
66. The ceremony took place on October 10.
67. After his recall as boundary commissioner, Weller moved to San Francisco and opened a law office. In 1851 he succeeded Frémont as U.S. senator from California. In 1857 Weller was defeated for reelection to the Senate but won the governorship (1858–60) and following that an appointment as minister to Mexico before retiring to private life (*DAB* 19:628–29).
68. George Bartlett, the brother of the next boundary commissioner, John Russell Bartlett, was accused of similar behavior.
69. We have found no evidence to substantiate this rumor.
70. GCG refers to the creation of California's first state government. Burnett was elected governor on November 13; Frémont and Gwinn were elected to the Senate by the California legislature, in joint session, on December 20.
71. Introduced in 1846 by David Wilmot of Pennsylvania as a rider to a House appropriations

bill, the famous proviso was often amended and never passed, but its central tenet, that slavery should not be extended to the new territory acquired from Mexico, was much debated by both political parties. California preempted discussion in Congress by presenting it with a fait accompli.

72. GCG's source for this information, Dr. E. Kirby Chamberlin, had been elected to one of two seats in the state senate for the district of San Diego and Los Angeles.
73. GCG's date on this letter "Sunday, July 28," is an obvious error. The events he described took place in late October. July 28 was not a Sunday in 1849; October 28 *was* a Sunday. Assuming he wrote the correct day of the month, this letter should be dated October 28. He dated his P.S. October 31.
74. GCG has more to say about Collier later in this letter.
75. This was a particularly galling report in light of the $50,000 appropriation made to Weller for the whole boundary survey (H.H.R. [Hiram H. Robinson], San Diego, November 1, 1849, in the *Cincinnati Enquirer*, December 13, 1849, reference courtesy of Charles Hughes).
76. Capt. Herman Thorn drowned on October 16, 1849. Lt. Cave Couts, who commanded the First Dragoons and with whom Thorn served, told the story much as GCG reported it, except that GCG had the wrong river. It happened on the Colorado, and the craft that capsized was a canoe (Couts to Emory, Camp Calhoun, October 17, 1849, copy in Emory's letterbook, box 1, f. 2, EP; Heitman, 1965, 2:958; Couts, 1932, 46).
77. Cave Johnson Couts came to southern California in the spring of 1849 with the army sent to occupy the region following the U.S.-Mexican War and was later assigned to the boundary commission. He decided to settle in San Diego and resigned his commission in 1851, married Ysidora Bandini (whom GCG mentions later in this chapter), and became a rancher prominent in local affairs (Scharf, 1976, 10).
78. Gray and his men came across Collier's party, which included a detachment of "semi-starved" dragoons, in the desert. Gray decided to send his assistant on to the Gila to do the surveying while he guided the men to San Diego himself. Because of the condition of their animals, Collier and his men believed that the dragoons would not make it by the wagon road, and Gray was not sure they would make it by the pass he had just come through unless he guided them: "I would not send my assistant [John Harris Forster] back to guide this party, because I wished to be there myself, in case of any accident or obstacle occurring, it being the first time that I believed [the pass's] ascent had even been attempted with packs" (Gray to Weller, San Diego, November 7, 1849, in SED 34, 45–46).
79. GCG says both men were drinking. James Nooney wrote in a private letter that Weller "struck and choked Gray and Gray very naturally got out his pistol and shot Weller" (Nooney to C. C. Parry, Camp Riley, November 9, 1849, PP). Hiram H. Robinson, a supporter of Weller's, says that Gray was intoxicated and that Weller was only making an innocent joke (H.R.R.

[Robinson], San Diego, November 3, 1849, in the *Cincinnati Enquirer*, December 12, 1849, reference courtesy of Charles Hughes). Weller regarded Gray's return with Collier's party as dereliction of duty: "Without entering into any discussion upon the subject, I must be permitted to say that your reasons for returning without making the personal examination of the mouth of the Gila River (for which your expedition started) are far from being satisfactory to my mind" (SED 34, 47). The fight between Gray and Weller, which must have been deeply embarrassing to both men, was not mentioned in the commission's official published reports or correspondence, but the press picked it up from Robinson's letter in the *Cincinnati Enquirer*, November 3, 1849, and the story appeared in the *New York Express* and the *(Washington, D.C.) Semi-Weekly Union,* December 20, 1849. Others, like GCG and Robert P. Effinger, described it in personal correspondence (Hughes, 2007, 146n41).

80. "Old Arquillo" was Santiago Argüello Moraga (1791–1862), who owned the sprawling Rancho de Tia Juana. He and his wife, Pilar Ortega, had twenty-two children; GCG was not exaggerating. Santiago's son, Santiago Emigdio, owned land to the north, where Emory established Camp Riley (Bancroft, 1884–1990, 2:702).
81. Effinger put it four miles from the ranch (Fireman, 2004, 51).
82. For a description of this surveying process, see Emory (1857–59, 1:4–5) and SED 34, 30–31. GCG's letter nicely complements Emory's more general report.
83. Emory says the flashing was done "for five nights in succession" (1851, 9).
84. GCG's disdain for the Indians seems to have been shared by other members of the party. Effinger, back from the same trip to the mountains, elaborated: "A more miserable race of beings never polluted God's creation than the Indians of this country. They are too lazy to work, too cowardly to fight! But simply have enough cunning to steal what will barely sustain them. They are no more account that the coyotes (wolves) that infest this mystic land" (Effinger to his mother, Camp Riley, October 3, 1849, in Fireman, 2004, 44). Such attitudes led to wholesale killing of Indians in California, a subject that historians have well documented. See, for example, Hurtado (1988) and Trafzer (1989). Had he stayed into the fall of 1851, Clint might have gained respect for San Diego's Indians. Led by Antonio Garra, apparently a Quechan, Cupeños and others revolted in San Diego's backcountry that November (Phillips, 1975, 71–94).
85. GCG has this just right, from a poem in blank verse, "Memory," by a then popular but now obscure English poet and moralist, Martin Farquhar Tupper (1851, 28).

CHAPTER 3
"Glorious State of Uncertainty"
November 15, 1849–July 2, 1850

Although the Mexican and American commissions had established the terminal points of the California survey on the Pacific and the Gila by October, they had made little progress in drawing a line between them, making detailed maps, or placing boundary markers. Then, in early December 1849, the American team shut down its operations entirely. The Mexican commission was still in the field, waiting at Yuma to continue work, but the American commission was out of money and would remain broke through the next summer.[1] The commission's remaining civilian staff, all owed back wages, continued to gravitate toward remunerative employment, and, as winter wore on, John Weller put the employees who had stuck with him on unpaid leave, or furlough.[2]

During this difficult time Clint held his place with the commission but went on leave and stayed in San Diego. On June 30, 1850, nearly a year after he first arrived in California, he told his parents that the surveyors in California remained in the "glorious state of uncertainty that we were when we first arrived."[3]

Clint had moved from Camp Riley into San Diego in early November 1849. A small town, much of whose male population had headed north to the goldfields, San Diego consisted of a Mexican town founded in 1769 and a port, La Playa. In town Clint found employment as a clerk with Capt. Julius Hayden, who had been "an old school mate" of Clint's brother Frank. Although the commission had no money to pay Clint's salary, the army did and the position was a godsend.— At some point that winter Clint moved into a room in Hayden's house, where the captain and his wife treated him "as one of the family." He continued to work for Hayden and live with him until June 1850.[4] Good that he did. With demand exceeding supply, rents in San Diego soared.[5]

In his free time Clint rode a horse, Old Whitey, waited for the mail, courted local ladies, and danced and drank late into the night—especially on Sundays, as was, he told family and friends, the local custom. Although the United States had conquered California, the elite Mexican families found it in their interest to accommodate the new order and to welcome and socialize with the Americans. As one of Clint's compatriots understood it, "most Californians hereabouts were on the American side in the war, and are proud of calling themselves *Americanos*, —'*No Mexican!*'"[6] Foremost among the welcoming party in San Diego was Juan Bandini, who served as alcalde, and his wife, María del Refugio. Juan Bandini, one astute historian has suggested, "was probably not pro-Yankee . . . but rather pro-winner."[7]

In San Diego, Clint took comfort in feeling superior to the Californios, whom he described to his mother as "indolent," and to Indians, who were beneath his contempt.[8] His sense of superiority did not stop him, however, from dancing with dark-skinned women and perhaps sleeping with them as well. As he wrote to a boyhood friend after a night of heavy drinking, "we danced until morning clear up till the sun came out after which we went home with galls some were black some were blacker some were the color of a chew tobaker [tobacco] or rather the color of a dark indian squaw." Dancing with dark-skinned women would have shocked his friends in Washington, who would have considered such women racially inferior, but Clint explained it away: "Such is the custom of the country."[9]

Clint's feelings toward San Diego seem to have changed during his first year there. In October he told his father that "this country . . . were it not for the Gold would be worthless."[10] The next April, though, he wrote to his mother: "I think it is one of the most delightful countries to live in that I ever was in."[11] He described the land as fertile, capable of growing a variety of fruit and grains.

Although San Diego had grown on Clint by April 1850, his relationship with Emory had soured. Clint had "several little spats with him" and was ready to quit.[12] Emory, it appears, had asked Clint to do more than he was prepared to do, and Clint thought Emory was in bad humor. Within weeks, however, the two reconciled.[13]

If Emory was in a bad humor that spring, it was for good reason. He found himself in charge of the penniless commission. The partisan attack on Commissioner Weller, which had begun right after his appointment, had continued in Washington. On December 19 Zachary Taylor's secretary of the interior, Thomas Ewing, a Whig from

Weller's home state of Ohio and Weller's political archrival, once again notified Weller of his dismissal.[14] A Democratic newspaper in Washington offered an explanation that was not far off the mark. Weller had done his work ably and economically, the paper said, but he was a Democrat and so a victim of Taylor's Whig administration. "We have no doubt the removal of Col. Weller will add thousands to the expense of the commission, which the people will have to pay; and all to gratify the political malice of the cabinet, instigated by Thomas Ewing. If they had not been stimulated by a most raging thirst for the political blood of Col. Weller, they would not have removed him before ascertaining whether or not his successor would accept the office thus vacated."[15]

A few weeks after dismissing Weller, Thomas Ewing appointed Major Emory to take temporary charge of the commission. Unlike John C. Frémont, Emory obliged. The previous autumn he had tried to resign because of the government's failure to fund the commission and its attempt to put Frémont, a political enemy, at its head.[16] Now, with Frémont elected to the Senate and out of the picture, Emory accepted, although his first instructions from the Department of the Interior were to cut back on his expenses and fire employees who were "not indispensable to the proper performance of the work."[17]

Emory had received notification of his appointment as acting commissioner sometime in the late winter of 1850, but he had inherited a commission wracked with debts, lacking credit, and without resources to finish its work. To make matters worse, although the California part of the survey awaited completion, the focus of the survey had shifted to the area of El Paso. At their penultimate meeting on the Pacific coast, on February 3, 1850, Weller and García Conde had agreed to meet in the Mexican town of El Paso del Norte in November to begin work on the other end of the line.[18] Continuing eastward from the Gila had become unrealistic as the demands created by the gold rush made men and materiel scarce in California and pushed wages and prices to astronomical heights—which had been clear from the beginning. As one observer noted in June 1849, "To proceed with the line survey beyond the mouth of the Gila will require a large number of laborers; they cannot be had for any price."[19]

With some surveyors en route to El Paso del Norte and others having gone off to the goldfields, Emory was stuck in San Diego with a handful of employees, all on leave and some unlikely to return. Indeed, with nothing else to do, even one of Emory's army officers sought his fortune elsewhere. Capt. Edmund Hardcastle, on leave in May and June, drove a herd of mules north to the mining region, hoping to sell them at $150 a head.[20] By April, Clint reckoned that, of Emory's group, only he, Hardcastle, Charles Parry, and Edward Ingraham remained. Surveyor Andrew Gray had one employee, the draftsman Henry Clayton. Clint characteristically overlooked the few laborers who remained, such as Frank Holley.[21]

Emory's frustrating circumstances were compounded by the strain that his assignment put on his family. While he languished in California on this stalled project, his wife implored him to come home. He had been gone the better part of six years. Rebuffed by the army in one attempt to leave California, Emory wrote to Secretary of the Interior Thomas Ewing and asked him to appeal directly to President Zachary Taylor to have him relieved of duty.[22]

In early April, before he received Emory's letter, Ewing sent word to Emory that an appropriations bill with $50,000 for the commission had passed the House but was bogged down in the Senate. Nonetheless, he anticipated its passage so that "funds will be sent to you . . . by the next [steamer] departure from New York."[23] By the end of June, however, two steamers had arrived in San Diego without the promised funds. Emory was furious. It appeared that his shipping agent in New York was right when he told Emory, "Congress has not done anything as yet for the Country it has only taken care of the pay for the members thereof and has discharged a great deal of gas."[24]

Emory wrote a blistering letter to Ewing at the end of June, observing that the secretary continued to send him instructions but no means to carry them out.[25] That same day Clint asked his father to use whatever influence he might have with friends in Congress to pass an appropriations bill for the commission: "You will not only oblige me, but a great many others who have not received any pay since January—besides a great many more who have not received any since last October—and also, some one or two have not received a single cent since they have been on the Commission."[26]

Congress did not make the appropriation until August, and even then the money never reached Emory in California. Instead, the $50,000 was sent to the survey's new commissioner, John R. Bartlett, whom Zachary Taylor had appointed in May (and Congress confirmed in June) to replace Emory. Emory, who still had to complete the California portion of the survey, later described his plight: "Not one cent of it was paid, as Congress designed, but it was improperly, if not illegally, diverted from its channel and given to the new commissioner, who expended it before he got on the ground, and incurred debts in addition far exceeding this sum. The persons for whom this money was intended, who had honorably sacrificed the certainty of private fortune to a sense of duty, were left in the field without pay and without subsistence."[27] Or, as Clint Gardner experienced it, "we have nothing to eat at present—nor no money to buy anything with."[28]

Clint found other sources of money and food during this first year in California. His job with the acting quartermaster, Julius Hayden, brought Clint a steady income beginning in November 1849. He also lived comfortably with Mr. and Mrs. Hayden but not without a sense of guilt. He was, as he told his parents, "one of the lucky ones of the flock."[29] His luck diminished somewhat, however, in June 1850 when the Haydens packed up and prepared to return to Washington and Clint lost his room. His surviving letters do not mention his new lodgings. At mealtime he believed that he remained one of the "lucky ones" because Major Emory included him at his "mess"—the small group that took meals with Emory.

In many of the thirteen letters in this chapter Clint provides rich descriptions of his social life in San Diego. The letters begin in the autumn of 1849, with Clint moving from Camp Riley into San Diego, and with John Weller still commissioner. They end the following summer, with Weller gone and the commission still broke but with Clint planning to stay on in California to finish the survey with Captain Hardcastle.

GCG to his father, San Diego, November 15, 1849

Dear Father

As the steamer is expected to pass down to morrow I thought I would write you a few lines—although I have but very little time to spare. I have been living here since last Monday as Clerk to Capt. J[ulius] Hayden, the Acting Quarter Master of the Escort—which was offered to me through Major Emory's influence: and after which he did not like to have me take it, for he said that it would be taking me away from that which I have made up my mind to make my profession—but as this is, as far as I can see, the only way in which I can get money enough to take me home; and now that is of the greatest importance to me, for as soon as I return to the States I can get in the [Pacific] Coast Survey—which the sooner I do the better. I do not wish you to understand by this that I am coming home soon—but what I am after is, getting money enough to take me at any time—so I will not be compelled to live here longer than I wish. I forgot to mention that I do not give up my place in the Commission, but am only on a leave of absence, and when ever there is any writing to be done (which is the only thing I had to do, as all the other work is done) the Major told me he would get a soldier to do it, and I am to pay him $1.00 per day, which I [can] well afford to do while getting $50.00 pr. month from the Commission—or rather $50.00 pr. month is running on—but I can't draw any of it—and $100. or $150. pr month from this place—I do not know which—but I know it is not less than $100, which will suit me, but I shall try for more if there is any chance of getting it.[30] In my last letter I told you that Frémont was coming down by land—but it seems that was only a false rumour which he had circulated, so as we would not expect him until the last of last month, at which time I suppose he did not get through electioneering: or perhaps he wanted to keep Col. Weller from coming up there to run against him, so he sent down by the steamer $5000. and placed $4000 more at his disposal up at San Francisco, from all of which I could only get $100 on my account: it seems that his [Weller's] friends from Ohio got the most of it—but never mind, it will be coming some time or other, so I will have to wait.

You said you would like to know how my accounts stand—there was due me on the 10th of this month $118.33 and I have about $80. of the 100 left—perhaps you may say, why what has become of the $20? Well I paid it for my washing, which has been running on since I have been here, and several other little things. I have had my old snuf[f] coloured coat, which mother or aunty knows well, cut off so as to get stuf[f] to make new sleeves and now it is a nice little coat like my gray one.

There is no news, so I will have to get you to tell Sarah and Alida that I will write to them next time before which I hope there will be a change of some kind. Give my love to *all* remember me to the neighbours:—tell Mother I have not seen a lump of gold larger than a five dollar piece.

Your Affectionate son Clint. Love to *all.*

GCG to his father, San Diego, November 30, 1849

Dear Father,

The Steamer arrived on the evening of the 23th. we all went down to the beach to get the news but were sadly disappointed, and worse than disappointed, for we received a mail that we had sent to the states on the 1st of October, it had been to Panama and returned here instead of going across the Isthmus, which I suppose was owing to the carelessness of the officers of the steamers, that accounts for Alida's telling me to be punctual in writing by every mail, for I have *always* written by every mail, but it seems the letters never reach their destination.[31] Notwithstanding the failure of the mails by the steamers, I was making myself quite easy and contented with the thoughts, that when a mail *did* come there was sure to be a letter for me, as you all have told me *often* that you write by every mail, yet on yesterday morning we were told that there was a mail from San Francisco, which came through by land and arrived the night before about one o'clock, so I immediately enquired for my letters, but found none from home, and had it not have been for the kindness of my *friend*, J. McCalla I should have been destitute of news from Washington. I was going to say my home but I don't live there any more or rather I say so to those around me, since I have seen some emigrants from there, who claim to be residents of that place and are rather hard customers, they laugh at me a great deal about the representatives of Washington that is among the young men of the Commission, but I am sorry to say they cannot boast of better from Ohio. I am rather unpopular among them, for the reason I do not make them my companions. I generally associate with young [William T.] de Bree, who is a son of Purser de Bree of the navy and is connected with Mr. Grays party, the reason I do not make others my companions are that they associate with the labours [laborers] and working hands of the commission, to be sure, as long as I mess at the commissioners table I have to eat with *them*, but that is all I intend to have to do with them, I treat them politely but that is all and I am glad to say that by to morrow I hope the remainder will start for San Francisco, as they intend going by land; the Commissioner and Surveyor with a good many of the party went up in the steamer, and I am sorry to say that the Commissioner [Weller] and Surveyor [Gray] have not come to good terms as yet.[32] Col. Frémont has not been heard of, what he intend[s] to do about accepting the commission no one knows.[33] Col. Weller says if he is not releaved [*sic*] he will return by the 1st of January or if he is not elected to the U.S. Senate, which I don't think there is any probability of if he is he will resign without saying any thing to Col. Frémont about it, but if he returns here he will try to get passage for those that wish to return, so I suppose those that dont wish to return will have to do without their transportation but however I think I shall return after I visit San Francisco and the intermediate places. I think it would be rather foolish to return without seeing the great attraction, "the Gold diggins", for what is a visit to California without seeing its gold. If I go to San Francisco it will be by land so as to see the country for the expence in the steamer is forty dollars and that in the steerage too, for it is impossible to get in the cabin they are so full and the expense from San Francisco to Panama is the same as from here to Panama, so there will be nothing lost in that way. When I get back to Washington I intend to join the [Pacific] coast survey, Major Emory told me he could get me a place about the same as I hold on this Commission.[34]

I enclose a letter to Alida which I wish you would send to her if she is still at school, the reason I send it to you is to avoid double postage. Tell mother I have commenced on my new socks and gloves and am about to get some new drawers, but they cant be flannel as she would have them. I was sorry to hear of the death of Mac Washington but glad to hear it produced such a good result.[35] Give my love to *all* and remember me to the neighbours. How does aunty get along with her needle work tell her I can beat her sewing. I have made a bed for myself and stuffed it with straw as good as an upholsterer and the only thing I dislike is that I have to make it up myself. Tell Sarah I think I can keep house better than her now, without she has improved a great deal since I left. I have but little time left, so I must bring my letter to a close so I will bid you adieu, repeating my love to *all*.

Your affectionate son,

G. Clinton Gardner.

Private

P.S. It would be well for me to tell you of an order that was issued, after I had accepted this place as Qr. M. [quartermaster] Clerk, which says, "no one must engage in any other business out of the Commission without first obtaining a leave of absence", so of course I had to get a leave of absence which was granted me; but I am now not entitled to subsistance, until it expires. Notwithstanding which I go to the Commissioners table, which I intend to do without taking any notice of my leave, as I think the order was issued for me alone, because it was after Doctor Chamberlin who was surgeon of the escort had resigned,[36] this is one of my reasons for thinking so, the other is because it came out just after I had taken this place which was held by one of Col. Weller's friends, but I am glad to say that friend was too much of a gentleman to say any thing about it, the reason Capt. [Julius] Hayden did not like him was because he was dis[s]ipated.

G. Clinton Gardner.

GCG to his sister Sarah, San Diego, January 4, 1850

Dear Sarah,

I thought of permitting the next steamer to pass without a letter, so as to be able to give you an idea of my future intentions, about which I can say nothing until Col. Weller returns from the north; but in my next I will tell you what I intend to do—so you will see that I send this merely to comply with the request which you all have so often made of me, that is, "to write by every mail," which I have done, up to this time, and will continue to do, although I do not receive anything in return; which I would say nothing about, had I not have received letters from some of my friends of much later dates than any from home; the last date from home, I think was Alida's note of the 12th of August, in company with the Daguerreotype of the family—with which I have passed many pleasant moments, and was thinking this morning as I took a peep at it, how you all had spent your Christmas week, which to me was rather dull, although I had done every thing in my power to make it gay. On Christmas eve I had a little supper in my room, which, by the by, I have not given you a description of,—it is small but pleasant. I have it ornamented with guns, pistols, swords, bookshelves, &c. but I must not forget the *Old Gold Watch*, for which I have made a case, and it stands up against the wall, like an old clock and, if anything, ticks much louder; it and the daguerreotype are the only things that keep me in mind of my home. (I think I will draw a card so as to number the heads in the likeness, that I may not forget the names of any one in the family.) I was asked the other day how many brothers I had. I stopped for a while, counted, and said three, but upon looking at the likeness I found that I had four, counting John, who is cut out, whom I counted before, then I commenced to think *who* could be the fourth one and upon looking over my *old* letters I found one signed Charles, which soon reminded me that I had such a brother, so you see from the fact of his not having written to me for so long a time, I had forgotten him (over the *left*); but this is running off the track—I was describing my room—which is ornamented in such a war like manner, as the Old Priest said the other day, "it is more like war than devotion;" he came in on Newyear's day to take a glass of wine, as I keep open house, but as I was not at home he took one with my friend and drank my health, the consequence of which is my bowels have been lo[o]se ever since, so I don't think I shall permit any more persons to drink my health up in that way, for if I do I won't have any; but now I am too far ahead, I have not told you how I passed my Christmas, which was so dull—I must go back further still—to my supper on Christmas eve, after which we all went out to see the devel [devil], as it was said he was about, not knowing where to find him we went in the church, (we did not expect to see him there) but there we found so many pretty young ladies that we went inside, and as soon as I got in the door, the Old Priest asked me back in the church—thinks I, now he is going to give me a glass of wine, & I was cogitating whether to drink before all those beauties, when he assigned me to a place in one corner, for which *politeness* I thought I would appear a little devout, so I kneeled down when they went to pray, and they kept me there so long that my knees became quite sick of the office they held that night, and have complained a great deal, so I have come to the conclusion, after deliberate consideration, never to kneel down in church again; well after coming out of that place I heard that the devil was to be seen down in front of the Alcade's [alcalde's] house, so I immediately went down there and found they had a large canvass erected all ready for the great performance, at first, seven quite handsome young ladies, dressed in white, with red shawls, and a gypsy hat swung on their shoulders, each holding an immense quantity of flowers on a staff, made their appearance, and after dancing and singing a while, two old men with white beards came along and joined them, and then came the devils, who would every now and then drag one of these old men out, I supposed in the infernal regions from the way they sung out; but what astonished me was to see the devil represented with wings instead of a long tail, and a sword in place of a wooden iron shovel, so after dancing and singing they went in the house one by one all except the devils "who could not come in" to pay their respects to an image which represented Christ (I did not go in myself for fear they would get me in a corner, as they had done once before that night in the church) after which the performance closed, and the devils, for two or three days, could be seen running about on horseback:—this is the amusement of a Christmas, would you not like to spend one here?[37] New year's day we had *nothing* to amuse ourselves with except visiting each other, until the evening came, then we all went

house one

Gardner's sketch of a man thumbing his nose.

to a great baylé [*baile*, or dance] and I had the pleasure of dancing with one of the prettiest girls in the room, after which I felt quite contented; so I went home and had pleasant dreams about home, my native home.

Well as my old friend against the wall tells me, it is much after the time to retire. I will have to bid you good bye, hoping soon to enjoy your company as well as that of the other members of the family Stop, let me see if I can give you all the members of the family. Father, Mother, Aunty, John, Jacob,—no, stop, I forgot Emma, Jacob, Franklin, Sarah,—let me think,—Charlie, Clinton, and Alida, hurah! hurah! hurah! I do know them all don't I? Now give my love to them all, including Uncles John [McLean] & Clint [McLean], also to all inquisitive neighbours, and that will be the last request I will ask of you tonight, so good night, and believe me

Your affectionate brother

G. Clinton Gardner

The Steamer I hope will take this tomorrow.

GCG to John M. McCalla, San Diego, February 3, 1850[38]

My dearest friend

I have several times commenced a letter to you but some thing has always stoped me, but now I will endeavour to get through one although I am at a great loss to know what to write about, should I write about the principal topic of the day my subject would be *mules*, which I know would be of little interest to you; almost every one about here now is preparing to take a drove up to San Francisco to speculate on and it is as a gentleman said the other day, "you can't ask a question now days but what they say mule to you", and I am sick of hearing about them so I will say no more on that subject, but will give you a short description of some of our dances, or baylés [*bailes*] as they call them, we have them here every night, sundays not excepted. just think of me some sunday night, while you are in Mr [Clement M.] Butler's church, dancing away with some pretty girl.[39] The other day they had a wedding in town, and as I was invited I went, they were married in the church by the old Priest after which the bride went out the back door, while the groom with his party came out the front door, and as it was very muddy, we having just had a rain, the bride with the bridemaids got into a wagon, not quite as good as some of the milk carts that run about Washington, for it had no top, and drawn to the house of the groom by six dirty rag[g]ed and almost naked indians, who I suppose were hired for the occasion,[40] where the groom met her and as it was about 9 or ten o'clock we had breakfast, after which the tables were moved and dancing commenced which lasted until about two o'clock, when they fixed the tables again for dinner at which they drank quite freely and gave several toasts one of which I will mention. One of the Young Officer[s] of our army was called upon he got [up] and said in very good Spanish as follows, "I have travelled over a good portion of the world but have never seen a more beautiful set of young ladies as there are in San Diego, so I propose the health of the young ladies of San Diego", to which a young lady from San Francisco got up and replied as follows "I not being a resident of San Diego do not consider myself included in that toast, but, in behalf of the ladies of San Diego will reply to it. I have seen a small portion of the world and travelled a little, but have never seen a more handsome young gentleman than yourself, so I propose your good health." What do you think of that, was that flattery for you? The young lady made the reply in such a sweet manner that one could not help taking notice of it. Well after the dinner was over which was about sun down the rooms were cleared again for a dance which lasted until about two or three o'clock in the morning then I retired and all the next day the fiddle keep agoing and I supposed it passed off the same as the day [before]. towards night [I] went in again and found them dancing in which I joined and danced until about one or two o'clock when the upper crust, as some of the young men call them, that is the first people of the place left and then we had the one horse concern which is an other expression of the young gents I will tell you how this originated; the only large room in the town is in the house of one of the first familys, the Bandinis, and as several balls had been given to the Americans we wish[ed] to return them so we got the room but it seemed some of them disliked the Americans and for that reason would not come so after a while no one would come except the Bandinis and a few others[.][41] then even they slighted some of the young Officers so they went [and] got up a ball and invited every body except the up[p]er crust and called it the one horse baylé and said the others were the two horse this was the origin of the one horse so after the up[p]er crust left there was no one there but these on[e] horse concerns so I thought I would [go] and take one or two dances with them and then retire but before [I] could [get?] around the doors were locked and no one could go out so we danced until morning clear up till the sun came out after which we went home with galls some were black some were blacker some were the color of a chew tobaker [tobacco] or rather the color of a dark indian squaw I often thought of what you all would say to see me dancing with such a piece in the states but such is the custom of the country. I was amused to see the fellows get tight one came came [*sic*] up when he was quite far gone and said "hollo Gardner we must have that gun fired off," refer[r]ing to ~~the~~ an old spanish piece that was taken in the war and given to the town, ~~dam it all the people are asleep in the town and by god the Sun is just commencing to light up the world why God dam it you have no idea of the expense he goes to to get the light, it costs fifteen hundred thousand million a minute and by god the people ain't up to see it why dam it all this money is thrown away, the people talk about the __ls but by God they are no body but the people that scrape the tar barrels and~~ [illegible] ~~lights up they only~~ [illegible] ~~the fire and when it gets burning God almighty~~ [illegible] ~~up himself but by god~~ [illegible] ~~off that~~ [illegible][42] so off he starts the canon had been fired about every fifteen minutes all night; the people came up from the beach or landing which is five miles from here to know what in hell was the matter. this is a great community I live in, every night a dance, drunk or something else~~, and you talk about your private __ps but they are nothing to these I have~~, you have no idea how I live, I live like a fighting cock a nice little room all to myself oh! it is sweet this is a sketch of it,[43] a view from the door, just over my bed is a map of the states, at the head of my bed is a looking glass and I g[u]ess I had better tell you that just over my table is a window and on the left side of my room is my gun, my knife and dressing case the latter is the thing on the shelf and shelf is that thing held up by a stick now dont you know which it is, I remember when I used to draw a horse I had to write horse under it so as to let them know what it was, same way with this, well now I think I have run on far enough with trifles so I will close without reading what I have written so you please not read it to any one or let any one see it. I expect you will though. I have not got over the brandy yet.

John McCalla Papers. Courtesy of the Historical Society of Washington, D.C.

Well now dear John, when you write, tell me about the young ladies and permit me to congratulate you on having acquired such a price [piece?], I think she is a very sweet pleasant, and agreeable young lady and is worthy of a better character then you, not intending to slight you at all. How are all the young ladies around town remember me to all that enquire after me and also to your family papa mama & last but not least miss Maria. I have not heard of though I should like to see very much what has become of A M. John Nigh, &c I wrote to John Nigh some time ago but have not heard of him yet.[44]

Well write soon old fellow and remember me to every body [illegible] your affectionate friend

G. Clinton Gardner

GCG to his father, San Diego, February 9, 1850

Dear Father,

I write in great haste to say, Capt. Hardcastle will enclose in this a draft for $500. Uncle Clint is here[45] and leaves this evening with the Capt. H[ardcastle] for San Francisco;

Love to All in great haste,

Your Affectionate Son

G. Clinton Gardner

Tell Mother box all arrived safe Much obliged &c.

G. C. G.

GCG to his sister Sarah, San Diego, February 19, and to his father, February 23, 1850

Dear Sister Sarah,

This morning the steamer was announced some time before breakfast—so, after making out an account for Capt. [Julius] Hayden, I got on my Old Whity, (as I call him) he belongs to the Qr. Master's dept; and took a gallop down to the beach,—went on board of the steamer, was introduced to Capt. Patterson,[46] who invited me to breakfast; after which I bid Col. Weller and Lt. Whipple, the bearer of this, good by. The Colonel has given us all a leave of absence for three months, at the end of which time we are to report to the commissioner by letter and wait further orders.[47] I was glad to hear from several of the passengers, that the Col. would not be removed—for I have now taken quite a fancy for him, notwithstanding there is no money to be had from him.[48] I am sorry to say, in answer to your letter of the 30th of December, which has just come to hand, that the $200 @ month is all a mistake—and I think you must have used Mothers eyeglass to have seen it in my letter of the 15th Nov. Genl [Bennet] Riley will not allow more than $100 @ month to Clerks,[49] with which I am quite content, as it is better than nothing: but I am sorry to find so little of it to show, at present: never mind—before my "own good judgment" turns my face homewards, I hope to have at least enough to dress myself with, of my own—which I am quite confident my friend Willie Jesup has not,[50] notwithstanding his waltzes and Polkas:—but you astonish me when you tell me of the convertion [conversion] of J. McCalla, as from his last letter I should judge that any thing else would happen—so, the next I expect will be quite a sermon: tell him he had better get Mr. [Clement M.] Butler to select the text,[51] but not to get one that will admit of the subject of dancing, as we have dances every sunday night, and it would be of no effect. Last sunday we had quite a large Pic Nic, which I should like very much to describe—but as it was something that was handsome, gay, and beautiful beyond description, you will have to excuse me: but, however, this being to you, I will oblige you with just a sketch of the manner in which it was gotten up. Mrs. Bandini, one of the first ladies of the town[52]—or as ex vice [president] [George M.] Dallas would say of the "City" of San Diego"—gave out word that there would be a small pic nic on the following sunday; and as the gentlemen, always on such an occasion are expected to furnish them selves with horses, I had my Old Whitey up and ready at the appointed hour, galloped over to the house, where I found all the young ladies ready—five vehicles held in readiness to convey them to the place appointed—besides which, two horses with ladies saddles on, one an American saddle, the other a Native saddle. Miss Ysidora Bandini rode on the American saddle,[53] and Mrs. Fitch, on the other[54]—by the by I must not forget to tell you what a California saddle for a lady, is—it is nothing more than a mexican saddle, with the stir[r]ups taken off, and the lady sitts on the right side of the horse—which perhaps you would say was the wrong side: well, we all got under way—the first vehicle (I won't say what kind) took the young ladies—the next, some young and some old—then some ladies and some gentlemen,—then came the Music car—after which followed the eatables,—we on horse back of course in the lead: upon starting out, Mrs. Fitch got a fall, but being rather *fat*, got up without being hurt. Miss Ysidora rode beautifully: I think she rides *almost* as well as yourself: well, after galloping four or five miles, we came to a halt, in quite a pretty looking valley, by a brook, where the trees or rather bushes formed quite a shade. While we were enjoying ourselves together, the wagons—Oh! I forgot—carriages were drawn up in a row, and a tablecloth laid down on the grass, upon which quite a fine dinner was sett out. Roasted Sheep, ducks, ham, eggs, sallad and all kinds of things,—I have never seen such a show before, in San Diego—after which puddings &c—then the wines—then the coffee—after which the gentlemen took a smoke: and reseated the ladies in their wa—carriages, and were ready to mount their horses—which we all soon did—and arrived in town by dark; then went in at Mrs. Bandini's,[55] where we had quite a pleasant baylé [*baile*], until three o'clock in the morning: that is what I call doing it up brown on a Sunday. I never thought what you all were doing, about that time, in Washington,—perhaps, just as I was dancing the Polka with Miss Ysidora, Mr. [Clement M.] Butler was praying—who knows but what, for John McC[alla]:—well, I only hope so, or some other one who is just as good-by your affectionate brother

Clint.

Dear Father

Feb. 23rd Since writing this to Sarah I have heard a great deal of news: when the Steamer came in, the other day, Major E[mory] was down at the Punta—near our old encampment[56]—with the mexican surveyor, and upon his return he found several letters for him—among them an order to act as commissioner and receive the books from Col. Weller; but the Col. left in the same Steamer for San Francisco—so the Major had not a chance to receive them, and has requested me to do some writing for the commission at night, after my Office duties with Capt. H. [Hayden] are over. I have not obeyed them strictly, nor don't intend to, until he issues an order taking away my leave of absence,—for if I do duty for the commission I am entitled to subsistance, which while on leave of absence I am not entitled to, so therefore I cannot subsist *myself* and do duty also.[57]

I have received all of your letters I believe, and papers also, for which I am much obliged to you *all:* tell Mother I am very much obliged [to] her for having exerted herself so much in furnishing me with so much news—for I have no doubt it must have been quite an exertion to get Charlie to write—although now since he is away from home he feels the want of a letter once in a while to entertain him, for, if he is any thing like myself, he soon gets tired of his companions, and seeks enjoyment from reading Old back letters, received from friends: if I have read once, I have read a dozen times, all my old letters and every time with more pleasure. I have received but one from him as yet, although, I anticipate having him as a regular correspondent—and that one I have endeavoured several times to answer, but it has always been my fate to have the steamer start out before I could get down to her—and that is the reason why you have not received any letters from me lately; it is not because I have not written, on the contrary, because I wrote too much, and did not give myself time to get it on board of the Steamer: but this one I will have on board, or lest all the instruments will be left—which are to be put on board in charge of Mr. Whipple,[58] who I intend getting to take charge of this to you, together with some money—I can't tell how much at present. You were speaking of the wages of this country being so high, but you forget that other things are in proportion—and I am almost certain that all the fortunes in this country are made mostly by gambling, which I do detest: even in this little place, there are one or two Monte tables, and the spanish people every cent they can get is put down on one, but is always raked in by the bank.

My paper warns of the time to close—so I will bid you goodby, for the present, your affectionate son, Clint

I send with this a report which Dr. Chamberlin, the Senator from this place [to the California Senate] has been kind enough to favour me with. G.C.G.

GCG to his mother, San Diego, April 6, 9, and 20, 1850

Dear Mother

I have written by every steamer, but have several times closed them too late for the mail, which will in some measure account for the supposed neglect on my part, and I will endeavour to follow your advise and write during my leasure moments, but will always mail my letters hereafter at least a day before the steamer comes, so as to be in time, and perhaps here it would be well to mention, that I have received no letters since January, except one from Mathilde for which I am much obliged to *her*,[59] but could have received several more from the same part of the world and not have been under half the obligations, as I have written to all of Emma's family with the exception of Mathilde—and she is the only one who thinks it worth while to write to a friend; and as for Frank he has entirely given me the slip, but however since I live in such a delightful climate why should I destroy its affects by thinking over such trifles; letters give one but few hours pleasure and then they are laid aside, and if it were not for the steamers I should never think of receiving one; when they come in, and I see other persons reading letters, to be sure, I feel like receiving one also, but since I go to the Post Office so often without receiving anything, I think I will stay away and mount my horse to take a ride over the hills, until the letter excitement is over, and there will be so much postage saved also, so take it all in all I believe I am satisfied, but should I run on much more with this stuff, I am afraid *you* will not be satisfied, so suppose I give you a little idea of things in general of this, as some people would say, barren country, but I think it is one of the most delightful countries to live in that I ever was in. I have no doubt but that almost every kind of fruit can be raised here with a little trouble, in some of the vallies the ground is very good indeed, and you find wild oats growing all over them, some two or three feet high; and the wheat of this country is also very fine. I think the heads are much larger and contain much more wheat than ours, but the people are very indolent and raise no more than sufficient for their own wants. I shall be glad when the state commences to tax them, for I expect then they will bring some of their produce in to sell, in fact they will have to do it, in order to pay their taxes or else their lands will be sold; I think when the tax gatherers come around some of these people will be for giving up some of their lands, which they pretend to own; some of them claim whole vallies some twenty and thirty miles long, which produce nothing at present—but with a little labour could be made valuable.[60] For several days past some of the Americans and a few others who own land want to have this southern part of California, south of Los Angelos, made a territory, but they cant raise a meeting large enough to draw up a petition: the people of Los Angelos have already made a petition, but the most of the Californians down here think the Americans are up to something that might injure them, so they wont come to the meetings[61]: then again I believe the Mormons claim this as being within the limits of their state, and are going very soon to send a great number of their people down here to settle, they have made several offers to buy any ranch between Los Angelos and this place—and I believe have bought one; they want San Diego as their port of entry, and I have no doubt if they do come here, they will improve this place very much, for I have always heard that they were an enterprising and industrious set of people—and I think it wants some such people to make this what it ought to be.[62]

April 9th. We, this evening, had quite an excitement in town—it seems that there has been a party of Texians out on the Gila who established a ferry, and have been noted for having commit[t]ed a murder—there was an Irishman who was coming to this country, and having got as far as the Gila thought he would raise a little money—so he built a raft and established an opposition ferry, for which this party tied him up and shot him[63]; so we have been told by several emigrants who have passed that way; this party came in town in order to get some provisions—and during their stay, one of the party took the liberty of shooting one of the soldiers, just for the sake of seeing if his pistol would go off, which he found went off very well, and shot the fellow through the lungs, the ball going in his back and coming out just below the heart; he attempted an escape, but was caught just one mile from town, and is now confined in irons waiting his trial.[64] I presume he thought he could do as a great many of these rascals have done in Los Angelos, kill any one without having any notice taken of it—but I think *he* will be strung up by the neck, if he is not shot before they have a chance, there are a great many who feel indignant enough to do it if they get an opportunity. General [Perisfor F.] Smith is now in town staying in the same house with Major [E. H.] Fitzgerald and myself. I believe I have written about the manner in which I live with Capt. Hayden and Mrs. Hayden, having entirely *cut* the iron forks and tables without table cloths[65]; but am only afraid it will not last long, for I expect they will order the Escort to be broken up, then I will have to go back, but not as far as I was before, as Major Emory has charge of the Commission now, of which I suppose you all have heard, you being at head quarters [i.e., Washington, D.C.]. I have been trying to get a place above [in northern California] but have not received an answer yet—should I succeed I shall go up there and see what I can do. I hear Fred Emory has made $20,000 and is very likely to make as much more—owing entirely to his surveying, which gave him a chance to speculate.

April 20th. I am quite disgusted and have now come to the conclusion to leave the commission without Maj. Emory is relieved. I have had several little spats with him of which I am getting tired, for it seems that the more I put up with, the more

I am expected to put up with; I have never written home the duty which I have been doing, or if I have I have always placed it in a good light, but now I will give you it, as it really is, and also my opinion—when I first came to Panama I went to draw some money, being without funds and desirous of returning the fifty which I got from Father. Charles Weller made out my account which I took to Maj. Emory for approval, *he* approved it but drew his pen through the word "computer," which took two hundred dollars a year from my pay, and set me back as junior assistant with the same pay as the Laborers of the time,—that was corrected through the kindness of Col. Weller—so my pay from the *start* was $600 per year and from the first of August $700 per year[66]; and not only that, but several times while in Panama, he made such remarks to me that would have induced any one else to have left, but I said not a word in return: and when I came to San Diego, I was not only junior Assistant—but private secretary to copy his letters in reference to the Escort—all of which I did in the best humor possible; and since I came to live with Capt. Hayden—during which time I have been on leave of absence—he has called me always to copy letters for him all of which I have done, except when it interfered with my duties as Qr. Ms Clerk, then I refused—and once during my leave of absence I hired a man to record for me and paid him $1.00 per day for a month, merely to oblige Maj. Emory; and now this morning he comes to my window before I am out of bed and tells me I have neglected my duty, because I did not bring to him some papers which I had written for him out of *kindness* and sent to him by a gentleman who had forgot them. He seems to think that I am entirely under his control whether on or off leave of absence, and now I have come to the conclusion to tell him, out and out, that I will not submit to it any longer. I know my duty as a member of the Boundary Survey—I will do it as long as I stay with it but as to doing any other duty, I *wont,* I will resign, as others have done before, for the same reason, so you may know I am not the only one who complains. Mr. [James] Nooney who was his first computer left him, I am certain from what he said, because he was of that disposition[67]; and Mr. Whipple, who I presume you have seen before this, has advised me to go to San Francisco—which I think I will do as soon as Capt. Hayden goes away, he expects to get orders by the next steamer from head quarters which will deprive me of my place, then I think, I shall go to San Francisco—if I can't get my leave extended to go, I shall resign, without Maj.

Emory is relieved, for I won't serve under him. I should like to hear from Father and see what his views are; but I expect before I can hear I will have acted, but however I think I can get along very well out here without being subjected to such abuses. Give my love to *all*, including *every body* and not omitting *any thing.*

Your affectionate

Son Clint

GCG to his father, n.p., n.d.

Dear Father,

Will you be so kind as to subscribe to the weekly inteligencer[68] for Capt. Hardcastle and have it sent to him directed, Capt. E. L. F. Hardcastle U.S. Topl. Engs. San Diego, Cal. and call upon Charlie for the pay, if so you will oblige me much, send me a receipt for the subscription also so as I can give it to the Capt.

Your affect

Son Clint

GCG to his brother Charles, San Diego, April 20, 1850

No. 11 East 11th Street New York, N.Y.

Dear Charlie

You must excuse me for neglecting you so but really I have written so little lately that I have almost forgotten how owing to my correspondence, with *home* even, having stopped for I have not received a single letter for the Lord only knows how long, and so I have written none in consequence but as Lt. [William Alloway] Slaughter who goes in the next Steamer has offered to carry a letter to you I thought I would accept his offer and get you to make some purchases for me. I will send by him $50.00, half of which I wish you to send to father the rest I should like you to spend as follows, viz; Get me some good large drawing paper and have it put up in a round tin case, I think the best—is called Elephanti Antiquarian, then get me a small case of pencils, I believe the different shades come in a small box—about two inches wide and the length of the pencil, these can be put in this tin case also I think, after which I wish you would try to get a good sketch book. I believe there are some made with all the leaves seperate those I think are the best I should like two of them, then get me some tracing paper which will also go in the tin case, and I should like also two or three knives, as I have none, now show your taste in selecting them and send them out by some private hands, for if you don't I shall have to pay quite a sum for them and then run the risk of losing them, you must direct them to me at San Diego care of the Post Master. I forgot to mention some two or three pieces of india ink and also india rubber. How do you get along with your new establishment, does New York agree with you?[69] I should judge not for there is not the enjoyment to be had there that there is in Washington, without you should get in some lucky circle of friends. Since I have been here I have got into a regular mode of living, I get up in the mornings, make my bed, sweep out my room and dress myself, after which I go to my breakfast, then return and sometime work at my Qr. M. papers or loaf about town until dinner, then after dinner I take a ride on horse back until dark then dress myself again and go to see the girls until 12 or 1 o'clock, dancing all the time this is the order of the day sundays not excepted for we dance more on sunday than any other time. I think this would be a delightful place to live if one could only get servants for as it is I have to take care of my own things and room, it is not expensive living down in this part of California one can live here quite moderate to what he can further north but then there is very little to do every thing is the same once in [a]while we get up a little excitement in the way of a ball or picnic which answers very well for a variety.

Upon receiving this you please call and see Lt. Slaughter he will write the name of the place he is staying at on the back of the letter so as you can find him. he is a very clever young man and will tell you more about the young ladies that I would write and also about the line as he was one of the officers of the escort.

Your affectionate brother,

G. Clinton Gardner

P.S. Your letter of Feb. 12th has just arrived but where [are] the three that you talk about. [It] is a decided case of quien sabe [*¿quién sabe?*—who knows?] for I have not seen but one before this and doubt very much whether you have for since you have been at New York the mails have always come quite regular.

You talk of coming to California yourself but I should advise you if you can get along at home to stay there, for this is not much of a place with out you have a mint of money or some one to pay your expenses, we pay 17 cents per lb. for potatoes and cant get them at that only from the Commissary Department of the Army and every thing else is in proportion rooms even rent here for $60 per month and at San Francisco for twice that. G. C. G.

Dr Charlie,

I have two *new* knives. I will send you as a present to [Uncle?] Clinton, so you need purchase but *one*. I have sketch books, I think, also. See my letter herewith.

By Lt. Slaughter on outside of letter:

You will find me at the Astor House at 3 o'clock.
Room 84. I leave this evening at five for Albany.
W. A. Slaughter.

GCG to his father, n.p., April 20 and May 3, 1850

Dear Father

Your letter of the 10th of February has just come to hand—and although I have just closed a letter to Mother, I will open it again, and give you these few lines in answer.

It seems, although you think it can't be done, that Col. Weller has broken up the commission, or rather adjourned it to meet at Paso del Norte, the line having been finished to the mouth of the Gila some time since—and only remains to have the monuments placed, which Capt. Hardcastle and myself are to be left to do, as soon as money is received to do it with, for I don't think I will go in the field again until some money is received. I still hold my place as Qr. M's. Clerk, and will continue to do so until the middle of next month, by which time I think the escort will be entirely broken up. Major [Edward H.] Fitzgerald has already taken command of the Dragoons and moved to San Luis Rey by order of Gen. Smith, and if all of Gen. Smith's order is carried out the same, all the Infantry here will be sent up to the Mission of San Diego, about six miles from this place, leaving no troops here—he says there is none necessary for this place, and if there is the Mission is near enough; I believe his object is to get hold of the Missions until Congress decides to whom they belong.[70]

You seem to think that the Astronomical Party is too far advanced for such a youth, but I much prefer it, as I there have a chance to see both surveying and astronomy, and for that reason, besides others, am glad I did not go in Mr. Gray's party. Should Major E[mory] leave, which is not at all probable, he would not have much influence in getting me a place with *Gray*, as they are very near, although I have no doubt but what I could get a place with him—and am sure I can always get some thing to do out here, as I believe I am not destitute of *friends*.

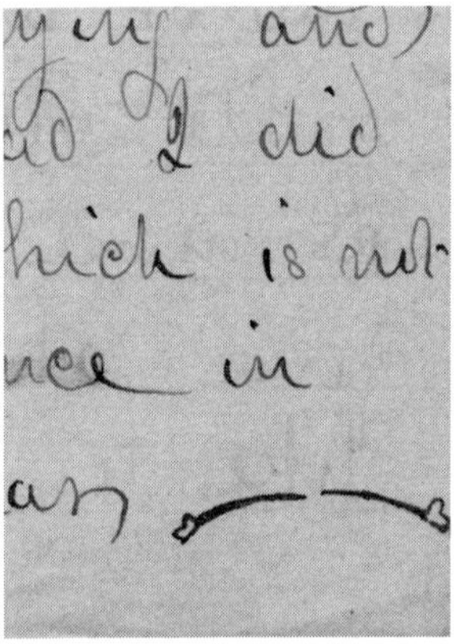

Gardner's sketch of two swords.

There are now five persons out in this country who still belong to Major E's party; Capt. H[ardcastle] who seems to be quite a pleasant gentleman, but I can not say much for his scientific knowledge; Dr. Parry who is a naturalist and botanist, is now at Monterey on leave,[71] he is a very fine young man indeed, but is one from whose appearance no one would take to be as smart as he is, for he is a very quiet and unassuming person; Mr. Ingraham, a nephew of Major [James Duncan] Graham who is now here on duty with the commission is quite a nice fellow,[72] but according to my notions is rather independant for one who is so much dependant on others; then comes your son, for whose character you can ask his mama: these together with Major E. compose the astronomical party;—there is now but one belonging to Mr. Gray's Party—he is the draughtsman—Mr. [Henry] Clayton of Ohio,[73] and a very nice English gentleman, but it is a pity he drinks: where the others are I can't say, and of Col. W's [Weller's] party there is but one left, he is the Act. Qr. M., Mr. Rust of New Orleans, and seems to be quite an agreeable person.[74]

The reason of Fred Emory leaving the commission I think was on account of the jealousy existing between the Major [Emory] and Mr. G[ray]. He went to San Francisco and is now quite rich—I intend to rob uncle Clint of all his good clothes as mine are almost gone; if Mother could send me some fine shirts—I would be glad—about one dozen; I should like to have them just like the best shirts I took from home, and just about the size of those she had made without any bosom, with one exception, make the bosoms a little longer with standing collars; and should it be convenient to get some made with ruffle bosoms and cuffs I should like them also—don't think I am joking, for I want to cut a dash and perhaps catch a fortune—who knows?

Give my love to *all.* In haste,

Your affectionate

Son Clint

---

on same page—entry of May 3, 1850

May 3 1850 The Steamer is expected to morrow morning so I mail this letter, before which I would say that I have again got on good terms with Major E; he abused me since I commenced this letter and told me I would have to leave Capt. H. [Julius Hayden] or leave the Commission—to which I replied I should leave the commission: so after he got over his pet,[75] he told me that he spoke in haste and that he did not intend what he said—so we are again I believe on good terms: that is his way when he gets a little excited and makes one feel rather awkward: but take him all in all he is quite a clever person so Good day. G. Clinton Gardner

GCG to his father, San Diego, May 14, 1850

Dear Father

I enclose here with an account of our wagon and forage master which will be good to you for ninety dollars as soon as presented to any quarter Masters. I purchased it because it is as good as a draft and is much cheaper, as I only paid eighty one for it being ten per cent off.[76]

Capt. [Julius] Hayden who I expect will start for the states in the next steamer has offered to take this and mail it in the states, and should you hear of his arrival in Washington I should be much obliged to you if you would call on him and his lady, both of whom you will find to be very agreeable persons indeed and who have treated me with the greatest kindness, for which I hope you will thank them.[77]

I will try and get time to write a long letter before the steamer comes which I have no doubt you will get before this.

Your Affectionate Son

G. Clinton Gardner

GCG to his mother and father, San Diego, June 30 and July 2, 1850

My Dear Father and Mother,

Permit me to answer your letters in one, those received in the May steamer were dated April 11th 1850 and those by the last steamer (June) were dated one from father and Sarah April 25th and one from Alida May 11th 1850.

I am much obliged to you, dear Mother, for the black silk neck-handkerchief, it came just in time—or I had worn out the one I had and have this one now, a very *fine* substitute. We are still here in this glorious state of uncertainty that we were when we first arrived: about one year has passed since our arrival here, and we are now no better off than we were the first night; and I think you may say not so well off, for we have nothing to eat at present—nor no money to buy anything with, and we had something when we first came. By the by, dear father, I would request that if you have any friends in Congress, whom you can get to push the bill for this boundary through, you will not only oblige me, but a great many others who have not received any pay since January—besides a great many more who have not received any since last October—and also, some one or two have not received a single cent since they have been on the Commission: not that I need pay myself, but that others who have been unable to get employment out of the Commission are much in want, and it is owing more to that than any thing else, that I have lived so much up to my pay in the Quartermaster's Department. I think that what I have sent home, together with a little I have now, is all that I have from five months pay;[78] to be sure my pay from the Commission is still due me—but when I think of the way in which I lived, when others around me did not spend a cent, I feel a little ashamed; yet I made myself comfortable and that was all, while others did not; and I may say now that I live more comfortable than any one of the young men of the commission—to be sure my wash bill is larger, because I have more pieces; I have sheets, pillow cases, night shirts, &c. that they have not, so my living costs more yet it does not come from my pay in the Commission, for I have as much due me now as any of the others who receive the same pay.[79]

I see by your last letter dear father of April 25th that you have not received my letter dated some time in January, in which I stated that the note sent by me against Charles L. Weller for $50 had been settled by Col. Weller. I hope you have not done as you say you will do, that is—write to the Colonel, for he will think it very strange that I did not have the note destroyed as I said I would, yet it is not my fault that my letters are not received: but however, should this come to hand, you will please destroy that note, and I think when this is received you will have received also a letter from me by Capt. Hayden (who I expect will go in the next steamer to the states on a sick leave) in which I sent $90 or rather a bill which the Quartermaster General or any Quartermaster will pay, amounting to that, the reason I sent this was because it cost but $81 which was ten per cent off and I knew it to be good according to Par. 1029, Army regulations of 1821.

Should you see Capt. H[ayden] and his lady when they visit Washington, I wish you would thank them for me for their kindnesses, which I am sure I can not easily repay; for while I was at their house I was treated as one of their family, and lived better than any other person in the town—for I can assure you she had everything that was to be had; It (as I have often said before) reminded me of home, when I took my seat at the table with silver forks, napkins, &c, but alas I have fallen a button hole since they broke up house keeping, for I have now come down to the iron forks and spoons. I now mess with Major Emory, and am as it were one of the favoured two, for Dr. Parry and myself are the only ones that mess with him—the others have been living at a little eating house, as destitute of eatables, as it was dirty, which the other day was compelled to close doors for want of funds, so they are now eating here, there, or any where—so I still think, as I have often written you both before, that I am one of the lucky ones of the flock, yet it was once, I did not think so, when I received the enclosed letter from Fred Emory which came just too late, as it arrived *from the states* in the May steamer[80]; but however I was about to go up in the June steamer which has just passed when the Major who came down on the 3rd of this month told me that Fred had got some one to fill the place which he offered to me—was that not a pity? but you know "everything is for the best" so I am content to remain with the commission.

But what seems strange to me is, that the Sect. of the Interior should write to the Major and say, his plans had been approved of, and that money would be sent out in the next steamer (which was the June steamer—but the money did not come) and at the same time send Col. [John] McClellan to the Paso del Norte, if I understand your letter dear father right.[81] I think there must be some mistake there—for I think I heard Major Emory say that his wife had sent him a copy of an order, which was sent to him by the same mail from the Department but which he had not received, telling him *either* to go on with the line himself as far as the Gila, or leave a party here to do so, or return himself with his party—just as he thinks proper. I will not be certain of this, so therefore I wish it to go no farther than yourselves, for fear of a little trouble.

Perhaps you both would wish to know what remains to be done on the line between this place and the Gila, as one would judge from the news papers that it had been finished—so I will save the trouble of the question, by giving the answer. The line, as far as the Gila, has only been run and marked in the ground for about thirty miles on this side,[82] and about eight or ten on the Gila side; so you will see besides finishing this, there are monuments, which are to come from the states, which we have heard from the Department have been ordered, but are not on the way yet. I expect to be one of the party left here, with Capt Hardcastle, to put up these monuments—but it will be some time yet before it is finished, if things move on as they do now: however I hope the next steamer will bring us some news concerning the affairs of this *industrious* party. I am in quite good health, and hope to hear the same from you all. Give my love to all, including the neighbours, too numerous to mention.

Your affectionate son

G. Clinton Gardner.

Tell Sarah, Alida, Luke and Charlie I will write by the next steamer to them. G.C.G.

Dear Father, July 2nd. After mailing this I found that I had omitted to mention even the subject I wished to write about, so I open it again for that purpose which is this—I have heard that there is to be a new regiment of dragoons formed for California: as I am in this country I should like to get a commission in that Regt. and I have no doubt but what you have some friends in congress who could do it—so if you would apply to them and get me such a commission, I would be much obliged to you—for I know it would be of service to me.[83]

G.C.G.

GCG to his uncle John McLean, San Diego, July 2, 1850

My dear Uncle John,

Permit me to occupy a few moments of your time, while going through the Capitol from day to day, which I have no doubt you do at present, by asking some of your friends in Congress, Members or Senators, to use their influence in getting for me a commission in the new Regiment of dragoons which is to be formed for California, and should you be a good friend of the President, or Sect. of War, perhaps you could assist me by giving me the use of, or using for me, your influence.

I beg of you, if there is a chance at all—try for me: of course, I want as high a commission as I can get—but will be content with a 2nd: Lieut:'s commission but I want you to apply for a 1st: Lieut:'s commission, then perhaps if I don't get the one, I will the other. I would ask also that you keep father reminded of it—and you will much oblige,

Your affectionate nephew,

G. Clinton Gardner

---

1. Emory notes that he stopped work on December 12 (Emory to Thomas Ewing, Sec. of the Interior, San Diego, March 1, 1850, in SED 34, pt. 2, 9; see also Werne, 2007, 37–38).
2. GCG was put on formal furlough as of February 17, 1850 (Weller to Emory, San Francisco, April 1, 1850, box 4, f. 25, EP).
3. GCG to his mother and father, San Diego, California, June 30, 1850.
4. For the schoolmate connection see GCG to his sister Sarah, Camp Riley, California, July 23, 1849. For lodging see GCG to his mother, San Diego, April 9, 1849. The 1850 census counted 650 in the town and another 798 in the county. The absence of men was noted by H.H.R. [Hiram H. Robinson], San Diego, Panama, June 16, 1849, in the *Cincinnati Enquirer*, August 16, 1849, reference courtesy of Charles Hughes. The quotation is from GCG to his mother and father, San Diego, California, June 30, 1850.
5. Hayden reported that "the rent of building has almost doubled within the last week" (Hayden to Emory, San Diego, November 26, 1849, box 4, f. 21, EP).
6. H.H.R. [Hiram H. Robinson], San Diego, Panama, June 16, 1849, in the *Cincinnati Enquirer*, August 16, 1849, reference courtesy of Charles Hughes.
7. Harlow, 1982, 145.
8. GCG to his mother, San Diego, April 9, 1849.

9. GCG to John M. McCalla, San Diego, February 3, 1850.
10. GCG to his father, n.p., Sunday, October 28, 1849.
11. GCG to his mother, San Diego, California, April 6, 9, and 20, 1850 (the quotation is from April 6).
12. Ibid. (the quotation is from April 20).
13. GCG to his father, n.p., April 20 and May 3, 1850 (GCG noted the reconciliation on May 3).
14. Thomas Ewing, Secretary, Dept. of the Interior, Washington, December 19, 1849, to Weller, in SED 34, 15). Supervision of the boundary commission was transferred earlier that month from the secretary of state's office to the Department of the Interior. Werne explains the unseemly political context (1987, 334–41, 344).
15. *(Washington, D.C.) Semi-Weekly Union,* December 20, 1849.
16. In writing to Thomas Clayton from Camp Riley on September 15, 1849, Emory did not specify his reasons (Emory, 1857–59, 6–7).
17. Thomas Ewing, Secretary, Dept. of the Interior, to Emory, January 8, 1850, in Emory, 1857–59, 1:8.
18. Weller to Clayton, San Diego, February 3, 1850, in SED 34, pt. 2, 2.
19. The observer was Hiram H. Robinson. See H.H.R. [Hiram H. Robinson], San Diego, Panama, June 16, 1849, in the *Cincinnati Enquirer*, August 16, 1849, reference courtesy of Charles Hughes. Informed observers like Emory agreed that manpower could not be found in California (Emory to Thomas Ewing, Sec. of the Interior, San Diego, March 1, 1850, in SED 34, pt. 2, 9).
20. Hardcastle to Emory, Santa Barbara, May 29, 1850, and Sacramento City, June 29, 1850, and Nooney to Emory, Monterey, July 1, 1850, both in box 4, f. 26, EP.
21. GCG to his father, n.p., April 20, 1850. Emory's list, dated April 3, is similar. See Emory to Thomas Ewing, Sec. of the Interior, San Diego, April 3, 1850, in SED 34, pt. 2, 16.
22. In December 1849 the newly formed Department of the Interior assumed supervision of the boundary survey from the Department of State. Emory's first draft of this angry letter, dated San Diego, March 30, 1850, is in box 4, f. 25, EP, but a copy of the final version, dated April 2, is in box 4, f. 28. The political and personal quarrels intended to stop Emory from returning home are explained in Norris, Milligan, and Faulk (1998, 93–102).
23. Thomas Ewing to Maj. W. H. Emory, Washington, D.C., April 10, 1850, in Emory, 1857–59, 1:9.
24. E. & W. Blunt to Emory, New York, May 15, 1850, in box 4, f. 26, EP.
25. Drafts of Emory to Ewing, San Diego, June 30 and July 4, 1850, in box 4, f. 26, EP.
26. GCG to his mother and father, San Diego, California, June 30, 1850.
27. Emory, 1857–59, 1:9.
28. GCG to his mother and father, San Diego, June 30, 1850.
29. Ibid.
30. Like other civilian members of the commission, GCG remained unpaid and needed to find

a dependable source of income. Clerking for Hayden carried a salary of $100 a month. The military escort attached to the survey was being paid regularly, since it did not depend on a congressional appropriation. Neither Weller nor Gray had received funds since beginning work in San Diego. By January 3, 1850, the government owed the employees of the commission more than $11,000 in salary (SED 34, 38).

31. When the United States contracted with the government of New Granada, to which Panama belonged, to carry mail over the isthmus in 1848, before news of the gold discoveries was generally known, New Granada got more than it bargained for. A San Francisco weekly estimated the volume of mail sent from that city between January 1 and February 1, 1850, as sixty-seven thousand letters and five thousand newspapers. Incoming mail for the same period, also routed through Panama, was reckoned at twenty thousand letters. Apparently, as the volume of mail grew throughout 1849, so too did the number of mailbags left to rot in the jungle. In March 1850 U.S. agents were sent to Panama to renegotiate the terms of payment to New Granada and to supervise the transport of the mails; see the *(Washington, D.C.) Semi-Weekly Union*, December 17, 1849; *(San Francisco) Weekly Pacific News*, March 1, 1850.

32. It might have been more surprising if Weller no longer harbored ill feelings toward an employee who had shot him.

33. The California legislature did not elect Frémont to the Senate until December 20.

34. Weller informed Washington of his intention to adjourn the commission on January 3, 1850; GCG was apparently anticipating it.

35. A family named Washington lived in Washington City, according to the 1850 census, but Mac Washington may have been a relative. GCG had an "Uncle Wash," who was a resident of Rahway, New Jersey. See GCG to his father, Mouth of the Rio Grande, July 3, 1853, and Sarah Gardner's notation at the bottom of GCG to his father, New Orleans, January 14, 1853. We cannot explain GCG's reference to a "good result."

36. Chamberlin, known as "Old Medicine," had been elected to the California Senate. As he entered the political arena, Chamberlin contracted with Dr. Charles C. Parry to serve as assistant surgeon to the military escort. After Chamberlin left, Emory put Parry in his place, leaving him with the title of assistant surgeon but tripling his pay from $40 to $120 a month. GCG witnessed Emory's signature. See Contract between Chamberlin and Parry, September 6, 1849; contract between Emory and Parry, Camp Riley, November 18, 1849; and Emory to Parry, Camp Riley, November 23, 1849, all in PP.

37. He apparently saw a local rendition of the Christmas folk play *Los Pastores*, in which the devil is a prominent character.

38. Incorrectly dated 1849, this letter comes from the John Moore McCalla [Jr.] Papers at the Historical Society of Washington, D.C., which granted us permission to publish it. Mark Herlong kindly called this letter to our attention and transcribed it for us.

39. Clement M. Butler served as rector of Trinity Episcopal Church.
40. The Indians may have been hired for the occasion, but forced labor of California Indians was not uncommon.
41. The Peruvian-born Juan Bandini married into the prominent Argüello family in 1835 and became a prominent rancher and community leader.
42. GCG, writing apparently in a state of inebriation, struck out the lines in this passage, but most of the words are still visible.
43. Presumably this was his room in the Haydens' house.
44. This is the only appearance of John Nigh in these letters.
45. In his capacity as U.S. Mail agent between New York and San Francisco. See the introduction.
46. We have not been able to locate Patterson's first name. Kemble also lists him merely as "Patterson" (1972, 68).
47. The joint commission adjourned on February 15, 1850.
48. News of Secretary of the Interior Thomas Ewing's dismissal of Weller on December 19, 1849, apparently had yet to reach San Diego.
49. As a clerk for the quartermaster, Capt. Julius Hayden, GCG was on the army payroll instead of the payroll of the boundary commission.
50. Another acquaintance of GCG's from his Washington boyhood and whose name appears only once in these letters. Willie Jesup may have been the son of Maj. Gen. Thomas S. Jessup, a Washington resident.
51. Butler was the rector of the Episcopal church that McCalla and GCG attended.
52. Born in 1817, María del Refugio Francisca Lugarda Argüello had married Juan Bandini in 1835 and became famous for organizing social events. After the American conquest of California the family fortunes began to decline rapidly, but Mrs. Bandini held "one elegant fiesta after another" (Pitt, 1966, 112).
53. One of Bandini's three daughters, Ysadora married an officer assigned to escort the commission, Lt. Cave Johnson Couts, in 1851. The Americans regarded her as the *"Belle of California"* by one account (H. H. R. [Hiram H. Robinson], San Diego, Panama, June 16, 1849, in the *Cincinnati Enquirer*, August 16, 1849, reference courtesy of Charles Hughes).
54. Josefa Carrillo, daughter of the prominent Joaquín Carrillo. She famously eloped to Chile in 1829 with Henry Delano Fitch, a Massachusetts ship's master. She became a widow in 1849.
55. The Casa de Bandini, built in 1829, still stands on the plaza of San Diego's Old Town.
56. La Punta was the local name for Santiago Emilio Argüello's ranch (Pourade, 1965, 65).
57. Emory had good reason to ask GCG to do double duty, working for Capt. Hardcastle during the day and writing for the commission at night. Emory had received instructions to reduce "the persons employed on the work . . . to the lowest number consistent with the proper though economical management of the business confided to you" (Emory, 1857–59, 1:8).

58. Emory had ordered Lieutenant Whipple to transport the astronomical instruments by way of Panama to El Paso in order to arrive by the time designated for the joint commission to reconvene on Monday, November 4, 1850 (SED 34, pt. 2, 9).
59. Mathilde was the oldest stepdaughter of GCG's sister Emma, who lived in Louisiana. Later that year Mathilde would marry GCG's brother Frank. See the introduction.
60. GCG's views regarding the untapped potential of the land in California, as well as his judgment regarding the Californios who owned it, was common among newcomers from the United States. They resented California's landed class as owning far too much and doing far too little with it. See Pitt (1966, 83–103).
61. Ranch owners in southern California rightly feared that the state legislature, dominated by the more populous north, would impose confiscatory taxes aimed at breaking up large ranches, to make land available for immigrants (Cleland, 1941, 122–23).
62. Mormons did claim San Diego and much of southern California below Los Angeles as part of the State of Deseret. In 1850 they abandoned that claim but proceeded to construct communities along a "Mormon corridor" that was designed to connect Salt Lake with San Diego (Arrington, 1966, 85–88). For a man of his time and place GCG takes a surprisingly positive view of the Mormons.
63. GCG knew only fragments of this story of rival ferry operators, involving the notorious gang of John A. Glanton and the murder of an Irishman named Callahan. For fuller accounts see Forbes (1965, 312) and Woodward (1955, 132–37). Cormac McCarthy's celebrated novel *Blood Meridian* contains an indelible description of the Glanton gang.
64. One of Glanton's men, named Brown, was arrested for killing the soldier but bribed his guard with $500 and escaped. Brown eventually settled in Los Angeles and became a deputy sheriff. The source of this account is Woodward (1955, 27), who obtained the details from the *(San Francisco) Daily Alta* of January 8, 1851.
65. In contrast to Major Emory, Mrs. Hayden used silver forks and tablecloths (GCG to his mother and father, San Diego, June 30 and July 2, 1850).
66. This is correct. See chap. 2, note 43.
67. James Nooney, a lifelong bachelor, was an 1838 graduate of Yale who had distinguished himself at mathematics. From 1844 to 1848 he was a professor at Western Reserve College. Like other civilian employees of the commission, Nooney had been put on leave with his pay in arrears and had gone in search of employment elsewhere. In his case it was San Francisco. If he left because of Emory's "disposition," he was adept at concealing it. He sent Emory at least three cordial, chatty letters from San Francisco on December 31, 1849, January 14, 1850, and March 28, 1850 (box 4, f. 22, 23, 25, EP). The first of these letters advised other members of the commission to avoid San Francisco unless they had employment there beforehand because "every thing is extremely expensive." Nooney continued to work as a surveyor in California until 1859 when

he became an astronomer on the 49th parallel survey. See *Graduates of Yale University Deceased during the Academic Year Ending in 1895* (New Haven, Conn.: Yale College, 1895), 286–87.

68. The *National Intelligencer*, published weekly in Washington.
69. We were not able to locate Charles in the 1850 census or identify his "new establishment."
70. U.S. authorities quickly came to understand that the ownership of the California mission lands was murky. They had become national property under Mexico, and so the U.S. military regarded them as public property. Harlow provides a cogent explanation (1982, 92, 331).
71. On February 17, 1850, Weller gave Parry three months' leave; two weeks later Emory appointed Parry as "Geologist and Botanist to the Boundary Commission" (Weller to Parry, San Diego, February 17, 1850, and Emory to Parry, Office Boundary Commission, March 2, 1850, both in PP). As the expedition's only naturalist, Parry had his work cut out for him in collecting plants and describing geological features. On leave from the commission that spring, he traveled up the coast to Monterey. In a letter to Emory from San Diego on September 20, 1850, Parry summarized his work, which he had done largely at his own expense since he had received only two months' pay since he joined the commission (box 4, f. 27, EP).
72. Maj. James Duncan Graham, a topographical engineer, was married to one of Ingraham's aunts—his mother's sister Charlotte (Cleaves, 1960, 355). Graham had asked Emory to find a position for his nephew, and Emory, having no positions left, asked Weller if he would hire Ingraham (Emory to John B. Weller, Washington, February 19, 1849, copy in Emory's letterbook, box 1, f. 2, EP).
73. Clayton was Gray's draftsman and first assistant.
74. On April 30 Emory discharged Richard Rust, who had been his secretary and was now serving as acting quartermaster. The commission had no need for a quartermaster since it had no property (Emory to Thomas Ewing, Sec. of the Interior, May 2, 1850, box 4, f. 26, EP).
75. A state of bad humor.
76. This account, which GCG compares to a draft, was a safe and easy means of sending money across the country to his father. In remote areas of the country people commonly purchased drafts from government agencies, in this case the forage master, that were redeemable in Washington. The price at which a draft could be redeemed fluctuated, depending on where it had been issued and by whom: the government, the military, or a bank. As a result drafts could be purchased and redeemed for more or less than their face value (Kingsbury, 1996, xxxiv–xxxv, 25n33; 47).
77. Not everyone found Hayden so kindly. One California immigrant thought Hayden's treatment of gold seekers passing through San Diego was reprehensible (Powell, 1931, 191).
78. From his position with Quartermaster Hayden.
79. The members of the survey party finally received their pay after Alexander H. H. Stuart took office as secretary of the interior in the fall of 1850. Major Emory returned to Washington that

November to find that "among the first acts of his administration, [Stuart] had sent relief to my party" (Emory, 1857–59, 1:10). GCG received his back pay in February 1851.

80. Fred Emory had gone north to the placers and was making money by surveying and speculating (GCG to his mother, San Diego, California, April 6, 9, and 20, 1850). Emory had written the following to GCG from Eliza, California, February 14, 1850: "I only received Capt. Hardcastles letter yesterday & hasten to answer by the steamer which sails on the 16th of this month, so I have only time to say come to San Francisco immediately. There you will find a letter from me. If not, come up to Sacramento City & from there to a Town on Feather River called Eliza, which point is my headquarters. I can give you $8.00 per day at first &c. Come up immediately my dear Gardner, or let me hear from you.
    My love to Hooper, Murray, Hayden, Slaughter &c. In haste, Fred Emory."
81. Brevet Lt. Col. John McClellan, the chief topographical engineer, was recalled from El Paso to Washington in November 1850. Commissioner John Russell Bartlett accused him of habitual drunkenness, but his real transgression was questioning Bartlett's judgment (Werne, 2007, 50–51; Goetzmann, 1991, 169, 171–72).
82. The work of Emory himself.
83. This is the first of many attempts GCG made to have his father or his uncle secure him an appointment in the military. Even as he seemed committed to advancing his career as a surveyor, the military beckoned.

View of the Eastern Slope of the Californian Cordilleras. From W. H. Emory, *Report on the United States and Mexican Boundary Survey* (1857–59). Courtesy of the DeGolyer Library, Southern Methodist University, Dallas, Texas.

CHAPTER 4

Finishing the California Line

July 23, 1850–June 1, 1851

In the summer of 1850 George Clinton Gardner's "glorious state of uncertainty" came to an end. The work of the commission resumed, and Clint went back to work, even though, as he told his parents, "all of us have been treated shamefully."[1] He would spend another year in California, completing the survey of the line and placing markers on it.

At one of the last meetings of the Mexican and American commissioners in San Diego, on January 29, 1850, before they agreed to reconvene in El Paso del Norte, they fashioned a plan for placing physical markers along the California boundary between the Pacific and the Gila River. The commissioners, Pedro García Conde and John Weller, had decided to erect seven monuments. In what turned out to be a stunningly bad prediction, the commissioners explained that a mere seven markers would suffice to cover the 150-mile California boundary "as a very large portion of the territory on both sides of the line, for many miles between the Pacific and Colorado, is barren and can never be cultivated by either party, this demarkation [demarcation] of the line is considered amply sufficient."[2]

For the American commission, placing the monuments fell to Capt. Edmund Hardcastle and his little team, including Clint, who became his first assistant.[3] The joint commission had specified that pyramids or obelisks of white marble should anchor the two ends of the California line and that the five intervening monuments be of cast iron (the marble Gila monument was soon jettisoned in favor of cast iron when the difficulty of hauling heavy marble over mountains and across deserts to the Gila became evident).[4] In February 1850, Hardcastle and his Mexican counterpart, Capt. Francisco Jiménez, agreed to have these monuments constructed in the United States and shipped to San

Diego by sea, with the countries sharing expenses equally. Since construction would take many months, the two men decided to wait until the first of the new year to meet again.[5]

Meanwhile, they had to complete the survey. In the summer of 1850 Emory stopped waiting for aid to come from Washington and took matters into his own hands. He obtained a military escort for his surveyors and permission to use the soldiers as laborers when they were off duty. To finance the work he borrowed money without authorization from Washington—first $500 and then $5,000 from the sympathetic collector of the port of San Francisco, James Collier.[6] The commission was back in business and surveyors took to the field again.

The financial crisis had passed but only temporarily. In the autumn Emory learned that John Russell Bartlett, who had been appointed boundary commissioner in May, had headed to Texas without sending any part of the congressional appropriation to California. In mid-October Emory left San Diego for Washington by way of Panama, hoping to accomplish in person what he had not been able to do by mail: get money to pay the commission's mounting California debts. He left Hardcastle in charge, assisted by Gardner and Ingraham.[7]

With the commission revitalized, even if in deeper debt, Clint's days of ease in San Diego came to an end. Fieldwork resumed in July, when Emory sent Hardcastle's team to complete the survey of the boundary line and select sites for the five monuments that would go between the starting and ending points. Their task, as Emory explained it, required computing "the direction of the line, making due allowance for the spheroidal figure of the earth, to mark it on the ground, and get the topography."[8] Hardcastle, however, was in San Francisco, apparently still not returned from a trip north to sell mules. Emory put Clint in charge until Hardcastle's return.

Clint, whose party included a small military escort, did not relish the responsibility. "I would prefer to have some one over me," he told his father.[9] Perhaps for good reason. When Hardcastle arrived at Clint's camp on August 6, at Tecate Mountain (today Tecate Peak), about thirty miles from San Diego, he found that "Mr. Gardner had already measured his base and the necessary angles for the purpose of connecting his last point (of the line measured from the initial point) with Tecate Mtn. and he was about changing his camp to the same place." Hardcastle, however, was not entirely pleased. Clint had not selected the best spot for a base of measurement, so Hardcastle chose a new one and redid the work.[10] Whatever Clint's shortcomings, Emory had confidence in him. In mid-September, when Emory summoned Hardcastle to testify in a court-martial in San Diego, he told him to "put things in Gardner's hands with specific instructions to push on until you return."[11]

Clint regarded this fieldwork as a way to gain "practical knowledge of my profession . . . as well as a knowledge of the world," and he described the surveying expeditions as "our pleasant tramps."[12] His private correspondence reveals little about the day-to-day operations, which often required strenuous work. Hardcastle, for example, described hauling a heavy instrument to the top of Tecate Mountain (the planned site, apparently, of monument number 4) with the help of "four stout & willing men."[13]

With the exception of a letter that he wrote to his parents on October 2, when he was temporarily back in San Diego, there is a long gap in Clint's correspondence from August 1 to November 18, 1850. Some letters may have been lost or perhaps he did not find the surveying noteworthy. As he wrote to his uncle Clint, from the oasis of Vallecito on the edge of the Imperial Valley, "all I could say about this place and our march would be but little as it has been very dull and disagreeable and this is the most barren spot on the face of the earth."[14] So much for his "pleasant little tramps."

At Vallecito in mid-November, Clint got into a quarrel with his immediate superior, Captain Hardcastle. Early on Clint had pronounced Hardcastle "quite a pleasant gentleman, but I can not say much for his scientific knowledge," but on a trip to the Gila Clint believed that Hardcastle had behaved improperly.[15] Hardcastle, Gardner, and Ingraham had left San Diego to return to the field on October 23. For protection they traveled with Maj. Samuel Heintzelman and three companies that were headed to establish a military post at Yuma. The group reached Vallecito on November 3 and was unexpectedly detained. The large party required more forage than the local terrain could provide, and the survey party was stuck until December 1, waiting for forage and other supplies to arrive from San Diego.[16] In the midst of this prolonged stay, which probably had men on edge, Clint supposed himself insulted by Hardcastle and, still smarting from an earlier slight by Hardcastle back in San Diego, disobeyed an order. Hardcastle dismissed him from the commission but quickly relented and reinstated him. Meanwhile, unaware that Hardcastle had forgiven him, Clint had saddled up and ridden to San Diego. He did not return.[17] The episode apparently cleared the air and paved the way for a happy dénouement. By January, if Clint told it right, he was "getting on quite finely with the Capt. now his manner towards me has entirely changed."[18] Hardcastle, however, may have had little choice in letting Clint back into his good graces. He had lost one of his assistants, W. A. Taylor (whom Clint curiously never mentions in his letters), and depended on Clint more than ever.[19]

Clint's quarrel with Hardcastle and his retreat to San Diego spared him a difficult time on the Colorado River. Hardcastle, following Emory's orders, had continued to the Gila and tried to run the line eastward. He got no more than nine miles into this area of shifting sand before lack of water, grass, and forage weakened his animals and forced him to stop. The surveyors returned to San Diego leaving a large part of the line unmeasured.[20]

Captains Hardcastle and Salazar were supposed to meet in San Diego on January 1, 1851, and formulate a plan for placing the monuments, but by that time Salazar had moved on to El Paso del Norte and the monuments had yet to arrive. The much-reduced commission waited out the winter in San Diego, where Clint found employment surveying a new town development for a group of investors and continued to enjoy a lively social life.

In mid-March, Salazar's replacement, Dr. Ricardo Ramírez, arrived, as did the six cast-iron boundary markers.[21] Hardcastle and Ramírez worked out a plan for jointly erecting the markers, and their parties headed east. On March 24 Gardner and Ingraham started from San Diego with two of the markers but without Hardcastle, who stayed behind to await the imminent arrival of the marble monument and inspect it. He caught up with his party at Vallecito. From there the surveyors continued to the Colorado River, arriving on April 10 and, during the next two weeks, installed two of the heavy monuments at predetermined sites, one at the junction of the Colorado and the Gila and the second at a point where the new international boundary crossed the Colorado River. Ramírez, who had been delayed, reached the Colorado on April 26 with two additional monuments that the teams jointly installed on the Colorado River where the emigrant trail crossed it, and at New River.[22]

At New River on April 28 Clint wrote his last known descriptive letter from California, to his sister Alida. He sketched out the trip from San Diego to the Colorado, explained how Fort Defiance got its name, told her how they installed the monuments, and described the sand hills. He gave her no hint of what Hardcastle graphically described as "discouraging difficulties—almost impassible hills of drifting sands and the intense heat." It was 108 degrees in the shade when Hardcastle wrote a report to Emory a few days later.[23] Clint made it sound like a pleasure trip.

Hardcastle still had to place two more monuments, calculate the azimuth line between Tecate Mountain and New River that had earlier defeated him, and connect the line coming from the Gila to the line coming from the Pacific. His little party returned to San Diego to pick up two more monuments and began to work its way east. In late May the surveyors placed one monument at a predetermined point, where the main road connected California and Baja California, and another to the east of the main road, where a trail ran between Rancho Otay and Rancho Jesús María. The cast-iron monuments installed, the men returned to Camp Riley, apparently to pick up supplies, then set out again for Tecate Mountain.[24] From there Hardcastle, Ingraham, and Gardner painstakingly ran the line for about eighty miles, pausing occasionally to erect loose stone markers. Hardcastle's team took astronomical measurements and did flashings, while Ramírez's men painstakingly measured the line with chain. Rather than working independently and then verifying one another's work, the two teams, both short handed, complemented one another and worked harmoniously.

When they reached New River in early July, the line they had measured from the east did not meet the line they had marked from the Colorado River to the west. They found themselves 1,864 feet south of a monument they had placed on New River in May. Hardcastle thought that a "disagreement of a little more than 1/3 of a mile" was understandable, given the quality of some of his instruments and the difficult circumstances. He and Ramírez made adjustments that required moving two monuments, and they completed their work on July 8.[25] A few weeks later Hardcastle acknowledged in a letter that he could not have completed the work had he relied merely on appropriations to the commission. He had succeeded thanks to supplies from the U.S. Army's quartermaster.[26]

The end of the California phase of the survey came on Sunday, July 14, with a ceremony dedicating the marble monument at the initial point on the Pacific coast. The monument weighed more than eight tons, and a special masonry foundation had been built to support it.[27] Clint left no description of the ceremony, but he was present. His name appeared twice on a document that was deposited in the base of the monument, first as one of the members of the 1849 U.S. commission, and second as a witness to the day's event.[28]

Monument at Initial Point, Pacific.

From J. R. Bartlett, *Personal Narrative* (1854).
Courtesy of the DeGolyer Library, Southern Methodist University, Dallas, Texas.

The day after the ceremony Captain Hardcastle held a celebration at his "camp near the monument," leading the editor of the *San Diego Herald* to effuse: "We do not remember ever to have spent a day more pleasantly since we have been in California. The ladies—God Bless their sweet souls—were loveable and lovely—Capt Hardcastle and his assistants were polite and in good spirits—the gentlemen put on their kind killing smiles."[29]

With or without a killing smile, Clint must have been in his element. He had become comfortable in California, and as the work of the commission came to an end he told his sister Alida that "San Diego has become a second home to me and I feel sorry to leave it."[30]

It had taken two years to complete the survey between the Colorado River and San Diego Bay—the same amount of time that García Conde had estimated it would take to complete the entire survey from San Diego to the Gulf of Mexico. As one historian summed it up, "The disorganization of the United States commission, caused in part by the hysteria of the gold rush, but in much greater part by the failure of the government to provide adequate money and credit, is largely to blame for the delay."[31]

The nineteen letters in this chapter begin with Clint's asking his uncle Clinton to help him replace his wardrobe and to send news about the congressional appropriation for the commission. The letters end with a missive to Alida, written on June 1 from the initial boundary point. In early August, Gardner and Ingraham shipped out of San Diego.

GCG to his uncle Alexander Clinton McLean, agent, San Diego, July 23, 1850

My dear Uncle Clint,

I regret much that I am unable to meet you at the landing but owing to instructions received from Maj. Emory I am obliged to start to morrow morning with a party to survey the line commencing at the initial point on the coast perhaps I may see the boat on which you sail pass down the coast but from the nature of the duty assigned me I will be unable to get even a leave of absence for one day so of course it will be impossible for me to meet you.[32]

I received the papers and am much obliged to you for them. I also received a letter or rather a note from Father, who tells me that any thing you have which will be of service to me you will let me have and he will replace it when you return and as I will not be able to see you I will enumerate a few articles—a frock coat, cloth—one that is rather tight for you don't send me an old one for I have more old clothes now than I know what to do with—Black cloth pants—don't fancy any other color—A few nice looking shirts would be acceptable—Under clothes I am intirely destitute of such as socks, drawers, under shirts—of the shoe or boot kind no. 8 are very good for me—of the necessary articles I believe I have enumerated all but however any little fancy article which will take the eye of a pretty girl would answer very well for these parts and should you have a "Colts revolver" by all means let me have it with all its appertenances, for such a thing is very usefull. I would not give a snap for any other kind—by the by one of the necessary articles I forgot to mention is buckskin gloves—don't think my *few* articles are too numerous for you must know it is better than 1 year and four months since I left home since which time my truck has not be[en] replenished.

Write me as long a letter as you have time to do and let me know something about your route and at what place and time letters are most likely to meet you. When you return to New York should you see Charly ask him if he has sent me that box if not send it as soon as possible by some private hands or else it will go to San Francisco and I will never get it. When you get to Washington I wish you would ask what has become of that $50,000 which was appropriated for this commission for we are out here without a cent and would starve were it not for the little hard Bread and pork which we can beg from the commissary department of the Army give my love to them all and tell them I am [well?] sorry I have not time to write but business before pleasure and if Father is in a good humor you might ask him if letter is not as abundant as note paper for I *like* long letters. Tell Mother I wish she would start Sarah or Alida on a sheet of paper for I find letters are Few and far between.

You must excuse this scrawl for I am very sleepy it being now about 1 hour after midnight so I bid

you good night
your affectionate
nephew G. Clinton Gardner
P.S. Aug. 1st. Please send the enclosed to father.
G. C. G.

GCG to his father, Camp Odeide near U.S. & Mexican Bdy., August 1, 1850[33]

Dear Father

Having a few moments left before my wagon starts for town I thought I would write you a few lines; and I am sorry I have not time to write to Sarah, Alida, and Charlie to all, of whom I believe I owe letters—I am out here sketching the topography of the line or rather learning to sketch for I have never done such a thing before—The Major [Emory] has been kind enough to send me out in charge of a party with an escort of 8 men together with one sergt; which to one a little distance off might seem quite pleasant but I would prefer to have some one over me and am in hopes Capt. Hardcastle will come down in the next steamer he is up at San Francisco.[34] Uncle Clint will be down in the next steamer but I will not be able to see him as I can't give myself a leave of absence but I have written to him to let me have a few of his clothes which I hope he will comply with. My time is gone so I will have to come to a close. Give my love to Mother and Sally.

Your affectionate son,

G. Clint Gardner

P.S. I am well. GCG

---

in another handwriting
I have trimmed the letters to save postage. Alida.

GCG to his mother and father, San Diego, October 2, 1850

My dear Father & Mother

Although I have let some time pass since writing to you, yet but a very short time has passed without my thoughts being with you *all* at home; I have often wished to be with you all, but it seems the time when we shall meet is no nearer at hand now than it was when I wrote my last yet I think there may be some hopes of my making a visit to the states soon. I have seen an order which the Major gave to Capt. Hardcastle, who is to be left in charge, which directs him to send me to Washington[35] with the notes on our return from the Gila should there be money enough left of $1,000. which the Major borrowed to defray the expenses of the Party of which amount, I am very certain there will be none left, and as Mr. Bartlett has taken the $50,000. which was to have been sent out here, no one knows where more will come from without Major E. succeeds in having some sent out here. From the letter I received through Uncle Clint from Father I judge you *all* think I am on my way to El Paso—but when the Major arrives in town he will tell you all better, as he says he will call to see you. Should Capt. H[ardcastle] offer to send me home without insuring to me that by going home I will not be cut off from the commission, I would much prefer to stay here—or go to El Paso by the states without going to Washington: where they will be very apt to cut me off—saying "it makes no difference now he is at home." Although I have a great desire to go home, yet at present I have a much greater desire to keep on the line—as I have no doubt but that it will last for some years yet[36]—and at the end of that time I hope to have a little more knowledge of my Profession than at Present: and I don't think any thing but higher wages under a good engineer, or an opportunity to visit the world with a good pay, would induce me to leave the commission, although all of us have been treated shamefully—as the Major will tell you.

I see by the Papers that Frank has at last been married, and I think has made a good match—and permit me to congratulate you on having such a fine daughter in law: only hoping that you may soon have *another*.[37] If I could tell in what parts to address my letter I would write them a letter of congratulation—so I hope you will in your next letter to them remember me and tell them should I visit the states again, it would do me great pleasure to call on Captain and Mrs. Franklin Gardner.

I see also that Miss Ann Nickelson has been married and is coming out with the new commission a portion of the way, her husband being Lt. [J. B.] Sutherland of the Marine corps whose brother is now a resident of this place.[38]

Well as it is growing quite late I will close—for I expect the Major can tell you all that I could—so give my love to *All*, including the neighbours.

Your affectionate
son G. Clinton Gardner.

GCG to his uncle Clinton McLean, Camp at Vallacito (Vallecito) on Route to Gila and Colorado Rivers, November 8, 1850[39]

Dear Uncle Clint,

I remember you told me that the November steamer would take you by the Port of San Diego and am only sorry that I am again compelled to let you pass with out meeting you; it seems fate has decreed that we shall not meet on this coast as this will be the third time you have passed; but however we may have the pleasure of meeting in a far better country soon.

I should like you to write me all the news about home a short letter as you pass up would be quite agreeable and then you can find time as you go down to write me a longer one. If you will leave your letter with Lt. Thos. D. Johns the depot QrM [quartermaster] he will send it out with Capt. Hardcastle's by the Express which leaves San Diego after the arrival of each Steamer. We are now on our way to the Gila, the Troops to establish a military post[40] and we to figure for a while on the line. I believe father mother & all at home are of the opinion that we have started across the country to El Paso or rather were of that opinion untill Major Emory [told?]

them better which I have no doubt he has done before this as he arrived soon after you left.

By the by should you have any thing for me from home, or else where, Gardiner & Bleeker,[41] I have no doubt, would take charge of it or any of my friends in San Diego; letters and papers such as the Express can bring out I wish you would give to Lt. [Thomas D.] Johns so as I will get them soon.

You must excuse the brevity of my letter, but really the time it would take to lengthen it is worth far more than the letter as all I could say about this place and our march would be but little as it has been very dull and disagreeable and this is the most barren spot on the face of the earth.

In haste Your
Affectionate nephew
G. Clinton Gardner

GCG to Lt. E. Murray, Dr. H. Hewit, Asst. Surg. U.S.A., and Lt. Thos. Sweeny,[42] San Diego, November 18, 1850

Gentlemen,

It being necessary for me to communicate to my family the fact of my discharge from the U.S. Boundary Commission by Capt. Hardcastle, you will confer a favour upon me by expressing your opinions as to the propriety of my conduct, under the circumstances which led to such discharge.

I have the honor, Gentlemen,
to be Very Respectfully
Your Obt. Servt.
G. Clinton Gardner

Dr. Hewit et al. to GCG, Vallecito, California, November 21, 1850

Dear Gardner,

Lt. Murray received your letter addressed to him Lt. Sweeny and myself and not having time to answer it before leaving I will do so at his request and in the name of the other gentlemen.

We look upon your and Capt. Hardcastle's late disagreement as a matter entirely personal and in no way affecting your character as a gentleman or your reputation in your profession and we consequently feel that a statement such as you request would be superfluous.

Lt. Murray informs me that Capt. Hardcastle does not intend to take any steps with reference to your position as an Assistant in the Commission of the Boundary Survey prejudicial to your interests or honorable standing.

With great respect
I am, Sir
Yours Sincerely,
H. M. Hewit
Asst. Surg. U.S.A.

GCG to Bvt. Capt. E. L. F. Hardcastle, Corp. Topl. Engineers, San Diego, November 18, 1850

Sir

As you have seen proper to discharge me from my office of Junior Assistant in the U.S. Boundary Commission, I have to request of you that you will furnish me, by the earliest opportunity, my discharge in writing, in order that I may be able to inform my family and friends of the reasons of said discharge.

I have the honor to be
Very respectfully
Your Obt. Servt.
G. Clinton Gardner

Edmund Hardcastle to GCG, Vallecito, California, November 20, 1850

Sir, In reply to your communication dated San Diego, Nov. 18th 1850 asking of me your discharge in writing that you may be able to state to your family and friends the cause of said discharge, I have to state, that you have been guilty of the offence of knowingly and wilfully disobeying my orders, when called upon by me, in the discharge of your appropriate and assigned duties. This charge I can sustain, but I have no desire to inflict upon you the penalty it deserves, as this is the first occasion I have ever observed such disposition on your part.

I shall therefore with-hold your discharge for the present—hoping that you have lamented of your folly and that your future conduct may be, as it had formerly been, unexceptionable. But until you promise to recognize and obey my authority you cannot be received as one of my party. You are at liberty to seek and engage in any employment about San Diego until my return there in Dec. or Jany. next—when you will report to me for further orders.[43]

I am, Very Respectfully,
Your Obt. Servant
[Edmd. L. F. Hardcastle Bvt. Capt. Corp. Topl Engs.]

GCG to his father, San Diego, California, November 30, 1850

Having nothing to do at present, dear Father, I believe I will give a little account of myself, going as far back as the day Major E[mory] started for home or rather a few days before that.[44] When he was at the Plaza waiting for the steamer commenced the difference between Capt. Hardcastle and myself, I asked his permission to ride a horse belonging to the U.S. Bdy. Comm. down to the Plaza to see Major E off, to which he gave his concent [*sic*] but when the steamer came in he had the horse saddled for a gentleman, who was then staying with him, and they rode off together without saying a word to me, although I was present and the horse had been lent to me, this I considered as entirely ungentlemanly, although I should not have taken as much notice of it if it had not have been followed by similar occurances, after which I made up my mind to do only my duty and if a personal favour was asked of me to refuse it, as that was the only way I could take revenge; and it happened shortly after this that he wished me to make a map of the survey of the harbour of San Diego for his private purposes which I refused, and as he could not order me to do it he had to let it pass, but I noticed afterwards that his treatment towards me was quite different, he being much more dignified than before; and finally when we arrived at Vallecito, on our way to the Gila, being in company with the Officers who were on their way there to establish a Military Post,[45] he thought, I suppose, he would show a little of his authority which he did in this manner. He [Hardcastle] called out in a very rough manner indeed to Frank Holley, one of the laborers, to enquire about a sextant which he had missed, Frank not knowing anything about [it], he [Hardcastle] called upon me in the same manner, at which time I was seated in a tent of one of the Officers eating some preserves, and he having called out for Gardner in such a manner I asked him if he wanted me, to which he replied in the same tone "to be sure I want you if I didn't I would not have called you," to which I replied that I was eating and would come as soon as I was through, that I had not been accustomed to being spoken to in that manner nor I was not going to be, upon which he came to the tent door and said what do you mean? I told him I intended what I said, to which he replied that I would have to change my manner towards him or he would find a way of making me do it. I told him he nor no one else could do that, upon which he said he had treated me as a gentleman and he expected the same of me. I told him every [one] had their opinions on that subject and that I was of the opinion he had not treated me as a gentleman, (refering to the case I have mentioned as well as others) upon which he told me I must leave the place. I told him I was ready to go as soon as he furnished me with transportation which I think he said at first he would do, but afterwards refused to furnish me with either transportation or money, but thank God as long as I live in this world, where ever I am placed, I can get living and that too in an honest manner.

Well after treating me thus he attempted to drive me out of Camp and out of my tent, saying if I did not put my things out soon he would have them put out for me, but he soon found much to his regret that his authority did not extend as far, for the tent I was living in did not belong to the Commission nor was it borrowed by the Commission, for I borrowed [it] on my own responsibility. Well the next morning after this occurance, I purchased on my credit a mule and with my blanket, pistols, holsters, &c. I mounted and started for San Diego[46] at which place I arrived in the course of time and meet [met] my friend Ingraham in search of the sextant, through whom I addressed a letter to Capt. H[ardcastle] and also one to the Officers of the Mess, copies of which together with their answers I enclose to you so as you can judge for yourself of my conduct on the occasion and should you deem it necessary to give any advice on the subject it will be very thankfully received and attended to, although it will be rather late as it will all be settled before your letter can be received. I was quite astonished to receive a letter addressed to me as Asst. U.S. Bdy. Com. after that occurance, for I thought I was no longer a member of the Com. but I think he has come to the proper conclusion, after a little reflection that he has no power to discharge one whom he does neither pay nor appoint.

Hereafter I shall seek to do my duty as long as I am with him, although I have no respect for him at all, my opinion of him is but small, for I think for a 2nd. Lieut. of Topl Engineers he supports rather too much dignity and should it be possible to be transferred to Lt. Whipple's party at E1 Paso I should be glad, for I have no desire to be under a man I can't respect.[47] I intend writing to Mr. Whipple on the subject and I know if it is possible for him to do it he will.

I received your letter of Sept. 22, together with the news papers, neck handkerchief and pocket handkerchief for all of which I [am] much indebted but am sorry to say your advice to be careful to preserve the good opinion of the Officers, &c. came too late, for I received it after arriving here from Vallecito yet I have always had the good opinion of the Officers about me and even of those whom I have held a poor opinion of, as you can see from Capt. H's letter, where he says my conduct "had formerly been unexceptionable".

It is a very hard matter to get such books on engineering as I should like and as for the advantage of the superior instructors around me it is but small for I consider Capt. H. a very poor speciman of a scientific engineer if I were with Lt. Whipple or Prof. [James] Nooney then I should consider my advantages great, and I only hope I will be able to get with Lt. Whipple that I may take advantage of his good advice for I think he is really a scientific man not only that but he is [a] gentleman in every sense of the word and is not capable of doing an ungentlemanly act knowingly, which is more than can be said of—another of the same corps.[48]

Well as I have several other letters to write I will bring this to a close hoping my conduct will meet your entire approbation.

Your affectionate son,

G. Clinton Gardner.

GCG to his father, n.p. (San Diego), December 2, 1850

I see by reading some "Unions," that I got hold of that Father Ritchie is defending Col. Weller and denounces [Horace?] Greeley as a liar, for saying the Col. was on a bender at Chagres and also about the prostitute; which if he does not take care will be proved against him, not at Chagres but at Panama. You can tell him he had better not harp too much on that string, for it will soon break.[49]

GCG to his sister Sarah, San Diego, December 2, 1850

My dear Sister Sarah,

Your letter of Sept. 23rd was duly received, and I am almost ashamed to acknowledge that I am indebted to you several letters and not only you, but mother and Alida also, for which I could give several good reasons, but knowing the amiable dispositions of you all at home I know that my neglectfullness will be overlooked without the reasons being given. I was quite surprised on my arrival here the other day at not meeting Uncle Clint, who wrote me some time ago that he would be along in the November steamer, I had obtained a leave of absence and had made up my mind to ride in from Vallecito on purpose to see him, but just before starting, I had a little muss with my commanding officer which resulted, much to my satisfaction, in my coming in soon[er] than was intended. Since I have been here I have had a great many applications for my services as a surveyor, the [San Diego] town council at one time proposed turning out the city surveyor and placing me in his place, he being a dissipated man and one whom they could not get to lay out their streets, the only objections that could be raised to me was my age, I being a minor, but however since this threat of the council he has reformed, but I fear it will not last long, it is a great pity too, for he is a splendid engineer and draughtsman.[50] Shortly after arriving here I received an official letter placing me here under orders and at the same time giving me permission to engage in any employment I wish until his return (Capt. Hardcastle) in reply to which I requested the loan of an instrument,[51] so as I could attend to some of the *numerous* applications I have had, I have not as yet received an answer but in the mean time I have borrowed one from the city surveyor and am now engaged in [the] layout [of] a new town of San Diego, for a company of numerous citizens of this place; perhaps you have not heard that the city of San Diego includes four distinct places, the first at the entrance of the harbour is called the Playa, the next the city of San Diego or the old town, next the city of San Diego or Harrasityville, upon which I am at work now laying it out a[nd] dividing it into blocks; the next and last is New San Diego or Graysville all of which Mr. Gray will describe to you if, you ask him I expect he will be in Washington when this letter arrives, for he started from here some time ago.[52] Well to return to your letter I would say that it is not necessary to excuse yourself for any advice you can give, for it all will be received very thankfully. The box sent by Charley I believe was acknowledge[d] to have been received some time ago, and by the by should this arrive before the departure of Uncle Clint I would like father to send me such books as he may deem usefull to me after consulting with Mr. Gray and Major Emory.

Well dear Sarah you must excuse the brevity of this scrawl, but really I have written so much to father that I am tired remember me [to] all my friends and my love to each member of the family.

Your affectionate brother

G. Clinton Gardner

GCG to his father, San Diego, California, January 2, 1851[53]

Dear Father,

Your letter of Nov. 10th was received in company with Sarah's—and I am glad to hear you say I run no hazard in coming on to Washington, which I should like to do very much: yet I do not wish to be placed on duty in the Topl. bureau—to be sure it will gain me standing there, but I think practical knowledge of my profession which I will gain in the field, as well as a knowledge of the world, will be of more advantage; but as the saying is I am open for anything.

You seem to think I have never mentioned those silk handkerchiefs, but I think I have thanked my dear Mother twice for them—but however such kindnesses can not be too often thought of, so I return her my thanks again; and to change the subject I hope you or Uncle John will permit me before long to thank you for having obtained that commission in the new regiment, which I understand is certainly to be raised.

I would ask, dear father, if you will be so kind to enquire about the pre-emption laws, if any, over this country—for I have some idea of taking possession of some land, if possible.[54]

If the steamer which is hourly expected should bring any letters from home I will endeavour to answer them by this mail. When you write to Charley, should Sarah find any news in my letters, give him the benefit of it as I will not have time to write him by this mail.

Your affectionate son,

G. Clinton Gardner

GCG to his mother, San Diego, California, January 17, 1851

My dear Mother,

Having put off writing until the last moment I must hasten it in order to be in time for the Steamer. Every thing out this way seems to go on as usual, no money as yet for the Commission and very little work, we have finished the Gila Notes I believe or almost so, I am now one of those Gentlemen of leisure who [k]nock around making just enough by laying out lands &c. in and around the City to pay all expenses. I contemplate taking a contract for the survey of this City if nothing is received from head Quarters to prevent [it] by the coming Steamer. I am getting on quite finely with the Capt. [Hardcastle] now his manner towards me has entirely changed, I have the use of what instruments I deem necessary for laying out lands, &c. The other day I was calculating what was due me from the commission which I find to be almost a years pay a good portion of which I should like to send home, if I could get it but however perhaps it may come when it is more needed. Last night I was invited to a wedding of the beauty of the place. Miss Murìa [María] Doloris [Dolores] Bandini now Mrs. Charles R. Johnson, the former about 14 years old the latter 21 years of age the wedding [was] very handsome not like those I have described before, but more Americanfied the supper on the occasion was the best I have ever seen in Califòrnia.[55]

Jan. 18th. The Steamer has just arrived. I received father's letter of Dec. 9th and also the card of Mrs. [first name illegible] North. I wish you would give my respects to her and say I am sorry I was not there to pay my respects to the bride. My respects also [to] the bride and groom, the former for old acquaintance sake, the latter as the husband of an old friend.

Uncle Clint, I thought, would have been in this steamer, but I [will] certainly look for him in the next (Feb. 4th) and I hope it will not be in vain.

My love to *All* in
haste Your Affectionate
Son G. Clinton Gardner

GCG to his mother and father, San Diego, California, February 18, 1851

Dear Father & Mother,

Accept my sincere thanks for the box which contained the different articles as enumerated in father's letter, all of which I was in need of. Uncle Clint arrived here on the *California*, under sail (she having broke some of her machinery) and was compelled to stay here until the *Constitution* came down.[56] I expect him here again about the 4th of next month on his return, he wished me to say that he was delayed two days here, so as father might mention the fact to the Department, he was in good health and fine spirits—but he would not go to see the young ladies. He said he had not time to get me a coat in New York, so I have sent to San Francisco for one—where I believe they can be had as cheap as in New York.

Capt. Hardcastle has received orders as father mentions in his letter, but I expect it will be necessary for us both (Ingraham and myself) to remain here, and if so he will send the instruments alone—however should it be offered to me I will accept it very quick.[57] Capt. H[ardcastle] is now in San Francisco. Ingraham has gone to the Gila again with Genl. [Persifor?] Smith so as to show him the line, and I alone am here in charge. By the by Dr. Parry is going to Washington in the steamer, I expect, with Uncle Clint; he will call and see you and will answer all questions.[58]

I have put off writing so long that now I have not time to write a very long letter, so I will have to bring this to [a] close. Give my love to all, in haste.

Your Affectionate Son,

G. Clinton Gardner.

P.S. Father, don't forget to use all the influence that you can solicit for a commission in this new *Regt.* for me and Mother you keep Uncle John reminded of it also, don't let the least opportunity pass.

G. C. G.

GCG to his father, San Diego, February 28, 1851

Dear Father,

I enclose herewith the second bill of exchange for ($500) five hundred dollars payable sixty days after sight. The reason I send one payable sixty days after sight is because I did not wish to pay the discount on it but should you wish to use the money immediately you can get it by paying the discount which is only 1 per cent. The first bill has been mailed by Capt. Hardcastle at San Francisco but I expect you will receive it by the same Steamer that Uncle Clint arrives in; I intend sending this either by Uncle Clint or Doc. [Charles] Parry of our party who is ordered to report to Major Emory at Washington. You will find the Doctor to be a very quiet unassuming gentleman one who is liked by all that know him. You must invite him up to the house to see the family tell Sarah and Alida they must not set their caps for him for he is to[o] quiet and not sufficiently fashionable for them, but should they like a person who can appreciate and describe a handsome boquet [*sic*] he is the person.

Your letter of Jan. 9th has been received, read, re-read and considered. I hope you will succeed in getting me in Lt. Whipple's party but I would much rather receive a commission in the new Regt.[59] That little disturbance with the Capt. has blown over entirely every thing goes on as usual the Monuments have arrived at San Francisco and will be here about the 8th of next month—then we are for the field again. If it were not for a dance once in a while, I [would] like to be out all the time on our pleasant tramps.

Oh by the by Uncle Clint told me that Major E[mory] had got me a Place in the [Pacific] Coast Survey should I come home. I wish you would remember me to him and thank him kindly for me—for I appreciate his feelings towards me as those of the best friend I have he has always shown the greatest kindness towards me in every instance.

Well—as Doct. Parry can tell Mother all the news, I will draw this to a close, thanking Sarah for her long Post Script—give my love to All.

Your Affectionate Son,

G. Clinton Gardner

GCG to his father, San Diego, March 23, 1851

Dear Father

In closing my account with the U.S. Boundary Com. I find Col. Weller made a mistake in the computation of a pay Roll which he turned over to Major Emory, since which Major E turned over to Capt. Hardcastle to pay upon. The consequence of which is that it knocks me out of about $40 or $50 which if you can collect may be of service. I send here with the Voucher for it, which you see explains it-self, and if you take it up to the Department, I have no doubt it will be paid.

I wrote a letter to Col. Weller about it, he acknowledges there is a mistake and tells me how to get the exact amount due me; I will enclose his answer to you.

You will find the Voucher (Voucher No. 4) refer[r]ed to in the account among

Capt. Hardcastle's Papers for the quarter end March 31st—which will leave in the same mail with this letter.

Remember me to Maj: & Mrs. Emory when you see them: also to Doctor Parry, with whom of course you are acquainted, before this.

I am, Your Affectionate

Son,

G. Clinton Gardner

P.S. The same mistake occurs in my friend Ingraham's account a voucher for which I send you also if you collect it, write and let me know, so as I can pay him the same amount.

G. C. G.

Emory to Charles K. Gardner, Washington, March 15, 1851

Major Emory, regards to Mr. Gardner and informs him he has sent a letter from Capt. Hardcastle (official), saying that in consequence of unforeseen circumstances he will be obliged to retain Mr. Clinton Gardner in California for some months.

Under all the circumstances, which I will explain, this is not unfavourable to your son's interest—quite the contrary.

GCG to his sister Alida, Camp at the little Lagoon on the desert,[60] April 28, 1851

My dear Alida,

It has been some time since I have written you a letter, but you must not think it is neglect on my part for it has always happened that something or other prevented just before the departure of the mail, and knowing that you hear from me always when I write home, I did not desire to send you an old letter or rather one that had laid in the Post Office for two or three weeks.

Well since I have commenced, perhaps you would like to know how I came in this part of the world, so I will give you a little account of our travels from San Diego, which place I left on, the evening of the 25th of March last, in charge of a party to place the Monuments on the line; Captain Hardcastle remaining behind to see the Marble Monument, which is to be placed at the Initial Point on the Pacific, landed; intending to overtake us at San Isabel, which is a small Indian Village in one of the finest vallies in the lower part of the State and has been one of the best ranches in the Country—there are still remains of vineyards &c. which show it, some of them by the by are still very productive.[61] But as the Steamer did not arrive in time—we got on as far as Vallecito, at which place is the Government Depot (between San Diego and the Military Post at the Mouth of the Gila [Camp Yuma]), where the Captain joined us, and in seven days more, we were encamped on the banks of the Rio Colorado, near Fort Defiance, which is quite a high point at the bend of the river about seven miles from the mouth of the Gila.[62] Perhaps it would be interesting to know where it got its name from; well, supposing you to have heard of the Murder of the ferry company that had established themselves on the river just below the junction, I will proceed.[63] As there is a great deal of emigration over land, and every one in California is bent on making a fortune, two or three parties were organized to take the place of those whom the Indians had killed and two of the parties actually started from San Francisco. The first party upon arriving at the junction commenced operation, but the Indians warned them of their danger and ordered them to leave, they not being disposed to obey, yet thinking their party was not sufficient to cope with the red men, concluded to go down the river to this point and there fortify themselves, and bid defiance to any attack the indians could make. Thus the point got its name. After the second party joined them, they moved up the river and commenced operations jointly, but finding the emigration not as much as they expected, they seperated and left only a part of them there, making it more profitable for those that remained—as there is not much danger anticipated since the Post has been established at the junction.[64]

Well, after leaving the Monument, which is placed on the left bank of the Colorado, the Captain and myself rode on, up to the Post, the same day—leaving the wagons to come on the next day in charge of Mr. Ingraham. After being encamped there two or three days, we placed the Monument, across the river where the stone pier, as represented on the map, stood—with due ceremony:[65] then moved down the river and placed the one we left near Ft. Defiance, from which place the Captain sent Mr. Ingraham, with pack-mules to carry water, to produce the line on to New River, across the sand hills; and after we had finished what little work we had to do there, we came on to this place in three days, traveling a part of the night of the second day, and met Mr. Ingraham at New River.[66] And now all that remains is to place the two Monuments, which are to go some where near this place, and connect the two ends of the line, which I think we will do in about two or three weeks—if the Mexican commission returns from the Gila in time, then we start for San Diego. We have four Monuments to place at the other end—the last of which will be the Marble one at the Initial Point: then our work will come to a close with a dinner, Picknick [*sic*], or something of the kind, in honor of the occasion.[67]

Should you see me at home on my Birth Day you need not be surprised for we expect to get off in the first July steamer. I look forward to it with a great deal of pleasure, when I shall meet Father, Mother, Aunty, brothers and sisters once more. Yet there is one thing that derogates from that pleasure [and] that is, parting from friends; San Diego has become a second home to me—and I feel sorry to leave it.

Well dear Alida you must excuse me from writing any more, for I have a visitor this evening I can't get rid of, I have turned him out the door of my tent, but he will come in, under the walls, and he is no less than a frog—to annoy me so. When you write home acknowledge the receipt of Sarah's and Father's letters of March 17th and give my love to them *all;* believe as ever,

Your affectionate brother,

G. Clinton Gardner

GCG to his sister Alida, Camp near Initial Point of U.S. & Mexican Boundary, June 1, 1851

Dear Alida,

It is with the greatest pleasure I commence an answer to your most interesting letter of the 25th of April for it has been some time since I have had the pleasure of answering a letter of yours although it has not been long since I wrote to you.

I am sorry to hear of the death of Mrs. McVeigh for the last recollections of her were very pleasant indeed. I expected to hear that Miss Lizzie was married before this for she must be quite "*ould*" by this time but however present my best respects to her—Is Miss Garner still at Patapsco[68] if so don't forget to remember me most kindly to her although I have never had the pleasure of seeing her yet from hearing you talk of her I can imagine myself quite well acquainted with her and should it not be too much trouble give a little discription of her. It will give me great pleasure to visit your school and should I return in time to take you home for your August

Vacation I shall not fail to do it yet my hopes of return[ing] home are rather dampened by the last mail from Washington.

I wish I could be present at another May ball and particularly as Miss Julia Hobbie was queen for I should like to have had the pleasure of a Waltz or two with her.[69] What a great shame I can't be about when all these beauties come out in society. I expect they will all be married before I return for even my sweetheart is to [be] married but never mind my day will soon come.

And so you did not find Dr. Parry very entertaining, you must tell Sarah that it was all her fault for he is very entertaining, if she had have only thought for a moment that no scientific man can be interested by trifling conversation, but if she had have commenced the subject of botany his eyes would have brightened up and then you could have seen what he was.[70] I expect Aunty could have entertained him more with her orange tree than you and Sarah could by a week's talk. But now dear Alida I must come to a close for I am in great haste preparing to start to the Mountains and the man is waiting to take my letter to town, so adios.

Believe me as ever your

Affectionate Brother,

Clint

Excuse this scrawl. Another time I will write you a letter.

G. C. G.

---

1. GCG to his mother and father, San Diego, October 2, 1850.
2. García Conde and Weller, Meeting of the Joint Commission, January 29, 1850, in SED 119, 62.
3. For GCG's title see Emory to Thomas Ewing, Sec. of the Interior, San Diego, March 1, 1850, in SED 34, pt. 2, 9. When Hardcastle was away from San Diego in May and June 1850, he left GCG in charge of his affairs (Hardcastle to Emory, Santa Barbara, May 29, 1850, box 4, f. 26, EP).
4. García Conde and Weller, Meeting of the Joint Commission, January 30, 1850, in SED 119, 62–63; Emory to Hardcastle, San Diego, April 3, 1850, HCMSA; and Hardcastle, San Diego, to E. & G. W. Blunt, New York, April 3, 1850, HCMSA.
5. Hardcastle and Jiménez, San Diego, February 23, 1850, in SED 119, 72–73.
6. Draft of Emory to Maj. Samuel P. Heintzelman, San Diego, July 22, 1850, and Heintzelman to Emory, San Diego, July 23, 1850, both in box 4, f. 26, EP. Emory paid the soldiers "from $1 to $2 per day" for manual labor (Emory, draft of letter to Thomas Ewing, September 20,

1850, box 4, f. 27, EP; Collier to Emory, San Francisco, August 31, 1850, box 4, f. 27, EP; Norris, Milligan, and Faulk, 1998, 102–3). Although Emory did not know it, a change in regime in Washington began to free up funds. D. C. Goodard, interim interior secretary, predicted that Congress would pass the appropriation "in a few days" and authorized Emory to issue drafts on the Department of the Interior in the meantime (Goodard to Emory, Washington, August 10, 1850, box 4, f. 27, EP).

7. A letter from Emory to Secretary of the Interior Stuart, purportedly dated September 25, 1851, New Orleans, appears in SED 119, 77–78. The date is wrong. Emory was still in San Diego. Hardcastle wished Emory "safe and speedy passage homewards" in early October (Hardcastle to Emory, San Diego, October 3, 1850, box 4, f. 27, EP). Emory left San Diego in mid-October on the *California*, arriving at Panama on October 22 (for certification of his ticket, see A. B. Forbes to Emory, Panama, October 22, 1850, box 4, f. 27, EP). By November he was home. See too Norris, Milligan, and Faulk (1998, 102–3).
8. Emory to A. H. H. Stuart, Secretary of the Interior, New Orleans, September 25, 1851, in SED 119, 77–78. See note 7, for the explanation about the misdating of this letter, but that does not change Emory's description of the work.
9. GCG to his father, Camp Odeide, August 1, 1850.
10. Hardcastle to Emory, Camp South Tecate, August 14, 1850, HCMSA. See also note 34 about GCG's letter to his father, Camp Odeide, August 1, 1850. The thirty-mile distance to Tecate Mountain is Emory's calculation (Emory to Thomas Ewing, Sec. of the Interior, San Diego, April 3, 1850, in SED 34, pt. 2, 14).
11. Emory to Hardcastle, San Diego, September 15, 1850, HCMSA. Thus Emory implicitly chose GCG over his friend Edward Ingraham, Hardcastle's other assistant. Emory notes that the party was still in the field in Emory, draft of a letter to Thomas Ewing, September 20, 1850, box 4, f. 27, EP.
12. GCG to his father, San Diego, January 2 and February 28, 1851.
13. Hardcastle to Emory, Camp South Tecate, August 14, 1850, HCMSA.
14. GCG to his uncle Alexander Clinton McLean, Camp at Vallacito [Vallecito] on Route to Gila and Colorado Rivers, November 8, 1850.
15. GCG to his father, n.p., April 20 and May 3, 1850.
16. Woodward, 1956, 49–50.
17. GCG to his father, San Diego, November 30, 1849.
18. GCG to his mother, San Diego, California, January 17, 1851.
19. Hardcastle to Emory, San Diego, January 18, 1851, box 4, f. 28, EP.
20. Ibid.
21. The cast-iron monuments arrived in San Diego on March 14, Ramírez on March 18 (Hardcastle to Emory, San Diego, March 28, 1851, box 4, f. 28, EP).

22. Hardcastle to Emory, Camp Stuart on New River, May 2, 1851, box 4, f. 30, EP.
23. Ibid.
24. We infer the party's return to Camp Riley from GCG to his sister Alida, Camp near Initial Point of U.S. & Mexican Boundary, June 1, 1851, and Hardcastle to Emory, Camp at Initial Point of Bdy. on Pacific, June 3, 1857, box 4, f. 31, EP.
25. Hardcastle to Emory, San Diego, June 3 and July 17, 1851 (quotation), box 4, f. 31, EP. For good overviews of this operation see Brown (1969, 160–68) and Rebert (2001, 100–7).
26. Hardcastle, San Diego, to Maj. Gen. Thomas Jessup, San Diego, August 1, 1851, HCMSA.
27. Hardcastle had left GCG and Ingraham to run the line while he stayed in San Diego to oversee the work (Hardcastle to Emory, Camp at Initial Point of Bdy. on Pacific, June 3, 1857, box 4, f. 31, EP).
28. Hughes, 2007, 142. See also Minutes of the Meetings between Capt. Edmund L. F. Hardcastle of the United States and Ricardo Ramírez of Mexico . . . , March 19–July 30, 1851, box 501, f. 59, HCMSA, and *San Diego Herald*, July 24, 1851, both references courtesy of Charles Hughes.
29. *San Diego Herald*, July 17, 1851, from Winifred Davidson's notes, San Diego Historical Society. The editor and chief writer for the paper was John Judson Ames (Pourade, 1965, 167).
30. GCG to his sister Alida, Camp at the little Lagoon on the desert, April 28, 1851.
31. Werne, 2007, 35–37.
32. As U.S. Mail agent between New York and San Francisco, Uncle Clinton made frequent trips along the Pacific.
33. This letter is in the collection of the Smithsonian Institution. The date is not clear on its face. The month is clear, and the date appears to be August 1 but could be August 17, and there is no year, but it was clearly written in 1850. A letter from Emory to Hardcastle (San Diego, August 2, 1850, HCMSA), ordering Hardcastle "to join the survey party, on the line, at or near the Tecate Mountain and take charge of the same," makes clear that the letter was dated August 1 and not August 17. Emory told Hardcastle that he would "find in the hands of Ast Gardner a mem. [memorandum?] to which I desire you will adhere." GCG was charge of this party and had the opportunity to name his own camp; he dubbed it "Odeide" for his sister Emma's teenage stepdaughter, with whom he seems to have been infatuated. Its location was "on an arroyo that flows to Tija Juana and about 5 miles distant from the base of Tecate Mountain" (Hardcastle to Emory, Camp South Tecate, August 14, 1850, HCMSA).
34. Hardcastle arrived at the camp on August 6 (Hardcastle to Emory, Camp South Tecate, August 14, 1850, HCMSA).
35. Hardcastle did not follow that order. On returning from the Gila he found that one of his assistants, W. A. Taylor, had left San Diego "without any authority from me." Hardcastle was reduced to two assistants, GCG and Ingraham, and would need both to erect the monuments.

Thus, he told Emory, he was sending his notes back to Washington with Dr. Parry "instead of Ast. G. C. Gardner as you desire" (Hardcastle to Emory, San Diego, January 18, 1851, box 4, f. 28, EP).

36. In predicting the length of time the survey would take, GCG was prescient.
37. Franklin Gardner married eighteen-year-old Mathilde Mouton. See the introduction.
38. The brother was Thomas H. Sutherland, who was elected county attorney in 1850 and 1851 and involved in land speculation in "Haraszthyville," or Middletown, with Emory and other influential people (Harrison, 2005, 46, 51, 56; Sutherland to Emory, San Diego, March 3, 1851, box 4, f. 29, EP).
39. This letter is in the Smithsonian Institution.
40. Maj. Samuel P. Heintzelman, at the head of three companies of soldiers, established Camp Yuma at a site overlooking the junctions of the Gila and the Colorado to protect the overland emigrants to California. Lieutenant Couts and his party had occupied a post on this site, called Camp Calhoun, in 1849 (Derby, 1969, 8; Forbes, 1965, 322; Sweeny, 1956, 49).
41. Gardiner (not Gardner) and Bleeker were apparently forwarding agents.
42. Lt. Edward Murry, Dr. Henry Stewart Hewit, and Thomas "Fighting Tom" Sweeny, a one-armed veteran of the U.S.-Mexican War. Maj. Samuel P. Heintzelman had sent them to the Colorado River to establish a fort at Yuma.
43. Emory had left Hardcastle in charge of the members of the commission remaining in California, so GCG worked directly under his orders.
44. Emory left San Diego in mid-October.
45. Camp Yuma.
46. A lengthy ride. GCG's ride must have taken him into the mountains about twenty-eight miles to Warner's Ranch; then he rode about sixty miles more from Warner's Ranch into San Diego.
47. The main body of the commission, under Bartlett, had already taken the field in Texas. Emory had sent Whipple by way of Panama, New Orleans, and Texas with "a suitable supply of instruments" to meet the main body in El Paso del Norte (Emory, 1857–59, 1:10).
48. Both Whipple and Hardcastle were graduates of the U.S. Military Academy (Whipple was fifth in the class of 1837 and Hardcastle was fifth in the class of 1842). GCG's preference for Whipple over Hardcastle is probably based on Whipple's deportment rather than either officer's engineering skill (Heitman, 1965, 1:499, 1025; Sweeny, 1956, 220n17).
49. The "Unions" to which GCG refers were issues of the *Union*, a newspaper established in Washington in 1845 by the influential journalist Thomas "Father" Ritchie at the request of James K. Polk. Before that, for forty-one years Ritchie had successfully operated the Democratic *Richmond Enquirer* and through it lent his support to politicians such as Van Buren and Polk while opposing the likes of Henry Clay and John C. Calhoun (*DAB*, 15:628).
50. GCG could be referring to former commission member Henry Clayton, who had been elected

county surveyor for San Diego in 1850 (Pourade, 1965,166). Speaking of Clayton, GCG had told his father in an earlier letter of April 12, 1850: "It is a pity he drinks."

51. There was precedent for GCG's request to borrow instruments from the commission: Lt. Cave Johnson Couts had borrowed Major Emory's "compass, chain, etc., to run a few lines more" when he surveyed a large section of Old Town and the town lots close to the beach at La Playa the year before (Scharf, 1976, 19n41).
52. La Playa was a commercial district at the water's edge, on the flatlands of Point Loma, five miles west of Old Town. It consisted of some stores, the customhouse, three houses where hides were stored, one owned by Bandini's son-in-law Charles Johnson, and the Pacific Mail Steamship Company's passenger and freight depot. Around this time the army had two companies living in a tent village at La Playa. Old Town consisted of homes and commercial buildings clustered around a plaza. Prominent Californios such as Juan Bandini, José Antonio Estudillo, and Josefa Carillo de Fitch owned property there. The "Harrasityville," or Haraszthyville, that GCG was surveying was a tract of 687 acres named for one of its developers, Agoston Haraszthy, a Hungarian native, the first city marshal, and the son of one of its first council members, Charles Haraszthy. It was adjacent to the bay between Old and New Towns and eventually became known as Middletown. The syndicate that purchased this land included two speculators associated with the survey, Cave Johnson Couts and William H. Emory. New Town, which GCG also called Graysville, was built in 1850 at the site of modern downtown San Diego. It was the brainchild of Surveyor Andrew Gray, who had camped at Punta de los Muertos, at the foot of present Market Street, in 1849, and who built a house there. Gray formed a partnership with several prominent local men, including William Heath Davis of San Francisco, who provided most of the capital. They bought 160 acres in March 1850. New Town thrived briefly when Davis diverted a planned army depot and barracks from La Playa to his town site across the bay by giving an army supply officer, Lt. Thomas D. Johns, one of the town's eighteen shares in return for shipping the army's lumber to the new location, but the town collapsed by 1852 when funds ran low (Pourade, 1965, 160–61, 163, 170–72; for Gray's role see Davis, 1967, 258).
53. GCG dated this letter 1850, but he clearly wrote it in 1851.
54. GCG was not the only Yankee in California interested in obtaining land. Thousands of argonauts, fed up with the drudgery of the mines, began to focus their attention on the agricultural potential of the vast tracts of "unused" land surrounding them. Even as GCG made this request of his father, the U.S. Congress debated the question of land tenure in California in an attempt to shift property out of the hands of the "contented" Californios and into the hands of "enterprising" Yankees. The result was the passage of the Land Law of 1851, proposed by California senator William Gwinn, which set up a three-man commission in San Francisco to determine the validity of the Californios' land titles "as defined by Spanish and Mexican law,

the provisions of the Treaty of Guadalupe Hidalgo, the principles of equity, and the precedents of the United States Supreme Court" (Pitt, 1966, 83–86).

55. Johnson was a former cattle auctioneer from San Francisco (Pourade, 1965, 170).
56. The *California*'s cross-tail and connecting links had broken one hundred miles south of San Diego on February 1, 1851 (*Steamer Pacific News*, February 15, 1851).
57. On December 23, 1850, the Department of the Interior ordered Hardcastle to forward to New York all the commission's instruments he did not need to run the line between San Diego and the mouth of the Gila. Although Emory had been ordered earlier to send the instruments to El Paso del Norte with Whipple, Emory sent only a portion since his orders had not made it clear whether the survey would continue east from California or recommence at El Paso. Therefore, roughly 150 instruments remained behind in San Diego, consisting of equipment Graham classified as "the main body of the surveying apparatus *proper*" (C. M. Conrad, *Report of the Secretary of War, Communicating, in Compliance with a Resolution of the Senate, the Report of Lieutenant Colonel Graham [June 18, 1852] on the Subject of the Boundary Line Between the United States and Mexico*, 32nd Cong., 1st sess., Senate Executive Document 121 [1852], 3, 5, 7–8).
58. Parry had resigned from the commission on September 20, but Emory asked him to withhold his resignation and instead travel to Washington and report to Emory there. The commission lacked funds, however, to send Parry back to Washington. Emory left and Parry remained in California (draft, Parry to Emory, San Diego, September 20, 1850; Emory to Parry, San Diego, September 24, 1850; Hardcastle to Parry, San Diego, October 15, 1850; Alexander H. H. Stuart to Parry, Washington, December 23, 1850, all in PP). Parry finally left San Diego on a steamer on March 4, 1851, with his botanical collections, a copy of Hardcastle's notes, and some instruments (Hardcastle to Emory, San Diego, March 28, 1851, box 4, f. 27 and 28, EP).
59. Lieutenant Whipple was with Bartlett's group near El Paso.
60. They were camped at what apparently was a seasonal lagoon near New River.
61. San Diego Mission had maintained a small chapel, or *asistencia*, at Santa Ysabel, which exists there yet today.
62. Four miles below Camp Yuma (Heitman, 1965, 2:494).
63. GCG had described this a year earlier in the April 20 entry in a letter to his mother, San Diego, April 6, 9, and 20, 1850, and apparently presumed Alida knew the contents of that letter or had read of the episode in the press.
64. Forbes provides a detailed account of this episode (1965, 310–24).
65. This on April 15. Beneath the monument they placed a piece of paper that contained the names of the members of the U.S. and Mexican commissions as of July 9, 1849 (GCG appears as a "Junior Assistant to the Astronomical Party"), and the names of those present on April 15, which included Hardcastle, Edward Ingraham, and GCG and a small military escort (Hardcastle, April 15, 1851, HCMSA).

66. New River was a channel halfway between Vallecito and the Colorado River that sometimes carried overflow from the Colorado back into the desert, toward the sinkhole that became today's Salton Sea (Pourade, 1965, 148; Forbes, 1965, 299, 322; Sweeny, 1956, 50, 115–16). GCG first heard of the New River shortly after he arrived in California (GCG to his sister Sarah, Camp Riley, California, July 23, 1849).
67. The placing of the marble obelisk at the initial point on the Pacific took place on July 14, 1851, with Captain Hardcastle providing what the *San Diego Herald* described as an "entertainment" (July 24, 1851). GCG was present. A document was placed in the base of the obelisk that contained the names of members of the 1849 Mexican and American commissions, including GCG's, and GCG also was among those in attendance who signed the document as a witness. The *Herald,* conscious of the historical importance of the moment, reproduced the contents of the document in full. Reference and text courtesy of Charles Hughes.
68. Alida attended the Patapsco Female Institute, located in what today is Ellicott City, Maryland. Miss Garner was apparently the sister of George Garner, who joined the survey in its Texas phase (GCG to his sister Alida, Frontera, Texas, May 8, 1852).
69. We identify Julia Hobbie in chap. 1, note 96.
70. The *Dictionary of American Biography* credits Parry with a warm and engaging personality, describing him as "beloved by all" (14:261–62). This stands in contrast to Sarah Gardner's apparent reaction to him.

CHAPTER 5

Home for a Day and Off to El Paso del Norte

September 7–October 20, 1851

On August 1, after they had packed the instruments and completed their work in San Diego, Clint Gardner and Edward Ingraham received orders to "proceed by the quickest and cheapest route" to join the new boundary commissioner, John Russell Bartlett, "in the vicinity" of the Mexican town of El Paso del Norte (present-day El Paso, on the U.S. side, did not yet exist). They would have to use their own funds to get to El Paso. Capt. Edmund Hardcastle, who transmitted the order "in accordance with instructions from Washington," told his two assistant engineers that he was not empowered to furnish them with any public money for the journey (perhaps they were to be reimbursed on arrival). If Gardner and Ingraham wanted to end their service with the commission, Hardcastle was authorized to pay their passage home. Hardcastle took the occasion to praise the men for their "zeal, efficiency, and general good conduct . . . in the laborious service that has been performed."[1]

Gardner and Ingraham decided to stay with the commission, but before reporting to El Paso they would enjoy a short home leave. They shipped out of San Diego some time after August 1, crossed Panama, and sailed into New York Harbor on September 7. Had events followed an earlier plan, they would not have made it home. Initially, Secretary of the Interior Thomas Ewing had ordered Hardcastle to proceed from San Diego to El Paso by land once he finished his work on the California line. The secretary had dispatched guides from Santa Fe to meet Hardcastle on the Colorado River at the mouth of the Gila. Hardcastle was to survey the Gila River with instruments he had used in California and, on arriving in El Paso, deliver those instruments to Commissioner Bartlett.[2] Gardner and Ingraham expected to make the overland journey with Hardcastle, but plans changed. After Secretary Ewing dispatched the guides from Santa Fe, he decided to postpone the Gila survey and focus on surveying the line west from El Paso to the Gila. Ewing's change of plans gave Hardcastle and his assistants the opportunity to make their way from San Diego to the Texas coast by sea rather than by land, and so they did, taking a detour from Panama to New York.

From New York, Ingraham headed to Philadelphia to visit friends (his family's home was in Louisiana). Clint traveled about sixty-five miles up the Hudson from New York to visit family in Newburgh. His father and mother had both grown up in Newburgh. Cousins, uncles, and aunts remained there, and his mother was visiting there.[3] Clint went to see her, his uncle John McLean, and a cousin, Eliza, at the nearby village of Little Britain. From there he returned to New York City to visit relatives, the well-to-do Sam and Mary Butterworth.[4] One of the Butterworths' nieces, also named Mary Butterworth, caught Clint's eye. She was a classmate of his sister Alida's at Patapsco Female Institute and, as Clint told his sister, "I think she is really a pretty girl and should like to have seen more of her than I did."[5] Although she was his cousin, he saw "no impropriety" in a relationship with her.[6] Alida tried to promote the relationship but without success.[7] Whatever attraction Mary held for Clint apparently faded. On a subsequent trip home he thought about visiting her but then "forgot."[8]

After visiting his mother and some of his extended family, Clint returned home to Washington, D.C. By then he had received orders to join Major Emory in New Orleans before the end of the month. Since his return to Washington the previous November, Emory had worked in an office preparing the official report of his work in California. In September 1851, however, the secretary of the interior, who was now Alexander H. H. Stuart, ordered Emory back into the field. He would assume his old position, chief astronomer of the boundary commission and, in addition, serve as "head of the Scientific Corps." He replaced Col. James Duncan Graham, who had been working as chief astronomer for nearly a year under the direction of Commissioner John Russell Bartlett. Stuart acceded to Emory's request that he be given an escort of dragoons since, as Emory had told Stuart, he would be traveling through territory "occupied by wild Indians." Emory had also asked to be put in command of army officers attached to the commission. His experience in California had taught him that the aid of the army was absolutely essential. Stuart honored that request as well.[9]

On September 13, the same day he received orders to return to the field, Emory asked Captain Hardcastle to instruct Gardner and Ingraham to meet him in New Orleans "if

Mr Gardner and Mr Ingraham elect to stay in the Commission."[10] Hardcastle, however, would no longer supervise their work. Emory ordered him to remain in Washington to finish preparing the commission's report.[11]

Clint had received advance warning that Emory would be sent to El Paso, but Emory's order to meet him in New Orleans at the end of the month took Clint by surprise.[12] He spent just a day at his parents' home in Washington before heading to New Orleans with Ingraham, who had joined him in Washington.[13] Clint's remaining correspondence contains no details about the first part of their journey, but on September 24, just eighteen days after their return to New York City from San Diego, Gardner and Ingraham were in Montgomery, Alabama. They had planned to take a boat down the Alabama River to the Gulf and proceed from there to New Orleans. Instead, they sent a telegram to Emory: "We cant get off until day after tomorrow at the earliest no boat no steam."[14] The next day, rather than wait, they changed transport and boarded a crowded stagecoach headed to Mobile, but that trip went badly. Top heavy, the stage tipped over, spilling its passengers and cargo. Gardner and Ingraham arrived in New Orleans on September 29, only to find that Emory had left the day before. Once again Clint had missed the boat and was stranded in New Orleans without money, at least initially.

At this point there is a large gap in Clint's remaining correspondence. He and Ingraham followed Emory to Indianola, then the largest seaport in Texas, and took a stage to San Antonio, where they caught up with Emory.[15] From there Emory's party made a forty-one-day trek to El Paso del Norte, which they reached on November 24.

The ten letters from this brief interlude in Clint's work with the boundary commission begin with Clint's informing his mother of his arrival at New York City on September 7, 1851, and end with another short note to his mother, written on October 20, as he traveled between San Antonio and El Paso. He would have more to say about the journey after he reached El Paso.

GCG to his mother, New York, September 7, 1851

Dear Mother,

Here I am at last in a perfect state. Uncle Clint insists that I should write, although I have nothing to say.

We arrived this morning. I have seen Maj. Emory, who tells me you are all well at home and I will see Mr Thompson this evening and find out in what part of the world I can find Uncle John and whoever is with him.[16] When I will arrive in Washington is a case of quien sabé [*¿quién sabe?*—who knows?].

My love to All,

Your Affect. son

Clint

GCG to his father, Little Britain, September 12, 1851

Dear Father,

Here I am at Cousin Eliza's.[17] I intended returning to New York to morrow but as the country air is so strength[en]ing, and my *health* so *poor*, I think I will remain until Sunday and be in New York on Monday.[18] Uncle John [McLean] told me there was no hurry for me to return to Washington so I will spend a little while here and then go to New Jersey if you do not send a copy of that last order bringing me home for I am waiting an opportunity to go to El Paso.

Mother wants very much to hear from home and particularly from that man, what brokes [*sic*] his bones which I think might be glued up by one of the carpenters. She has not seen a letter from home for a week.[19] Tell Alida I want her to give a ball to receive me when I arrive as she has charge of the Gardner Casa.[20]

Your affectionate son,

Clint.

The following was written on the back of the letter:

Dear Father,

Uncle John told me when I was in Newburgh that I need not be in a hurry to return to Washington if you explained the mistake I made in answering Major Emory. For I might as well remain here until I could find a way of getting to El Paso and see this part of the country. As I am entitled to my board while in New York I think I can draw the same while visiting out here, if so it will much over pay my expenses so I don't think my visit will take any thing out of my pocket.

I wish you would write and send your letter to New York care of Mr. Sam Butterworth giving me all the news concerning my position. Ingraham told me he wished to remain in Phil[adelphia] to see his friends and that he would wait for me to come there, so as we can both report together. I should like very much to visit New Jersey to see Aunty and your relations, write soon, my love to all.[21]

Your affectionate son,

G. Clinton Gardner.

C. K. Gardner to GCG, Washington, September 15, 1851

Dear Clinton,

I wrote you a hasty letter of the Orders of Major Emory and my conversation with him—care of Mr. Thompson, 11 East 11th Street. I learn that during my absence at the Dept. Capt. Hardcastle had sent Frank [Clint's brother?] up to the House to enquire whether you had arrived, and to say that he wished to see you—Whether you go with Maj. Emory or not, I think you might dispense with a trip to New Jersey. Charley said New Jersey was a dull place for him. I have not yet seen Capt. Hardcastle (tho' I have called 3 times) and do not know what his wishes are.

Maj. Emory will be at the Astor House and will embark on Wednesday the 17th in the Winfield Scott.[22]

Your mother wishes to show you off and enjoy the flattering attentions of her aunts, cousins &c. In fact, she is vetted *[quizzed] wherever she goes, but tell her I will vet her at home in all what is reasonable.*

Write, at large, your views & conclusions.

Yours truly affectionately,

C. K. Gardner

If purchance [perchance], you have left N. York when this arrives at Mr. Butterworths, I trust Sarah will get Mr. Butterworth to telegraph you to come in & see Maj. Emory.

GCG to his mother, Montgomery, Alabama, September 24, 1851

Dear Mother

I arrived here this morning about 3½ o'clock and will be compelled to remain until tomorrow to get passage in the stage. The river is so low that the boats cannot get up. There is one expected daily but should it come to day I would not go in it as there is no certainty as to its getting back so it will be impossible for me to get to New Orleans now before the evening of the 28th or morning of the 29th. Should

I have started from home the day before I would not have been any better off for the cars run off the track and did not connect so we overtook Frank H. [Holley?] at Richmond.

I will write on arriving at New Orleans my love to all in haste your affectionate son,

Clint.

Send my letters Via Santa Fee, and pre paid.[23]

GCG to his father (telegram), New Orleans, September 29, 1851

I arrived this morning Major E. left yesterday with word for us to proceed to San Antonio leaving neither funds or orders—What shall we do Answer quick.[24]

G. Clinton Gardiner [*sic*]

Col. Charles K. Gardner to GCG (telegram), Washington, October 1, 1851[25]

Wiring $100 via S. F. Butterworth.

GCG to his father, New Orleans, October 1, 1851

Dear Father,

I received your telegraph this morning but had made all my arrangements before. Major Emory did leave $50. for Ingraham & myself but we did not get it until yesterday evening.[26] He sent it to Capt. [Robert Emmet] Clary and we called on Capt. C. just after arriving here and no sooner did he see us than he told us he had no money to advance to the Bdy. Com. before we asked him for any so you may imagine how low the Commission is just about this time.[27] Ingraham made a loan of some two hundred and will defray my expenses until we join Major E. so I did not avail myself of the hundred you offered thinking it would only be a draw back to you. My outfit has cost about $50. and we will have plenty to take us to San Antonio so I have no doubt we will have a nice time, Should I ever start for an out of the way place again I would never go by Orleans for it seems to be my luck but stop I must not crow for am not out of it yet who can tell what may happen yet.

Give my love to *All.*

Your Affectionate Son

Clint.

GCG to his sister Alida, New Orleans, October 3, 1851

By way of encouraging you, dear Alida, to a correspondence, I will give a little idea of the pleasures of an upset in a stage coach.

On the evening of the 25th of last month I started from Montgomery Ala: in the stage for Mobile, for the first ten miles we had a plank road and got along very pleasant, altho' there were nine inside and one out with the driver besides quite a heavy mail, together with all our baggage, the amount of which you can imagine as mine was about one tenth: and as the baggage took up more than the rack behind the coach the remainder together with the mail had to be placed on the top, which made us quite top heavy, but notwithstanding all this we got along quite well until the night of the 26th, during which day no one had had anything more than an Alabama breakfast, which is composed of bread not quite the specific gravity of lead, meat that no one could masticate and butter capable of standing alone, so you can imagine that our appetites were quite keen for the anticipated supper to which every one was inquiring the distance, but lo' upon the last inquiry the coach began to c[a]reen first this side, then that, and it was then that the supper was forgotten. In a few moments more the coach was balancing itself on the two starboard wheels, and, as one of the gentlemen remarked "over she goes," then for the scrabbling to get out, but as luck would have it your most humble come tumble was on top, the first one out, so you see I had the pleasure of seeing all that was going on. The first thing that attracted my attention was as I thought an animal of some kind which had been caught under the stage in its fall and was creeping out, but much to my surprise I found it to be no less than a little Jew, one of the passengers, who it seems had got out of one of the under windows and was scratching his way from under the stage.

It was not long before all were taken out and the stage righted again. All the injury done was to two persons an old lady from the North, who was going to visit her daughters in the south and an Old gentleman of this place by the name of Pynchon, who had his bones a little bruised and his eye blacked.[28] These are the pleasures of a stage coach drive, it only took twenty minutes for the whole of this to be performed commencing ten minutes before ten p.m: and ending ten minutes after ten, by which time the coach was righted and every thing fixed.

It is hardly necessary for me to say that every one's appetite failed them, with the exception of Ingraham & myself both of whom I think got the worth of our fifty cents as it was the best house on the road.

Ask Father if he knows Mr. Pynchon, tell him I called on him after arriving here to see how he was getting on, and we were talking about my affairs and he told me if I had mentioned my case to him he would have been glad to have served Ingraham and myself and he only regrets he is not able to take us around the city, what do you think of that? When we were only travelling companions for two nights and a day.

Write soon and let me know all the news, since you can't tell me anything about the beauties of Patapsco [Female Institute] now, you can write about those of Washington. By the by has Mary Butterworth gone to Patapsco if so the beauties there are greatly increased, for I think she is really a pretty girl and should like to have seen more of her than I did, for I had hardly time to tell whether I thought her handsome because I had not seen any pretty girls for so long, or whether she really was handsome.[29] I suppose you correspond with her if so give my best respects.

Give my love to all including all the pretty girls. My respects to the neighbours, &c., &c., &c., &c. Your affectionate brother

G. Clinton Gardner.

GCG to Samuel Butterworth, New Orleans, October 3, 1851

Dear Sir,

I arrived here on the morning of the 29th of Sept. just too late to overtake Major Emory. I received your kind letter of introduction to Mr. F. H. Masterson, for which I am much obliged to you for I find him to be a very nice gentleman indeed.[30] I have almost finished the laborious task of making my outfit but almost every turn adds something to my list. We sail sunday morning for Indianola [Texas] and from there to San Antonio, where I hope we will succeed better in overtaking Major Emory.[31] Mr. Ingraham is here with me and has not yet seen home, nor will he have time, so I think myself lucky to have spent a day at home.[32]

Remember me to Mrs. B. [Butterfield] and the children also Miss Charlotte and Mary.

With great Respect I remain

Your Obt. Servt.

G. Clinton Gardner.

GCG to his mother, Camp near Ft. Inge,[33] Texas, October 20, 1851

Dear Mother,

We arrived here yesterday after having quite a pleasant March from San Antonio. As we stop here for a day or two to make some preparations before crossing the plains, I thought I would write these few lines and ask you, or father, to get for me a gold pen—just the size of the one enclosed—which I have just broken—you can enclose it to me just the same as I do this one. I see by the papers that a mail route has been established via San Antonio. I met Peter Brady in San Antonio and he told me he was connected with the contract—or something of the kind—but however, father can always find out the most direct route for letters.[34]

When I get to the end of my march I will write you all an account of it, so at present content yourselves with this. Tell Alida to tell Mary Ellen Brady that Peter is enjoying perfect health and enquired much about his home—and I was sorry I had not seen more of her and Mrs. [Margaret] French, so as I could have given him some news.

Give my love to *all* in haste,

Your Affectionate Son

G. Clinton Gardner.

---

1. Hardcastle to GCG and Ingraham, San Diego, August 1, 1851, HCMSA.
2. Thomas Ewing, Sec. of the Interior, to Emory, Washington, April 29, 1850, box 4, f. 25; draft of letter from Emory to Gray, San Diego, July 13, 1850, box 4, f. 26, both in EP. Ewing had sent "[John L.] Hatcher, Francisco and three other men of Santa Fé" to the Gila. Hatcher had guided the Collier party (see chap. 2) to the Gila in 1849 (Foreman, 1937, 23).
3. GCG to his father, Little Britain, September 12, 1851, makes it clear that his mother was there.
4. Samuel Butterworth is listed in the U.S. Census of 1850 as a thirty-eight-year-old resident of New York City, married to Mary, who was thirty-seven. The couple had four young children ranging in age from two to eight, and servants (U.S. Census, 1850, Ward 18, New York, N.Y., roll M432_557, p. 121).
5. GCG to his sister Alida, New Orleans, October 3, 1851. Mary Butterworth was not the daughter of Sam and Mary Butterworth. We know that because GCG told Sam Butterworth, "Remember me to Mrs. B. and the children also Miss Charlotte and Mary" (GCG to Saml. Butterworth, New Orleans, October 3, 1851). Mary was their niece. GCG told his sister Sarah: "Give my love to all the young ladies and when you write remember me to Mr. Sam

Butterworth & his kind lady also to their fair niece at Ellicotts mills [the location of Patpasco Institute]," a reference that is almost certainly to Mary Butterfield, who was GCG's cousin (GCG to his sister Sarah, Frontera, Texas, February 16, 1852). Mary Butterfield's nickname was Mamie; GCG used the names interchangeably. GCG explicitly identified both Mary and Charlotte as his cousins: GCG refers to Mary as his cousin in a letter his sister Alida, Frontera, Texas, May 8, 1852.

6. GCG to his sister Alida, Frontera, Texas, May 8, 1852.
7. GCG to his sister Sarah, Camp Sumner opposite Presidio del Norte, July 23, 1852.
8. GCG to his sister Alida, Brazos Santiago, May 17, 1853.
9. Emory to Alexander H. H. Stuart, Washington, September 13, 1851, and Stuart to Emory, Washington, September 13, 1851. This exchange followed a meeting with Emory on September 9. The letters may be found in box 4, f. 32, EP. See also Norris, Milligan, and Faulk (1998, 105–6).
10. Emory to Hardcastle, Washington, September 13, 1851, HCMSA; Norris, Milligan, and Faulk, 1998, 105–6.
11. Emory to Hardcastle, September 13, 1851, HCMSA.
12. See GCG to his father, September 12, 1851; Emory to Hardcastle, Washington, September 13, 1851, HCMSA.
13. GCG noted in a letter to Samuel Butterworth, October 3, 1851, "I think myself lucky to have spent a day at home." Emory had shipped out of New York for New Orleans on September 17 and expressed disappointment that Ingraham had not brought his "trunk" to join him. However, Emory ordered Ingraham to meet him in New Orleans after "passing by way of Washington," where Ingraham was to stop at Emory's house and pick up some books (draft of a letter from Emory to Ingraham, New York, September 17, 1851, box 4, f. 32, EP).
14. GCG & Ingraham, telegram to Emory, Montgomery, Alabama, September 24, 1851, box 4, f. 32, EP.
15. Emory had left instructions for them to take a steamer to Indianola or Galveston and the stage to San Antonio. Emory had reached Indianola by October 1, on a steamship called the *Mexico* (Emory to Capt. [Robert Emmet Clary], Steamship *Mexico*, Mississippi River, September 28, 1851, box 4, f. 32, EP).
16. Perhaps Emory traveled from Washington to New York specifically to meet his employees. At this point Emory apparently did not know he would be reappointed chief astronomer. Mr. Thompson was apparently a family friend. GCG's father sent a letter to his son care of Mr. Thompson (C. K. Gardner to GCG, Washington, September 15, 1851), and GCG visited "the Thompsons" when he was in New York in April 1853 (GCG to his father, New York, April 22, 1853).
17. This was apparently Eliza Welling, to whom GCG refers in a letter to his sister Sarah, El Paso,

November 27, 1851. Two Eliza Wellings lived in that area in 1850. One lived in Newburgh and the other in New Windsor, of which Little Britain was a part. Since GCG writes from Little Britain, the New Windsor Eliza Welling seems most likely to have been his cousin. She was Eliza A. Welling, born around 1800 in New York and married to Peter Welling, a farmer. They had several children: Elizabeth (fifteen), William (twelve) Joseph (nine), none of whom GCG mentions in his surviving letters. Three adult males also named Welling lived in the household (U.S. Census, 1850, New Windsor, Orange County, N.Y., roll M432_573, p. 260).

18. GCG wrote this letter on a Friday.
19. In large letters, shaped in a florid hand, GCG has written on the top of this letter, "Mother's Letter," apparently to remind his father to write to his wife. We cannot explain GCG's allusion to broken bones.
20. When he wrote to his mother from New York on September 7, GCG addressed the letter to her at Washington City. He did not know that his mother was absent. His letter of September 12 says that Alida was in charge of the household in Washington City, making it clear that his mother was not there. His mother had apparently traveled to Little Britain because GCG's letter suggests that she was there while he was and says that she wanted to hear from home.
21. GCG's father was born in Morris County, New Jersey, and the 1850 census shows a good number of Gardners in New Jersey, some in Morris County. GCG does not provide first names for the family members he wished to visit.
22. Emory did depart on the seventeenth on the *Winfield Scott* as planned.
23. The advantage of sending mail to the El Paso area by way of Santa Fe was that it often was faster. In the early 1850s mail could reach El Paso two ways: overland, on the Santa Fe Trail, or by steamer to Indianola and then overland to El Paso through San Antonio, a service operated by Henry Skillman and subsidized by the U.S. Post Office. By either route mail might be delayed or lost because of bad weather or Indian attacks. GCG apparently asked his family to send letters by both routes so some letters would get through to him. By 1851 mule-drawn mail wagons set out from Independence, Missouri, and from Santa Fe on the first of every month; on one record-breaking occasion in March 1852 the monthly mail from Santa Fe to Independence made it in seventeen days (Taylor, 1971, 28–32). Even the secretary of the interior sent copies of mail on both routes to ensure that Emory received at least one (Stuart to Emory, Washington, June 16, 1852, box 5, f. 40, EP).
24. In response to this telegram, GCG's father apparently appealed for help to Samuel Butterworth, who wired one hundred dollars to GCG in New Orleans, while GCG's father wired him to tell him where to call for it and that it was for GCG alone (Saml. F. Butterworth to Messrs. Buchanan, Carroll & Co., Gravier Street, New Orleans, Washington, October 1, 1851; C. K. Gardner to GCG, Washington, October 1, 1851). Emory had, however, provided for them, as GCG explained to his father on October 1.

25. This telegram appears to be missing from the Gardner letters at DeGolyer, but its contents are preserved in transcripts that were purchased with the original letters; in the transcripts it is followed by a telegram from Samuel F. Butterworth to Messrs. Buchanan, Carroll & Co., Gravier Street, New Orleans, that says: "Pay to Clinton Gardner, one hundred dollars, when called for." That is followed, in the transcript, by a note of the same date from his father: "Clinton Gardner Call on Buchanan & Carroll, Gravier Street, for one hundred for you alone."
26. Emory had asked steamship captain Robert Emmet Clary to look the two assistants up at the Verandah Hotel, "or elsewhere," give them the check, and tell them to meet him in San Antonio without delay (Emory to Capt. [Clary], Steamship *Mexico*, Mississippi River, September 28, 1851, box 4, f. 32, EP).
27. By the time GCG, under Emory's command, entered the field on the Texas end of the survey, Bartlett had been in command of the survey for a year and had spared no expense in assembling and outfitting a large party that included "a detachment of Topographical Engineers, a contingent of civilian surveyors, a force of fifty mechanics, a platoon of field scientists sponsored by the learned societies, a collection of [Bartlett's] personal friends and relatives, and a small navy" (Goetzmann, 1991, 169).
28. Probably G. A. Pynchon, listed in the 1852 New Orleans city directory as a "dealer in Liverpool Salt," at no. 4 New Levee. Courtesy of Mark Cave, archivist, Williams Research Center, Historic New Orleans Collection.
29. Samuel and Mary Butterworth's niece Mary.
30. The 1851 New Orleans city directory listed an F. X. Masterson working at the corner of Gravier and Tchoupitoulas [*sic*] and residing at the St. Charles Hotel (courtesy of Mark Cave). The city directory does not identify his occupation, but the 1850 U.S. Census (Representative District 3, Orleans, Louisiana, roll M432_234, p. 116) identifies him as a twenty-five-year-old merchant with the surnames of Francisco Xavier. GCG has clearly written *F. H.*, perhaps writing the initial for Xavier phonetically.
31. Indianola, a deepwater port on Matagorda Bay, no longer exists. It was founded in 1846, during the U.S.-Mexican War and served as an army depot and the starting point for the military road to San Antonio in GCG's day. Hurricanes of 1876 and 1886 destroyed the town.
32. Home for Ingraham was Grand Gulf, Mississippi. Before shipping out to Texas, Emory told Ingraham: "At New Orleans it will no doubt be in my power to arrange matters that you can visit your family at Grand Gulf" (draft of Emory to Ingraham, New York, September 17, 1851, box 4, f. 32, EP). On January 24 Ingraham took a packet from New Orleans to "home" and two days later wrote to Emory from Grand Gulf (box 5, f. 35, EP). Grand Gulf, on the Mississippi about sixty miles southwest of Jackson, is in Claiborne County. According to the 1850 census, Ingraham's father, Alfred, was forty-eight, a planter born in Pennsylvania, and his mother, Elizabeth, born in Spain, was forty-five. She was a sister of Gen. George Meade's; he too was

born in Spain (Cleaves, 1960, p. 355). Edward and two brothers were away when the census was taken, but the family had five daughters aged four to twenty-two living at home (U.S. Census, 1850, District 2, Claiborne County, September 5, roll M432_370, p. 122). By 1852 GCG had apparently met Ingraham's father. In a letter to Emory from Fort Inge Leona Station on September 14, 1852, Ingraham says: "I recd. A letter from Father to-day who begs to be remembered to you & Gardner" (box 5, f. 43, EP).

33. Fort Inge was established in Uvalde County, Texas, in 1849 as part of the first federal line of frontier forts; it was named for Lt. Zebulon M. P. Inge, a West Point graduate who died at the outset of the U.S.-Mexican War at the battle of Resaca de la Palma (Thomas T. Smith, "Fort Inge," *NHBT* 2:1105).

34. Peter Rainsford Brady, for whom Brady, Texas (seat of McCulloch County), was named, was a contemporary of GCG's, born in 1825 in Georgetown, Washington, D.C. Before this encounter with GCG in San Antonio in 1851, Brady had served in the U.S. Navy and the Coast Guard, surveyed land in Texas, patrolled the western frontier of Texas in 1847–48, joined the rush for gold in California, and then returned to Texas and joined Bigfoot Wallace's company of Texas Rangers. Between his stint with Wallace, which ended in 1851, and 1853, when he joined Andrew B. Gray to survey between Texas and California for the 32nd Parallel Railroad Company, Brady had a connection with a mail contract, as GCG suggests. In September 1851, the month before GCG's encounter with Brady, the postmaster general awarded a contract (that apparently superseded Brady's contract) to Henry Skillman to provide mail service between San Antonio, El Paso, and Santa Fe from November 1, 1851, until June 30, 1854, for $12,500 a year. Brady's later life remained peripatetic. He pioneered in Arizona; gathered intelligence information and supplies for the Union in Sonora, Mexico, during the Civil War; returned to Arizona, where he held public offices; and became involved in the mining and cattle industries. He remained in Arizona until his death in 1902 (Marilyn Carroll Smiland, "Peter Rainsford Brady," *NHBT* 1:696; Austerman, 1985, 23).

From W. H. Emory, *Report on the United States and Mexican Boundary Survey* (1857–59).
Courtesy of the DeGolyer Library, Southern Methodist University, Dallas, Texas.

CHAPTER 6
Frontera, Texas, a "Derned Mean" Country
November 27, 1851–May 16, 1852

On November 24, 1851, Maj. William Emory's small party reached the town of El Paso del Norte, an oasis in the Chihuahua Desert with irrigated pastures, gardens, and vineyards—"one extended vineyard," as Emory put it.[1] The Rio Grande, whose waters sustained this oasis, seemed little more than a creek to Clint when he encountered it at El Paso, but in this dry world what seemed a creek by eastern standards had seemed like a large river to the Spaniards who left the first written accounts.[2] To them it was, variously, a Rio Bravo—a wild river; a Rio Caudaloso—a river carrying much water; and a Rio Grande—a great river.[3] From El Paso to the Gulf of Mexico the Rio Grande served as the new international boundary under the Treaty of Guadalupe Hidalgo. The old town of El Paso del Norte, with about five thousand inhabitants, stood on the right bank of the river, in Mexican territory (the town today is called Ciudad Juárez).[4] The American side of the river, across from El Paso del Norte, was sparsely settled. Following the U.S.-Mexican War, Anglo settlements had begun to spring up on the river's left bank, bolstered by gold seekers who had poured through the region on their way to California. Some made purchases, adding to the local economy, and others had stayed to add to the local population. Among the new settlements on the U.S. side of the border was Magoffinsville, home of the wealthy Chihuahua trader James Magoffin, who would bankroll a penniless Emory at a critical juncture and who would befriend Clint.[5] The new commissioner, John Russell Bartlett, made his headquarters at Magoffinsville and described it as consisting of stores and warehouses built around a square. He predicted that it "will, no doubt, be the centre of the American settlements at El Paso."[6]

In Clint Gardner's day the only town called El Paso was El Paso del Norte on the

Mexican side of the river. Texas, which claimed that the region fell under its jurisdiction, had created El Paso County in 1849, but the present-day city of El Paso on the American side of the river did not take its name until 1859, when a surveyor platted the fledgling town of Franklin, directly across from El Paso del Norte, and called it El Paso (Franklin, then, rather than Magoffinsville, became the "centre of the American settlements at El Paso").[7] Nonetheless, before 1859 Americans sometimes applied the name El Paso to the whole region on both sides of the river. At one point Clint identified his mailing address as "El Paso," even though he was on the American side of the river in another community some distance from El Paso del Norte, and he referred to the U.S. mail's being delivered to El Paso, by which he surely meant delivery to the U.S. side of the border.[8] Clint, then, like other Americans of his day, used El Paso as shorthand not only for El Paso del Norte but for the surrounding area.

Another Anglo-American settlement was at White's Ranch, or Frontera, eight miles north of El Paso del Norte, where the American commission had rented land the previous autumn to build an observatory.[9] Frontera became Clint's home for the next six months, and it gave him a case of the blues—a complaint he never made to his family when he lived in San Diego.[10] In search of diversions he traveled frequently to El Paso del Norte, a journey made dangerous by Apaches, but El Paso was no San Diego. El Paso, he reported to his mother, contained "no society at all" and "nothing to interest a civilized person."[11] The women, he told his sister, were "without beauty, education or anything else which constitutes a lady."[12] Despite the discouraging prospects for romance, Clint continued to urge his family to send him fancy clothes, including "a black frock coat, pants & vest of cloth," and boots made of the "best English Calfskin . . . with broad toes the sole *projecting* about *¼ or ½ of a[n] inch all around.*"[13]

Emory's party had arrived during an unusually rainy spell that November, and the men spent their first two weeks camped near El Paso del Norte in deep mud before making the commission's headquarters at Frontera.[14] Mud, however, was the least of their worries. They had also stepped into what Clint, with considerable understatement, called a "snarl."[15] The principal cause of the snarl was Bartlett, who had arrived in the El Paso area a year before, in late November 1850, while Clint was still in California. Bartlett's inexperience and poor judgment had left the commission without money and with more problems than progress.

John Russell Bartlett was a man of parts—a skilled artist and an accomplished scholar with deep interests in ethnology, archaeology, and linguistics, but he was not an administrator. His Mexican counterpart, Gen. Pedro García Conde, sized Bartlett up as "a clever fellow, but unqualified for the labors that we must perform."[16] Bartlett, like Sevier, Weller, and Frémont before him, owed his appointment to his political connections. For much of his life Bartlett was closely associated with Providence, Rhode Island, where he had settled in 1824 after spending his first eighteen years in Canada on the north shores of Lake Ontario. In 1836 he moved to New York City, where his wide-ranging scholarly interests flowered (he published *The Progress of Ethnology* in 1847 and the highly regarded *Dictionary of Americanisms* in 1848), and he came to know some of the leading figures of his time and place. When Zachary Taylor's new Whig administration came into power, Bartlett hoped to become ambassador to Denmark so he could educate his children in Europe. Instead, he had to settle for commissioner of the U.S.-Mexico boundary survey—an appointment he accepted in June 1850. Later in life Bartlett remembered accepting the position with pleasure: "Although my life and pursuits had always been of a sedentary character I always had a great desire for travel, and particularly for exploring unknown regions. I had, also, ever felt a deep interest in the Indians and was glad of an opportunity to be thrown among the wild tribes of the interior."[17]

In northern Mexico and the American Southwest, Bartlett found ample opportunity to travel but demonstrated a woeful lack of administrative ability. He hired a large number of men, "many urged on me by their Congressional friends merely to get them away from Washington," he later explained.[18] These men were political appointees, as was Bartlett. Also like Bartlett, most turned out to be inexperienced and unqualified. Some were also quarrelsome or corrupt. Bartlett's teamsters committed at least three murders en route to El Paso. Bartlett and the chief astronomer, Col. John McClellan, got into a bitter feud, and Bartlett had him recalled. Colonel McClellan, in turn, charged Bartlett's brother George F. Bartlett, with fraud and the quartermaster, James Myres, with embezzlement.[19]

In the El Paso area the murders and contretemps continued, as Clint noted in his letters, and the incompetence of Bartlett's employees became manifest. Emory bewailed "the total and astounding incapacity of the assistants that have been turned over to me as surveyors. Of the whole number there is but one single one qualified and he is

[a] foreigner who scarcely speaks the language."[20] Bartlett was forced to admit that the Mexican commission, which had a fraction of the employees he did, could outperform his men.[21] The Mexican commissioner, García Conde, also outperformed Bartlett, talking him into a notorious "compromise" that gave Mexico much of what it wanted.

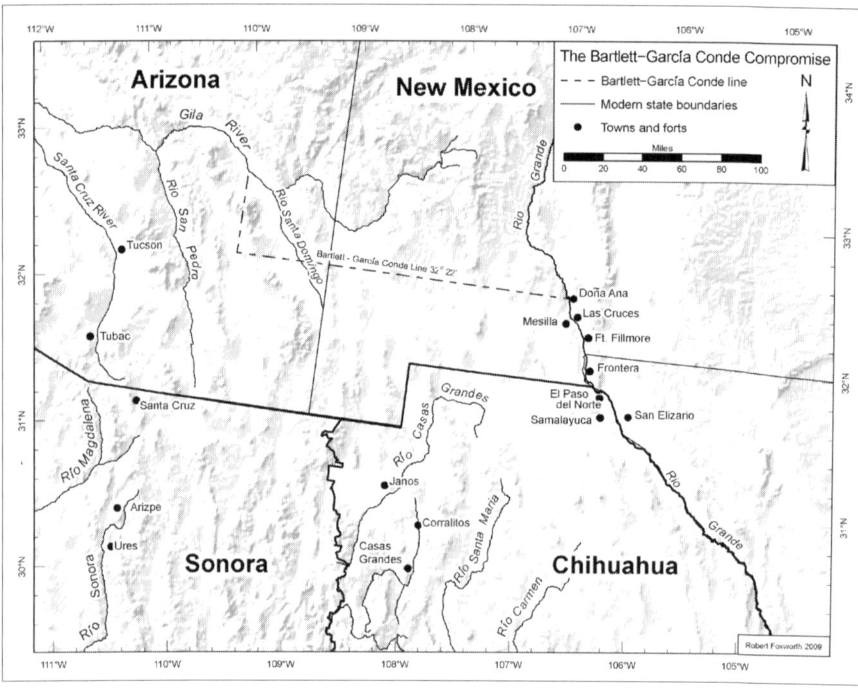

Article 5 of the Treaty of Guadalupe Hidalgo said that the new boundary should follow the Rio Grande's deepest channel "to the point where it strikes the southern boundary of New Mexico; thence, westwardly, along the whole southern boundary of New Mexico (which runs north of the town called Paso) to its western termination; thence, northward, along the western line of New Mexico, until it intersects the first branch of the river Gila." These instructions lacked precision, as the phrase "north of the town called Paso" suggests. The southern and western boundaries of New Mexico had never been surveyed, so in the interest of greater specificity the treaty makers agreed on a definition: "The southern and western limits of New Mexico mentioned in this Article, are those laid down in the Map, entitled 'Map of the United States . . . by J. Disturnell.'"

That additional language obscured more than it revealed. The Disturnell map showed the boundary running about eight miles north of El Paso del Norte, but El Paso del Norte was in the wrong place on the Disturnell map: forty-two miles farther north than its actual location and more than one hundred miles farther east than it was on the surface of the Earth. Should the line run north of actual El Paso del Norte or should it run north of where El Paso del Norte stood on Disturnell's map? Should the line start at the same longitude as the actual El Paso del Norte or one hundred miles farther east? The treaty left room for either interpretation. On Christmas Day 1850 Bartlett and García Conde agreed to split the difference. They would begin surveying a line west from an initial point forty-two miles north of El Paso, at 32° 22' north latitude, but they would recognize El Paso's actual longitude on Earth and start their survey on that line of longitude.[22] The compromise, which set the initial point at the village of Doña Ana a few miles north of the new town of Las Cruces, put the fertile Mesilla Valley and its two thousand residents in Mexico rather than in the United States, to García Condde's satisfaction.[23] For his part Bartlett was pleased that the United States had gained territory west of El Paso that held the Santa Rita copper mines, which Mexico might have justly claimed.[24]

In the spring of 1851, when the weather became favorable for surveying, the American and Mexican teams, led by Amiel W. Whipple and José Salazar Ilarregui, located the 32° 22' starting point on the Rio Grande. Then Bartlett sent Whipple to survey westward from that initial point along the agreed-upon line for three degrees of longitude before turning north toward the Gila along what they had decided was the western boundary of New Mexico. The survey underway, Bartlett took off on a sightseeing trip through northern Sonora, which he found had been devastated by Apache raiders.

Bartlett's return to the survey in June 1851 coincided with the arrival of the commission's civilian surveyor, Andrew Gray, who had been long delayed by illness, and Lt. Col. James Duncan Graham of the Corps of Topographical Engineers, who had come to replace Col. John McClellan as chief astronomer and head of the scientific corps. Both men protested the Bartlett–García Conde compromise, believing that Bartlett had given away too much. Most important, they believed he had surrendered the only practical route for a railroad across the Southwest to the Pacific.

Surveyor Gray refused to accept the compromise. He correctly pointed out that the treaty specified that such an arrangement required the signatures of the chief surveyors as well as of the commissioners from both countries, and he refused to sign or proceed with surveying until the question was resolved. Meanwhile, Graham ordered Lieutenant Whipple to stop surveying the Bartlett–García Conde line. Whipple also joined Gray and Graham in opposing the compromise. Although the three men united against Bartlett on the issue of the compromise, they also disagreed among themselves. Graham thought he outranked Gray, and Gray refused to recognize Graham's authority. Bartlett refused to recognize Graham as "head of the scientific corps," and Graham refused to issue scientific instruments to Bartlett. As in California, civilian-military discord grew so bitter that it paralyzed the work of the commission. As reports of the chaos reached Washington, they alarmed the Whig administration, which had appointed Bartlett and had already approved his compromise. The secretary of the interior, A. H. H. Stuart, fired Graham and Gray.[25]

At this juncture Secretary Stuart dispatched William H. Emory to Texas to replace Graham as "chief astronomer and head of the scientific commission." Emory was also to replace Gray as surveyor and was ordered to endorse the Bartlett–García Conde compromise. Emory's appointment, however, required presidential approval and the consent of the Senate. That would not come until the end of August. Emory had opponents who feared that he would follow orders to sign off on the Bartlett–García Conde agreement. His champions argued that Emory would refuse. He would find a way "to evade the question or resign." Emory, as it would turn out, did well at evading the question.[26]

The day he arrived at El Paso, November 25, 1851, Emory sent Clint to Frontera with dispatches for Lieutenant Colonel Graham, and Clint found Graham "quite delighted at returning to Washington."[27] For good reason. As Emory quickly discovered, the commission suffered a litany of woes.[28] First, it was broke and in debt to the wealthy Chihuahua merchant James Magoffin.[29] Second, Edwin Voss Sumner, the military commander of the Ninth Military Department in New Mexico, did not believe he had authorization to provide military support to the commission, and, as winter turned into spring, it turned out that he had no men to spare in any event.[30] This was no small matter. The astronomers and surveyors would be working, as Emory put it, "in a country in possession of the Indians who are now in an active state of hostility." Apache raiders in search of livestock and captives had devastated much of northern Chihuahua and Sonora, wreaking havoc on the survey's operations. Third, the Mexican commissioner, Pedro García Conde, fell ill and died at Arizpe on December 19.[31] Fourth, the American commissioner, John

Russell Bartlett, Emory lamented, was "absent and no intelligence of his whereabouts . . . within my reach." Bartlett had been gone since August, ostensibly in search of supplies. By one report he had headed to Guaymas and by another report to San Diego, and from San Diego he might go to Washington by way of Panama.[32] In August 1852, a year after his departure, Bartlett would eventually return to the El Paso area by way of the Gila River route, after many detours. By then Emory would be gone, surveying on the lower Rio Grande.

Initially, Bartlett's absence paralyzed the commission's work and left Emory in a quandary. Although Emory had responsibility for both the astronomy and the surveying, Commissioner Bartlett held the purse strings and Emory could not communicate with him. Desperate and angry, Emory planned to travel west in search of Bartlett.[33] Emory sent Clint to present his boss's credentials to the Mexican *prefecto* at El Paso del Norte, who agreed to give Emory an official pass to travel across northern Mexico with a military escort.[34] Emory hoped to leave in early January, but putting together a military escort and getting proper shoes and clothing for the men delayed him. He finally set out in search of Bartlett on January 25, 1852, but got just thirty-six miles west of El Paso del Norte when dispatches from Washington reached him that persuaded him he would not be able to overtake Bartlett. Emory returned to Frontera.

Clint had hoped to accompany Emory on his search for Bartlett. Instead, Emory had left Clint in charge of the observatory at Frontera, a brief responsibility, as it turned out, since Emory was back at Frontera by the end of the month. Before Emory's return, though, Clint's morale seems to have dropped to a new low. Emory's absence, the commission's unhappy situation, and the dismal social life in El Paso seemed to push Clint to ask his father, his uncle John McLean, and two prominent members of Congress to get him a commission in what he understood to be a newly forming cavalry regiment. Clint had made a similar request when he lived in San Diego, but his appeals from Frontera, first expressed on January 27, 1852, during Emory's absence, were more insistent and continuous.

Within days after his return to Frontera, Emory learned more about Bartlett's whereabouts. Mexican traders from Sonora delivered a letter from Bartlett, who reported that he had contracted typhoid in the Sonoran town of Ures, where he had been ill for two months. When he could travel, he would go directly to the coast and then sail north to San Diego.[35] Emory's heart must have sunk. Bartlett had left no funds, and Bartlett's employees were idle as well as unpaid. One, who had worked for Whipple as a "computer," complained to Emory that he was forced to live in a rude hut, or *jacal*, cooking for himself and "involuntarily accepting of shower baths, while it rains—a thing not very pleasant or agreeable in this season of the year."[36]

Before taking off for Sonora in the summer, Bartlett had tried to launch two surveys that seemed essential, regardless of the outcome of the controversy about the 32° 22′ starting point. He had sent Gray, the commission's surveyor, and Whipple, a topographical engineer, to survey the Gila toward its junction with the Colorado. Bartlett had also ordered Col. James Duncan Graham to survey the Rio Grande, but Graham had refused.

Emory got his topographical engineers and their crews back to work on the Rio Grande survey. As he told Senator James A. Pearce, a fellow Marylander and a supporter of Emory's, "I have taken a different course from Lieutenant Colonel Graham. He stood still, until he could get things . . . to his liking." Emory, in contrast, took "the means at hand" and made the most of them.[37] He set up an observatory on the Rio Grande about twenty miles below El Paso del Norte at the town of San Elizario, the site of an old Spanish presidio and the seat of El Paso County, and placed the observatory under the direction of Lt. William F. Smith, a topographical engineer.[38] Emory ordered another observatory built farther downstream at Fort Duncan near Eagle Pass under the direction of another member of the topographical corps, Lt. Nathaniel Michler.[39]

Although Emory put men back to work, the financial situation remained dire. The men lacked arms, ammunition, clothes, shoes, pay, credit, and a sufficient military presence to prevent Apaches from driving off their mules and horses. As he did in California, Emory improvised, borrowing money from the army or wherever he could.[40] If the government did not send money, the situation at El Paso would become, as Clint put it, "just as it was in California."[41] The money did not come. Congressional Whigs and Democrats argued about Bartlett's dismissal of Gray and Graham and his compromise with García Conde, and they withheld authorization to spend funds.

Through the winter and spring of 1852 Clint had an insider's view of events at Frontera and beyond. Sometime before or shortly after Emory's party reached El Paso the previous November, Emory had made Clint his secretary, or clerk. Clint retained the official title of assistant surveyor and added "clerk" to it.[42] Clint's previous experience in San Diego as clerk to the army quartermaster probably made him a logical choice for Emory, but Clint was restless in the position. In letters to his family he expressed interest in fighting Apaches or leading a surveying party, and he continued to pester his father about getting him a military commission. His father apparently did not cooperate, thinking it best that his son stick with surveying. Clint Gardner did. Except for a couple of weeks in April, when Emory entrusted him with leading a reconnaissance south of El Paso del Norte (of which Clint provides a rare account), Clint stayed with Emory at headquarters, writing and copying letters, assisting in the astronomical work, and learning firsthand the commission's astonishing range of challenges.[43]

The twenty-three letters in this chapter begin in November 1851 with Clint's brief account to his sister Sarah of his journey to El Paso del Norte and conditions there on arrival. The letters end with Clint still in the El Paso area, explaining to his father on May 16, 1852, that the surveyors were in "perfect hot water" and that he suspected a conspiracy against Major Emory.

GCG to his sister Sarah, El Paso, November 27 and December 11, 1851

Dear Sarah

I arrived here last Monday, with Major E. in advance of our train one day, after having a very agreeable march of forty one days on the road.[44] The Major and myself started from San Antonio the day after our wagons and the first march was twenty five miles which I found anything but pleasant for one who had not been in the saddle for several months. On our trip we had but one Norther[45] which met us about the head of the Limpia [Creek][46] after that the weather was quite agreeable when one got used to the cold the thermometer being below freezing. I could but laugh, when we were riding along in a rain to think how careful Mother and Cousin Eliza Welling were one evening to keep me in the house during a rain when I had made an engagement with John Jackson and there I was wet through with a cold north wind blowing on me and not even a good house to go in—Oh! that one could always have a Mother with him. After arriving here the Major sent me up to see Col. [Lt. Col. James Duncan] Graham on some business and I found the Col. quite delighted at returning to Washington to be sure he regrets being relieved yet he much prefers getting out of the business, as it is in such a snarl.[47] It seems the Commission has a much worse name here than it ever had in California as the Major says it seems to be oŭr luck to come in at the last moment when all the money is spent. I have heard that Bartlett has actually drawn on next years appropriation so of course he can't have much. We as yet have not heard what part of the world he inhabits, some say he has gone to California, at any rate we know he is not here.

The surveying party in Limpia Canyon. From W. H. Emory, *Report on the United States and Mexican Boundary Survey* (1857–59). Courtesy of the DeGolyer Library, Southern Methodist University, Dallas, Texas.

We are now enjoying a fine snow storm it reminds me much of home for it is the first time for many years that I have seen one. I think we are fortunate in getting here in good quarters before it commenced although we have had much colder weather in Camp. As yet I am unable to say where it will be my luck to be placed. We are now in Coon's hacienda[48] in Government Quarters but I expect when every thing is turned over we will stay at White's ranch,[49] about seven miles above this where Col. Graham now lives. The Major talks of starting across [the border to Mexico] to join Bartlett if such a thing can be done, but the people about here seem to think the most direct way to do that is to return to New York and wait for the Chagres Steamer as it is the report that he came out here only to write a book and for fear of being relieved before he could finish it he has gone right straight through on some pretex[t] or other.[50] Since the Commission came out here there has been several Murders and such like committed and if you remember the publications about the Murders &c. before the Com. arrived up here, you must know they do not say the Murderers were members of the Com. but I have been told that they were discharged members of the Commission.[51] From what I have heard Mr. Bartlett's party must have been rather a hard case.

Dec. 11th. I have been so busy since the major made me his Secty. that I have not had time until now to place this in an envelope which I now do in great haste as the mail left here for El Paso yesterday send much love &c. to all.[52]

Your Affectionate

brother,

G. Clint. Gardner.

GCG to his father, Near El Paso, December 11, 1851

Dear Father,

I believe I answered your telegraph respecting the draft while I was in New Orleans the reason I did not draw of course you understand as [Edward] Ingraham and I made a loan for some two or three hundred. Since I have been out here I have not drawn a single cent and supposing you wish to know how my finances stand I will give as follows

I owe different persons (in debt) $200.00

Amt. due from the Com. up to the

commencement of the 4th Quarter, 1851 $ *250.00*

So you see I stand up yet and after the Major appointed me his Secty. my pay is $1000 per annum which dates about 10th of Oct. some where and besides this I have my bill for transportation board &c. from New [Orleans] to San Antonio and perhaps I may be able to draw 10 cents per mile which my appointment allows me by the authority of the Secty. of the Interior yet it don't date back far enough some say but my duties date back as far as New York. How ever I will try it.

I wish you would write me and let me know if my accounts which I left home was paid if so I will send you some of my pay accounts certified to in great haste,

Your affectionate son,

G. Clint. Gardner

I owe Uncle Clint $5. I think or perhaps more which I wish you will pay for me also that 2 got from mother.

GCG to his mother, Frontera near El Paso, December 19, 1851[53]

Dear Mother

I have just sent my letter home by the mail via San Antonio yet in case that may fail I will write you the few lines via Santa Fé.[54]

Since we have been here we have had quite severe weather two or three snow storms, northers &c. something the people here say quite uncommon. Perhaps it is all for our benefit, who knows? Tell Alida I have not yet visited any of the young ladies in this part of the country so of course I can not give you a description of their manners and customs so what shall I take for the subject of this short letter the only object of it is to encourage you all to write and to state my health is as usual. The Major expects to start across the country in search of Mr. Bartlett in a few days perhaps I may go with him if so I hope he will continue on to San Diego yet it is not probable.[55] I should like [to] visit again some of my old friends in that part of the world and as for the country it is more desirable than this.

I wish you would keep Charlie reminded of my boots and shoes which I have placed in his charge in case the letter was not received I will repeat my wants. I should like to have Mr. Wheatly make for me of the best English Calfskin one pair of boots with broad toes the sole *projecting* about *¼ or ½ of a[n] inch all around* with

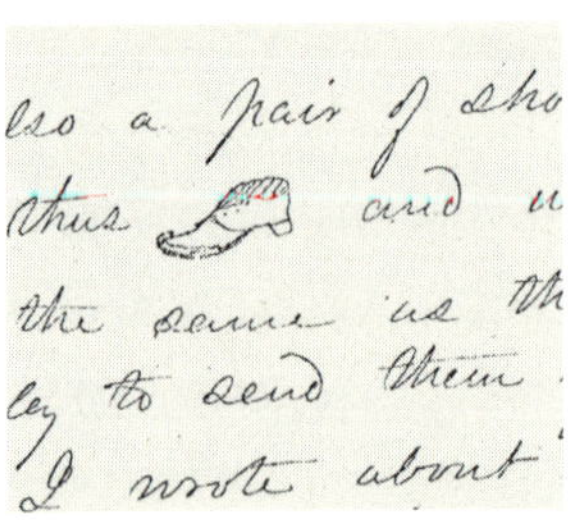
lso a pair of sho
thus and u
the same as th
ley to send them
I wrote about

the tops lined with red morocco so as when *turned over* to come up to the knee and when *turned up* to come some distance above the knee the distance I will leave to the taste of Charley I wish also a pair of shoes to tie up in front, thus and with the soles projecting the same as the boots. I wrote to Charley to send them by Peter Brady to whom I wrote about it,[56] but if he (Charley) don't hear from him tell him to send them addressed thus

G. Clinton Gardner

U.S. Bdy. Com

Care of Maj. W. H. Emory U.S.A.

El Paso

and write a letter to Mr. J. R. Sweet &co care of J. R. Sweet &co San Antonio asking him to forward them saying that I will pay all charges incurred.[57]

I am afraid you will think I am filling up my letter too much with my own business instead of trying to entertain you all at home but really I have nothing to say except that I am well and send much love to all.

believe me your affectionate

son

G. Clinton Gardner

GCG to his mother, Frontera, Texas, January 8, 1852

12 o'clock midnight

Dear Mother,

I can imagine you all now enjoying yourselves at the great Jackson ball[58] while I am waiting for an occultation, I would be asleep but the Major [Emory] will be out observing in about two hours more and I will watch until then.[59] My time for writing home is but limited now since I have been appointed Clerk to the Chief Astronomer and Head of the Scientific Corps,[60] which keeps me writing almost all day so I take this opportunity to write these few lines.

As yet we have not seen the Commissioner [Bartlett] and as for myself I do not expect to see any other than Major E. but of course, you all know better at home what is going on in the Department. My opinion is that Mr. Bartlett has gone over to San Diego and expects to be relieved before he will have time to return this way and in fact I know if he once goes to San Diego he will not be able to return on account of the difficulty of making an outfit.[61] I received Father's letter, dated Oct. 2nd directed to me at San Antonio a few days ago, but should have received it immediately on my arrival here for it passed us on the road in the mail which is now carried by Peter Brady who I suppose you know. If you wish to send me any thing direct it to San Antonio to him care of James R. Sweet & Co. being carefull to send it also care of Major E. nine miles above El Paso to this place be particular about the nine miles above E1 Paso for there is a Frontera in Mexico. I suppose tho' that you will have sent my shoes & boots before this reaches you.[62]

I find this not like San Diego for there is no society at all and the Commission being out of funds as usual it is hard work to live. The Mexicans in El Paso have had great doings there during the holy days. Bull fights &c. but nothing to interest a civilized person.

We live here at White's ranch as it were in a retired country seat and see nothing but the travellers passing by, except when we go to El Paso.[63] So far we have had quite a pleasant time, the Major gave a dinner the other day to the Mexican Commission or to as many of them as were here and not intending to flatter my Commanding Officer, it was said to be the finest dinner ever served up in this country *only* about 12 courses. The only difficulty was the change of Plates which were cleaned up in double quick time.[64]

How is Uncle John has he improved much Father says nothing of the family in his letter. Give my love to them *all*, and my respects to the neighbours.

Tell John I have obtained the Eliments of Lyell's Geology since I have been out here.[65]

Your affectionate Son,

G. Clinton Gardner

P.S. Jan. 11th, 1852. Tell Alida I have *just* received her letter of the 29th of Sept. via Santa Fee. You will see now the difference between the two routes, yet I wish you would write by both mails.[66] I will answer her letter in a few days.

G.C.G.

GCG to his father, Frontera, Texas, January 11, 1852

I received, dear Father, your letter of Oct. 2nd, 1852 [1851] & was much obliged to you for the prompt answer to my Telegraph from New Orleans.[67] I am also glad it arrived after Office hours for I would have been sorry if you had have gone to the Secretary about it as we found after a while th[at] the Major did leave money for us not sufficient I was compelled to borrow about $100 which I did before receiving your telegraph knowing that I had better get it if possible without calling on you. We not only lost the connection at Mobile but after or rather before we connected we were detained by as they would say out here, an "*up sot.*"[68]

You tell me there was a defensive account of Mr. Bartlett's operations out here and against our work in California in the Intelligencer.[69] Now altho' I am but small potatoes on the Boundary yet I think if I were in Washington I would let out some secrets and as it is I wish you would get Uncle John to say something in reply. As for assigning to Mr. Gray the credit of any of the work done on the Pacific side it is wrong. I can say that any such statement is *false* and *do say it* being responsible for the same. Mr. Gray never did anything but measure a portion or perhaps the whole marine league. Mr. [Charles J.] Whiting his first Asst. surveyed the harbour, for which I don't believe he got any credit, and from his survey Mr. Gray made a map taking in the Teajuana [Tijuana] Valley, which was I think surveyed by Fred. Emory, adding the topography around the initial point himself which is not correct and this is the map which if you remember was the first map you sent me in California and on which I plotted, as Capt. Hardcastle will tell you, the line and found his "approximate direction of the line" as he calls it out about one mile in five. What excuse has he for putting down the "approximate direction of the line" when he had the Latitude and Longitude of the two points except that he was unable to compute the azimuth. And as for his surveys at the Gila he never was within ninety miles of the place.[70] His Asst. Mr. Foster [John Harris Forster] went there and was there about three days and with the assistance of Mr. Whipple made a sketch which Mr. Gray completed afterwards from a map he got from the Mexican Commission which proved to be a copy of Lt. Whipple's Map.[71] Why if you remember you sent me two maps while I was in California the size of a leaf of a public document, which I received while on the desert in Company with Capt. H[ardcastle] and Dr. [Ricardo] Ramirez the Mexican Engineer and as I opened it the wind blew one of them towards the Doctor who caught it and at the same time remarked that it was their map and they had given Mr. Gray a copy. I afterwards learnt that they (the Mexican Commission) had taken Lt. Whipple's map as the official so you see the map made by Mr. Gray's triangulations proves to be no other than Mr. Whipple's map. An other proof is right on the face of it; Lt. Whipple established the initial point there by triangulation and as you know it was impossible to mark it in the river so he placed two marks in the same straight line equidistant from the initial point on each bank this straight line with its three points was converted into a triangle showing that it was a copy and an imperfect copy at that.

Now my opinion is that Gray, after Capt. H's attack on him for this false publication has now gone over there so as he will be prepared to say he has been there.[72] This is the man who claims to be Head of the Scientific Corps or rather that portion appertaining to surveys—well what qualifications has he? he makes a good looking note book and that is all saying nothing of its accuracy.

I wish you would send me all the publications connected with this work for I should like to criticize them not expecting of course to gain any credit myself, yet I think altho' not having ever written for the press, I would endeavour to correct things a little.

As for Bartletts operations out here I may ask, what has he done? why the Officers of the Topographical Engineers have run a portion of the para[l]lel of Latitude, about a degree, and have placed two parties on the Rio Grande and one on the Gila. But before I go any further let me tell you who belongs to the Commission or rather who did belong to the Commission why 127 young men who, I don't think ever were in the field before, except a few, and who I have heard knew nothing of Engineering; to give an instant of it one was asked how many right angles a triangle could have. He said it depended on its sides an other was asked the sum of the three angles of a triangle he said it depended how many of them were obtuse or acute; of course this is "entre nous." This is the material for the survey why Col. Graham will tell you [that he] had to place his draftsman in charge of a surveying party not for the want of surveyors but for the want of competent ones.[73]

All these people have been drawing pay and that is the way the appropriation goes without accomplishing any work but now Major E. is here and I only hope he may be made the Commissioner for if he is the whole work will be done this administration.[74] Since he has been here he has placed the whole force at work every man is now in the field except those on the sick report yet now Mr. Bartlett and his tribe or as I have heard them called "those boot lickers of his" are not here and I hope for the good of the country may never be here again; but now I am speaking of hear say when I talk of Mr. Bartlett's party or Commission yet what I have said in the Commencement concerning Mr. Gray are facts and I wish you would see Capt. H. [Hardcastle] and get Uncle John to write a little on the subject by way of an editorial for some paper and if necessary use my name, altho' it will not do me any good yet I claim the right as one of the party to correct any such *false* statement and if you will send me the different papers concerning this affair I will soon right the matter myself.

I thought Jake promised to send me the Intelligencer regular yet I suppose his marriage has made him forget all about a poor devil like I out in a wilderness.[75]

Write often my love to all.

Your Affectionate Son,

G. Clinton Gardner

GCG to his father, Frontera Near El Paso, January 27, 1852[76]

Dear Father,

I have just heard that there will be some new Regiments raised during this Congress and I wish to renew my frequent application for a Commission.

I send enclosed a letter to Dr. [William M.] Gwinn who I think will use his influence and also one for Gen. [Joseph] Lane of Oregon which you will be kind enough to deliver with such an application as you think best to make.[77] Now please don't treat the subject light for it has been my great desire to join the Army as you well know and now I think my long service in California under the Government entitles me to one.

I wish you would use all your influence to get it for me since you are in Washington right on hand and know to whom to apply.

Please not to drop the subject until accomplished as it is my ernest desire to get in the Army.

Your Affectionate

Son

G. Clinton Gardner

Please send the enclosed letter to Uncle John where ever he may be. G.C.G.

GCG to Honorable William M. Gwinn, U.S. Senator, Frontera, Texas, January 27, 1852

Dear Sir,

I take the liberty of addressing you these few lines to solicit in my behalf your influence to obtain for me a commission in one of the new regiments, which I have heard will be raised during this Congress.

It has long been my desire to enter the Army and I now hope if these regiments are to be raised, I will be able, with the aid of your influence, to obtain a Commission.

With great respect I remain

Your Obt. Servt.

G. Clinton Gardner

U.S. Bdy. Com[[m]]ission

GCG to General [Joseph] Lane, Frontera, Texas, January 27, 1852

General Lane of Oregon

Dear Sir,

Having heard that there will be raised during this Congress a new regiment as an increase to the Army I take the liberty of soliciting your influence in obtaining for me a commission.

I have now been in the service of the Government on the Frontier for the last three years, attached to the U.S. & Mexican Boundary Commission, and it has long been my desire to enter the Army, which I hope I will now be enabled to do, with the aid of your influence.

With respect I remain

Your Obt. Servt.

G. Clinton Gardner

GCG to his father, Frontera, Texas, January 27, 1852

Dear Father,

I send enclosed my pay account for $250. in due form, which perhaps you can get paid at the Dept. if not I expect Chubb & Schenck will cash it for a small discount as they are Mr. Bartlett's bankers.

Mr. William Smith[78] is a trader and goes on to New Orleans with Mr. [Marine Tyler Wickham] Chandler[79] expecting to leave there about the 1st of April to return, he can bring out for me those Boots & shoes which I requested to be made in my last two letters. If you call on Mr. Chandler he can tell you how to send them to New Orleans but don't ask Mr. Chandler to bring any thing for when he returns he comes in great haste as he is only on leave of absence. I wish you would also

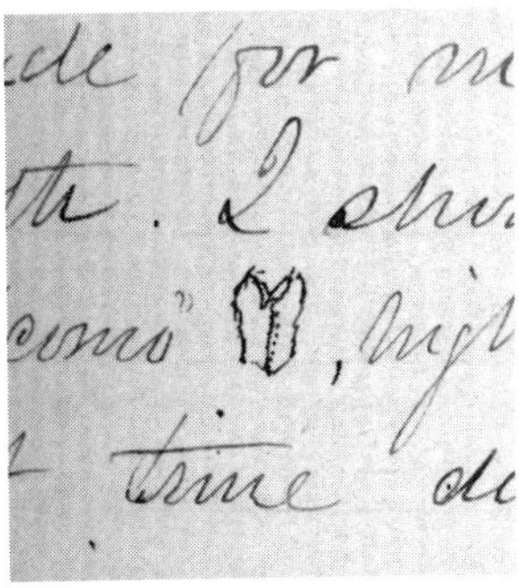

have made for me a black frock coat, pants & vest of cloth. I should like to have two vests to button up, "*como*" [i.e., as] high in the neck. But in case you have not time don't detain the Boots Shoes.

I wish you would write concerning the Commission for we are all in oblivion of the movements at Washington. Mr. Chandler can tell you somethings about the affairs out here.

The Major [Emory] has started for the Gila and left me in charge of the observatory here to car[r]y on the work but expect he will return before Chandler leaves as Radsminski [Charles Radziminski] has sent an express to overtake him.[80]

The Major says he would not be astonished if Uncle John was sent out here as Com[mission]er tell him he had better apply for it and if he comes out not to bring any protegés with him.

With much love to *all* I remain

your affection[ate] son,

G. Clinton Gardner.

In great haste.

GCG to Major Emory, Frontera near El Paso, February 1, 1852[81]

Sir,

I have the honor to report in obedience to your circular of January 8, 1851[2], that I have been engaged in my duties as Clerk during the past month.

In obedience to your letter of instructions of January 16th, I commenced observing all the culminations of the moon and moon culminating stars on the 29th of January and have observed on the moon and twelve stars; obtaining three transits of the moon and seventeen transits of the stars over the meridian of this place. I have also, in company with Mr. [John Henry] Clark,[82] made one set of observations for the determination of the values of each division of the level belonging to the bronzed transit.

I am very Respectfully

Your Obt. Servt.

G. Clinton Gardner. Clerk to the Chief Astronomer and Hd [Head] of the Scientific Corps U.S.B.C.

GCG to his father, Frontera, Texas, February 6, 1852

Dear Father,

My last letter home was sent by Mr. [M. T. W.] Chandler, Asst. U.S.B.C. [assistant U.S. boundary commissioner] and as I desire to write by every mail I will take this opportunity of giving you [a] little idea of the country in general.

Since my arrival here I have lived at this place which is about nine miles up the river from El Paso.[83] It is a barren country without grass sufficient for our animals in appearance it much resembles the valley of the Colorado except there is more timber and grass in the latter. The river is nothing more than what we would call in the States a creek winding its way down thro' the different chains of mountains and the town of El Paso is situated at the head of the opening of a pass just below this place. The ridge is about six miles thro' and it is from this it gets its name.

The country about here is greatly annoyed by the Apatches [Apaches] who are continually committing murders and thefts. I expect to see them come down on this place soon as they have done before, it was only this evening it was reported that nine had been seen just opposite here on a small rise as it were to inspect the Valley.[84] Mr. McGoffin [James Wiley Magoffin] quite a rich man has had his animals stolen frequently from his ranch, about three miles below El Paso and it was but the other day we had a false alarm.[85] He has offered twenty dollars for the scalp of Apatches and has sent out a herd of animals guarded by some twenty men for the purpose of killing them. I think if the Government were to do some thing of this kind it would soon exterminate the race. The best plan would be to send some of the Indians from the East to exterminate them giving so much for every scalp it is the only way that the safety of our citizens can be procured. As for attempting to kill them off by our troops it is impossible for they do not know how to hunt them yet if they give me an appointment in one of the New Regiments I will willingly service and do my best—yet I know it can't be done. It seems to me when an officer takes thirty days to get his company ready to make a campaign of two months as one in California said it would do the Army must be degenerating for what reason I can't tell without [i.e., unless] it is the meanness of Congress in its appropriations yet I hope now they will increase it and should Col. Weller be the next senator from California which I think is a certainty I wish you would apply to him to use his influence for me and put in for me an application signed by as many influential men as you can find in Congress.[86]

You will find Mr. Chandler to be a bag of gas if you give him a chance to explode. He can tell you more tales about the Commission than I know you will believe yet a great number of them I expect you will find to be so. He will tell you how you can send me any thing out from the States by Mr. William Smith the trader, but don't ask him to bring me anything without it is a letter.[87] I intended to get Peter Brady to bring out what I wanted but alas poor fellow, the Indians have either killed him or taken him prisoner which is not very likely or at any rate not he who is known as a man who has often taught them to fear him while he was a Lieut. in the Texas rangers.[88] I believe he has some property in the southern part of Texas and in fact he told me [he] owned a large track of land about which I will inquire when I go to the Paso next.

When you send me papers cut out the parts that will interest me, any thing concerning the Commission and in fact every thing for they drop all the newspapers on account of the want of transportation.

Should Col. Weller come to Washington don't fail to call on him and remember me to him by asking him for his influence.

Tell Mother not to let you forget to see all these influential people for I know if you have not some one to remind you of it, it will soon be forgotten.

Give my love to all and write often and about everything.

Your affectionate son

G. Clinton Gardner

The last news from Washington has cut down the Commutation for subsistence allowed to the Officers on the Commission[89] and as I am a little interested, just about $1500 per annum as Clerk to the Chief Astronomer &c. if you have any influence in the matter I wish you would use it to get it again. Tell Uncle John also to do every thing in his power to have it allowed again for I have not yet had a chance to draw it. My pay is $1000 per annum and with the Commutation of $4. per day if they had not have cut it down would have been an item.[90]

G. C. G.

February 8th, 1852. Please send the enclosed to Uncle John.

GCG to his father, Frontera, February 16, 1852

My dear Father,

I have been thinking that, your opinion of the Army during peace, which I have often heard you express, would influence you in making an application for a Commission for me.[91] But I assure you we are not at peace on this frontier, which I believe is the destination of the new regiment if raised, for the Indians are daily committing depredations upon the different settlements on this river and some most daring murders have been committed within almost gunshot of the posts which cannot be prevented, and for what reason? because there is no dragoons to give them chase. They have bid defiance to the troops and have come down on those stationed at the [Santa Rita] Copper Mines, taking all their animals and killing one or two men. It was but a few days ago they killed on the plains above this place some four dragoons while escorting a wagon, and the rest six in number together with the teamster made their escape by running. Yesterday while we were seated at breakfast we were startled by a man riding up full run to state the Apaches were in the valley just above here at the grazing Camp, carrying off our mules—The Major immediately ordered us all in the saddle to give chase, he at our head, but [we] were just too late they carried off eighteen together with all the animals at the ranch, nine miles above this place, killing one herder and taking a boy prisoner.[92] When I say there are no dragoons I mean none available for the purpose of hunting out these yellow wretches, for there are but few and mounted very indifferently, all of which are required to escort the different trains back and forth.[93]

It is foolish for me to commence to enumerate the different outrages committed for "every day brings something new". I wish you would mention these things at Washington and get Uncle John to use his influence in raising this new regiment for I assure you it is necessary for the prosperity of this country, (which is derned mean) as well as necessary to fullfil the treaty by which we are bound to protect this frontier against the Indians.[94]

It is my earnest desire to start out after these rascals and if Capt. [Abraham] Buford who commands the only dragoons in New Mexico makes a campaign and will take volunteers, I for one will go with him, with the Major's permission.

I hope dear father you will do all you can, and also Uncle John, to oblige your most

Affectionate Son,

G. Clinton Gardner

GCG to his sister Sarah, Frontera, Texas, February 16, 1852

Dear Sarah,

I expect you begin to think I have forgotten you by not writing, but you must know the mails are far between & I must answer those letters I have received, which answers, as they are received at home, makes but little difference to whom addressed. I have been anticipating a long letter from you telling me of the Doctor's wedding but who knows but what the Indians have received it.[95] Can you imagine one of your letters, which when put in the post Office you thought would entertain me for a month, opened, looked at, but not read, and burnt by the Indians. I never commence a letter now but what I think of poor Peter Brady who must have been killed, but he is but one out of almost a hundred that I have heard of. As it was remarked the other day we or those at Washington think nothing of it without it happens to be a friend, then it is only poor fellow, but if a relative, what fuss is kicked up.

When you write to Emma give my love to her and the family, a letter from them would be a curiosity. I was looking over my pile of letters and find from them but few & from Aunty none. I wish you would tell her so, for I should like to receive one from her in duplicate one sent by each mail.

Tell Mother I see that Alida puts two stamps on her letters which is unnecessary for one stamp is sufficient to these parts my only wish is she would write by both mails via Santa Fe & via San Antonio with one or no stamps.

Give my love to all the young ladies and when you write remember me to Mr. Sam Butterworth & his kind lady also to their fair niece at Ellicotts mills.[96]

My love to all in haste,

Your affectionate brother,

G. Clinton Gardner

GCG to his brother Charles, Frontera, Texas, March 18, 1852

It was my intention dear Charlie to write you a long letter but I have put it off so long that now the mail is about to start.

I take the liberty of enclosing a letter to John Ingle which I wrote a week ago and hope you fill up the direction and mail it for me or if possible give it to him in person.[97]

Congratulate the Doctor for me and remember me in the most affectionate manner to Sister Rosa.[98]

I should like very much to hear from home once a *year* if not oftener no letters except those that I have acknowledged have been received.

There is no society here at all it is very unlike San Diego I wish I were again there to see the beauties of that small but great place.

Remember me to all my friends, especially the fair beauties of the hill.

If it were not for the "mancebas"[99] one would not know how to live here but however I have found that those belonging to the Commission have not frequented the best circle so I made it my business to call on one of the old residents of El Paso and found the young ladies quite gentle in their appearance and very modest but they had not the beauty of the elite on the Pacific yet they had the accomplishments and they plaid [played] the harp and sang the whole evening for me.

Write soon and give me all the news.

Tell Mother I wish she would stir up some of my most affectionate sisters and get them to send a letter every week the first by the Santa Fe mail and then the next by the San Antonio mail alternate[ly] and I would then receive some as the mail only comes once a month when the Indians don't get it.

Tell Father not to forget my commission in a new regiment for this country or Oregon no matter which.

Give my love to All

Your affectionate brother,

G. Clint. Gardner

GCG to his father, n.p., March 1852

Dear Father,

I would be much obliged if you would return [to] Frank Holley that which I borrowed from him at Montgomery Ala. for he has written to me about it as for Ingraham's account I have nothing to say about it, it is not for me to meddle with other persons debts until asked to do so.

The Quarter Masters paper of his, I never had it was not left with me so I know nothing of it if Ingraham was here he perhaps could tell him about it as it was left with him. I wish you would write me about those pay accounts of mine for if they can be drawn I would like to hear. This Commission is the meanest affair that was ever got up under the Government. I only hope that after the whole affair has cost some $1,000,000 some thing will happen to make it of no account and compel them to do it over again. I think the meanness of the Dept. can't be excelled the idea of cutting down by [my?] pay in commutation for personal expenses $1460. a year, but I believe it was done on the recommendation of that man Bartlett if so I hope after he is paid some rascal a little sharper than he, will steal it from him.

Well do write me soon for I should like to hear what is going on in the Dept. as well as Congress about which no one else except you can give me any information. And tell me my chances for the New Regiment.

Believe me your affectionate Son

G. Clint. Gardner.

GCG to his sister Alida, Frontera, Texas, March 21, 1852

1 o'clock A.M.

Dear Alida,

I was quite delighted just now when I was woke by the rap at the door announcing to the Major the mail. I immediately got up and found your letter of the 14th

December in connection with Sarah's & Mother's for all of which I must express my sincerest thanks for altho' they are of not very late date any thing is very thankfully received & especially that as it contained two letters from San Diego.

You commence about my condescending to answer your letters but if that is the way you are going to talk I think I am a little ahead for I have received but one letter from you and I think if you look over the letters received from me you will find three or four. Well so be it I care not what you all say so you write me once in a while. I have become now quite reconciled to my fate and as Sarah says I have been very much troubled for the last two weeks by that disagreeable companion the blues. I went Yesterday to the Pass to see if I could not sneak off from him around some of the old adobe walls but "no puado" [*no puedo*—I wasn't able to] and Ingraham has gone to the Gila so of course he can not console me[100] well what is to be done? There are no ladies in this country except the Officer's wives and they don't live within forty miles and the Indians are so bad one runs a risk in going just the nine miles from here to the Pass.[101] I long much for the time when I shall meet you all again but then at other times I think it matters not whether I ever return for I think often there can be nothing to amuse or entertain me in fact a year spent in this country ruins any one for the pleasures of society all that one can do here is to go to the Pass once a week and see some of the members of the Commission for an hour or so or perhaps some of ladies without beauty, education or anything else which constitutes a lady. Quite unlike those on the Pacific who are in every sense of the word ladies yet they are not highly educated.

But why should I go on with a letter when I feel as if I could say nothing that would interest you so I think I will close it dear Alida and try to give some news by the next mail believe me your affectionate

brother, Clint.

Remember me to Miss Mary Butterworth and also to all the Girls and if Delia Conde is on the hill ask if she don't remember Charlie Snowden.[102] He is out here.

Dear Mother, I am in a perfect state at present with the exception of being a little sleepy[.] toe grande [his large toe] is very well I thank you and the pain in my brest ain't any worse. I am obliged to you, for I have not felt it since I left San

Antonio. My blood was in a poor state the other day for I had small "clons" [?] on my face so I took salts every night and they went away. No I thank you not nails but boils had.

your affectionate Son, Clint.

GCG to his father, Frontera, Texas, March 27, 1852

My dear father

Contrary to the instructions of the Secretary of the Interior to the Commissioner, I have sent you by the hands of Mr. [M. T. W.] Chandler my pay account for two hundred and fifty dollars,[103] which I suppose by this time you have received and presented and perhaps it has been paid, quien sabe [*¿quién sabe?*—who knows?], if not, I am sure it will be when the circumstances are known which are these.

When I left San Diego I was without funds and have never been paid any of my salary up to this time, directly, that is to say the amount paid down on my pay account; my necessary expenses were paid by Capt. Hardcastle for which I gave him my pay accounts, but I have received no more than that absolutely necessary to furnish me with the necessaries of life and now I think when it is impossible for us to draw our money out here we ought to be allowed the privilege of drawing at home, if Mr. Bartlett or his agent could have given me funds I would not have sent my pay account to you.

The Young men are now compelled to get their pay accounts cashed out here at fifty per cent discount,[104] in order to get their clothes and I am sorry to say that even those of the Commission, and no other than the brother of the Commissioner, is speculating off of others who are placed in this condition.[105] Is it not enough to disgrace any public work? when those attached to it, who if they are not in truth friends to each other, should at least respect the character of their countrymen sufficient to give a favourable impression to strangers instead of lowering themselves to such meanness.

Well I console myself with one thing that I as yet have not been compelled to make a sacrifice, nor will do it, I will live without funds before any of these thieving rascals shall plunder me. It is very astonishing to me that our government permits such things to be done the speculators have been reported to Washington and no notice has been taken of them and for what reason? because the removal of Mr. Bartlett might perhaps effect political influence a little. You all in Washington I know are inclined not to believe these reports that have been published thro' out the Union, but I assure you things can't be reported worse than they are and if money is not sent out here soon, the work must stop there is no help for it. The Commission is now in debt some $100,000 without having accomplished, you may say, any thing.

Since the death of the Mexican Commissioner [García Conde] the Mexican government have made the Astronomer & Surveyor the acting Commissioner,[106] thus uniting the three in one which is the most sensible idea & if our government will do the same the work will soon be at an end for all the time that has been lost in contentions of one kind and an other which will always be the case when a contemptible yankee is at the head.

Mr. McGoffin [James Wiley Magoffin] has adopted me as his son and tells me I shall always have what funds I want, not on my pay accounts at a discount as those with whom I am associated would do, but he is a liberal open hearted old gent and by the by an old friend of the Majors, for he was here when he passed along to California overland.[107]

The Indians are becoming worse, they come now and take mules from under ones nose with no fear, I expect some night to hear the rascals in the house as if [they] were taking our horses well let them come, if I can't kill them I won't miss my horse for I am bound to kill one and I would rather kill my horse than let them have him.

I wish you would tell Uncle John to use all his influence to have money sent out here and also the Major placed in charge of the whole work for it is the only chance they have of finishing it and if money is sent without he has the control over it, it will not benefit the work one iota for the debts will consume more than Congress will be willing to appropriate at one time.

Write me and let me know if you have drawn that difference due Ingraham & myself and also if there is any chance in the new regiment.

Your affectionate son

G. Clinton Gardner

GCG to his sister Sarah, Frontera, Texas, April 9, 1852

Dear Sarah,

I have put off writing until the last moment and now have nothing to say and no time to think of anything so if this does not prove as interesting as it should you must take the will for the deed.

Since my last letter nothing of importance has transpired the Indians have kept at home or they have gone in some other direction and I am glad to say in contradiction of what I wrote some time ago that Peter Brady was not taken by the Indians as near as I can learn they attacked them and stampeeded their animals but rather got the worse of it and in consequence of the delay he returned to San Antonio.

We are living here now quite comfortably and if money is not sent out soon it will be just as it was in California nothing to do and a great many to help us. I say comfortably because the last impression on my mind was that of comfort and that was made at dinner to day. While thinking of the hardships of camp life dinner was announced and the Major & I took our seats to eat, what do you suppose? salt beef hard bread &c., &c. no but a fine beef stake [*sic*] asparagus rice and every thing but potatoes after which the second course was served which was apple pie what do you think of camp life? all of this is cooked in earnest by Ernest (our cook) not the late king of Hanover but a frenchman who does things up brown.[108]

I am now going out, after this mail passes and I draw up all my letters on my first reconnaisance. It is not my duty but I am going to parts unexplored and am going to play astronomer solitaire for a while and perhaps soldat[109] if Indians come around my observatory.

I will write you about my trip by the next mail [110]

Tell Mother I enjoy perfect health with the blues once in a while about the colour of this ink. I enclose something here for your kind consideration as well as that of all the other members of the family who can make such things handsome and neat. I am in much want of something of the kind and I wish you or Aunty, Alida or some others would show some nice work.

Give my love to *all* including sister Rosa. I have not heard yet of the marriage except indirectly as I believe I hear of all the news of home.[111]

Write soon and often

Your affectionate brother G. Clint. Gardner

GCG to his father, Frontera, Texas, April 9, 1852

My Dear Father,

We have not as yet heard anything of Mr. Bartlett, but I see by the papers that Mr. Grey [Andrew B. Gray] has arrived in San Diego and I expect Mr. Bartlett was not far behind him altho he went by Guymas [Guaymas].[112]

However the work now goes on quite well without their presence the only want being that of money which supplies all wants I hope tho' [when] this reaches you an appropriation will have been made. It is my opinion that Mr. B. will go to Washington from San Diego, if so the appropriation will fall into his hands and if possible it should be prevented or at least a part of it, for this reason the debts of the Commission out here are large and pressing and will consume any amount that Congress will appropriate while the work will not be promoted one iota; it will be an easy thing to prevent this if it is not too late, if you will get some of the members of the house to attach a proviso to the appropriation, placing a certain portion in the hands of the Astronomer and Surveyor by doing which the work can be kept afloat, but if something of the kind is not done the whole appropriation will go to paying debts which have been contracted, and the work will stop.

Out of the two hundred thousand dollars together with eighteen months of time, for which the Commission is in debt, of the numerous persons employed there was the Initial Point on this river [the Rio Grande] determined and the parallel of Latitude run for about fifty eight miles west of the river and now the Commission is at an expense of some ten thousand dollars per month at this place with a great many doing nothing. The Major has sent all in the field over whom he has command and he has now some six parties actively engaged without one cent of money while Mr. Bartlett with his two hundred thousand dollars could do no more than what I have mention[ed] and at the same time has contracted more debts than the Major has to keep all these parties engaged.[113]

I think the Department should place the whole work in the hands of the Major or rather the money to do the work with, they have already placed the work in his hands but no funds to do it with and should it not be too late I wish you would get some such proviso attached to the appropriation for without it nothing can be done as I have stated above and a suggestion to some of the Members would save us perhaps a years longer service in the field, for with money we can finish and be [in] Washington this time next year.

Has any thing been done about the new Regiment yet? I wish you would keep a sharp look out and do your best for me for I consider it of the greatest importance to me to get a Commission if possible, and you know the promotion in a mounted regiment is rapid and will be more so if engaged on the frontier in active service as they will be if sent to this country.

I wish you would write me often and give me the different news afloat, my last dates from home are the 14th of Dec. 1851 while Major has letters from Mrs. E. as late as Feb. 12th—I have not as yet seen any of those papers the Doctor was to send me so regularly. I wish you would remind him of them for every thing of the kind is read and exchanged about for others. I am much obliged to you for sending me my California letters and hope you will continue to do so for they are all of great interest to me. You had better have them forwarded without taking them out of the Post Office for it will make no difference to me. Give my love to all and believe me,

Your affectionate son,

Clint

If you could get the commutation allowance allowed by a gentle hint to any of the big bugs about Washington it would be quite a raise for I am or was entitled to it and when they cut it off it was taking just fourteen hundred from me every year which would be some thing by the time I get home. However perhaps it is all for the best for if I got it my companions would envy me for no others except officers got it.

GCG to his family, Frontera, Texas, April 10, 1852

Gardner family

I take great pleasure in acknowledging the receipt of two letters one from Alida, Sarah & father of Jan. 11th—14th the other from Alida of Feb. 14th enclosing a diary from Mother for which I express my sincere thanks.

Next to the youngest

member,

G. Clint. Gardner

Love to All

GCG to his sister Alida, Frontera, Texas, May 8, 1852

Dear Alida,

This [is] the second piece of paper I have take[n] to write you a letter if my pen will do its duty but some how or other the paper gets greasey and refuses to take the ink and one as lonely as myself at present soon becomes vexed then wo[e] to pen and paper. Here I am alone with no one but my servant & horse for the Major, Ingraham & Dr. [Charles] Parry have gone to Fort Fillmore and the others are out collecting weeds or something else.[114] I am or ought to be at this time packing up the instruments to send to San Eliazario [Elizario][115] but when the Cat is away the mice will play what the odds I will write you a few lines and then finish them.

I wrote in my last letter that I was about starting for Lake Guzeman [Guzmán],[116] which place I visited making a reconnaisance of the country and returned without seeing a single Indian altho' they were watching us and if I had not have keep my party so close we would have had a little trouble as I had only ten men in all and twelve mules to take charge of.[117] I would describe the country but it would be useless for just imagine a barren and desolate plain with here and there sand hills and you have the country between this valley and the valley of the lake about ninety miles without water. In going out I went down on the road through El Paso to Cerralitas[118] but my object was to explore the country between this and the lake, or the Major's orders rather, so I returned direct across to the place as near as I could by the compass and found no water for 90 miles, which would have been very severe on my animals if divine Providence had not have ordained that it should rain one day before I left the Membris [Mimbres] river[119] and the day following which soaked the mules so that they required but little altho' at the time I said it was just my luck to fail in my first sextant observation on account of the clouds and rain and was no little vexed when I found Indian camp fires all around me at night which precluded the possibility of my remaining an other day without getting in a fight which I would have been very glad to have done if all my arms had not have been wet notwithstanding my orders were contra for I think my party would have stood a pretty sharp fight. I found Lake Guzeman and also Lake Santa Maria rather brackish I was going to say but it tasted of Potash and we had to encamp on the rivers to get water.[120]

The Membris [Mimbres] empties into Lake Guzeman[121] and the Santa Maria into Lake Santa Maria; the rivers are over grown with cotton wood and willow which makes them quite pretty I regretted at first that I had not my sketch book with me but after the rain I was glad for it would have got very wet and the sketches spoiled. I have not taken a pencil in hand to draw since I left home and I fear I have forgotten how however I will draw you our ranch and enclose it before the mail leaves.

May 16th. I just happened to think that I had commenced a letter to you and Mother both of which I have to finish before I leave these parts or else they never will be sent. Day after to morrow we move down the river where to God only knows for I think this is about the most uncertain time of year with us however when you write again direct your letters to the *Presidio del Norte*[122] via San Antonio or else I may never get them and put U.S. Boundary Commission out in bold relief and perhaps they may be picked up by some of us.

I have sent all our instruments to San Eliazario which is some 40 miles down the river so our next halt will be there but it will be but for a short time if the Indians or some other wild animals don't stampede our mules.

I delivered your message to [George G.] Garner about his Mother which he was glad to hear as he received no letter from her what has become of his sister[123] I think he told me she was threatened with the consumption which I was sorry to hear for I believe it is seldom cured but I know two instances of its being cured or rather will be cured soon for the way the Verduras [vegetables] as the greazers[124] call them fly at dinner & supper one would think we had the consumption and I know after we get on the road it will soon be cured.

Well dear Alida I must close this nonsencical [*sic*] scrawl and bid you good night.

Give my love to *All* including the girls of the neighbourhood and not forgetting my two cousins, Charlotte and Mamie [Mary Butterworth], and particularly to the latter since I decide with the former that there is no impropriety in it.

Your Affectionate brother,

G. Clint. Gardner.

GCG to his mother, Frontera, Texas, May 11 and 16, 1852

My dear Mother,

I was delighted yesterday to receive by the mail from Santa Fe Alida's long and interesting letter of March 13th together with Sarah's of the same date.

I had commenced in anticipation an answer to Alida's which I will finish and send with this in answer to Sarah's which I address to you as my last letter was to her. It would give me great pleasure to clear that Black Rock property which I believe was promised me when I was quite a youngster and if you will get the Doctor or Father or any one to advance the funds I will be responsible for it with interest until the appropriation is made.[125] I would send my pay accounts home and you could have them discounted but there is an express order of the Secretary of the Interior forbid[d]ing it, so I doubt whether the one I sent home by Mr. Chandler would be paid even if there was money, but however it should be as it was sent before the receipt of the order at this place.

I think before the conclusion of the work I would be able to clear the whole amount and it would be with the greatest ease if the commutation allowance was allowed again so as I could draw the $4 per day and purchase my own subsistence but however you can tell father that if it can be arranged he can calculate on me for six hundred dollars by this time next year and if my pay accounts will be paid at Washington, which he can easily find out, I will send them at intervals as they come due.

We are now preparing to move down the River but as yet—

May 16th. What I was going to say when I left off writing the other day I can't tell but however I will continue by telling you we are off day after tomorrow without something turns up; and altho' we stop at one or two points I desire my letters after the receipt of this to be directed to Presidio del Norte via San Antonio and quizas [*quizas*—maybe] I may get them and write me immediately about the money doings I have mentioned so as I may know how to act for you know every move I make now brings me close to the settlements and by the by I wish you would have me a suit of black made. Coat, vest & pants of the finest black cloth frock coat and will write you by the next mail how to send them. Tell Matlock & Griffen[126] to exert themselves for I want them [the clothes] to visit the ladies in on my return to the States or as I return;

I have been calculating the time it would take Uncle John to get to New Orleans and I have come to the result which is this, the mail from San Antonio comes up in 2 or 3 days and if they bring my boots they will be in time for my trip down the river and the answer will be + and on the contrary—, so, of course, at present it is ?

Give my love to all reminding them all to remind you to remind the tailor to make my coat a frock coat and vest fashionable and the pants according to the latest style with spring bottoms large enough to cover a good size hoof.[127]

Your Affectionate

Son,

G. Clint. Gardner

GCG to his father, Frontera, Texas, May 16 and 18, 1852

Dear Father,

I have just heard that Colonel Weller has been elected to the Senate, so I wish you would urge through him my appointment in the Army and if you ask him I have no doubt he will use his influence in getting the appropriation through for this work, as well as the restoration of the commutation allowance; he told Ingraham in San Francisco if either of us wish any thing he could do for us, to call him, so I wish you would urge these points and my commission in particular.[128]

We are now out here in perfect hot water, it seems to me there is a general conspiracy against the Major and I think it has been all along the same for I noticed when we arrived there was no one that offered us quarters or any thing else the Major had to *request* every thing. for two weeks while we were laying in camp in the rain surrounded with mud four or five inches deep; there were young men in quarters doing nothing but playing cards from morning until night. And when the Major came in authority he soon put all to work which they seemed to think something strange, especially amoung [*sic*] the Officers. Lt. N[athaniel] Michler who came up with me from San Antonio the Major sent back in charge of a party on the Lower Rio Grande because he was well acquainted with that portion of the river[129] and Lt. [William F.] Smith, who is well acquainted with this portion which is the most difficult and on which more credit can be gained, the Major put at San Eliazario [Elizario] in charge of a party, which he did not like as he was desirous of getting down in Texas so he commenced in al[l] sorts of ways to raise a fuss and finally did by his impertinence to the Major; first he complained of the responsibility which he had in reference to his Surveyor[.][130] well that was removed then it was something else, until now he is relieved, but in arrest with Charges against him,[131] and after Charges were preferred against Smith, he out of spite commenced to urge

Lt. [O. H.] Tillinghast, Qr. Mr. & Comsy. [quartermaster and commissary] on who commenced disobeying orders himself well as counter-manding the Major's orders and telling the Major his orders were absurd &c. threatening [to] starve us out arm the teamsters &c. to prevent [the] Major's orders from being executed, so of course he got in arrest with Charges against him and out of retaliation prefers charges against the Major of the Most scandalous and false kind bringing up things that occurred some five months ago, just after our arrival here, at a dinner given by the Major at which he was an invited guest and was as gay or as he terms it "drunk" as any one around the table, and it is then he accused the Major of being drunk and traducing his Government, all of which is false for I can remember as well as he & perhaps better what was said.[132]

Can you imagine anything more mean low and degrading as that? a man who pretends to be a gentleman to accept the hospitality of an other and then attempt, for it is nothing more than an attempt and can result only in his own disgrace, to disgrace him by any conversation that might have occured as this was, at a Bachelor's dinner, or the same thing [as] there were no ladies present. But why discuss the character of a man who has been caught in one or two *lies*? he surely is not worth it.

May 18th. I have just time to close my letter before we start so I will bid you good o. P. H.

Your Affectionate

Son G. C. Gardner.

---

1. Emory, 1857–59, 1:91.
2. GCG to his father, Frontera, Texas, February 6, 1852.
3. Horgan, 1954, 2:981.
4. The population figure comes from Bartlett (1854, 1:192).
5. For more on Magoffinsville see Timmons (1999, 47–56). Emory credited Magoffin with rescuing the commission by advancing Emory goods on credit and by helping him establish observatories and get surveying parties into the field (Emory, 1857–59, 1:11; GCG to his father, Frontera, Texas, March 27, 1852).
6. Bartlett, 1854, 1:193.
7. Timmons, 1990, 144.

8. GCG to his mother, Frontera near El Paso, December 19, 1851; GCG to his sister Sarah, Camp on Las Moras Creek, September 21, 1852. See too GCG to his father, Camp near Fort Duncan, Texas, October 8 and 10, 1852.
9. Timmons (1990, 108–9) put it eight miles north of El Paso, as did Emory in a letter to A. H. H. Stuart, secretary of the interior (Frontera, Texas, 8 miles north of El Paso, January 8, 1852, box 5, f. 35, EP). In a letter to his mother from Frontera on January 8, 1852, Clint put the distance to El Paso del Norte at nine miles.
10. GCG to his sister Alida, Frontera, Texas, March 21, 1852; GCG to his sister Sarah, Frontera, Texas, April 9, 1852.
11. GCG to his mother, Frontera, Texas, January 8, 1852.
12. GCG to his sister Alida, Frontera, Texas, March 21, 1852.
13. GCG to his father, Frontera, Texas, January 27, 1852, and GCG to his mother, Frontera near El Paso, December 19, 1851.
14. Months later GCG mentioned this to his father in a letter from Frontera, Texas, May 16, 1852.
15. GCG to his sister Sarah, El Paso, November 27, 1851.
16. García Conde to Máximo Yánez, March 11, 1851, quoted in Werne (2007, 54). García Conde arrived on December 1, shortly after Bartlett.
17. Bartlett, 2006, 37. See too the delightful work by Hine (1968).
18. Bartlett, 2006, 38.
19. The story is ably summarized in secondary sources, including Faulk (1967), Goetzmann (1991, 170–71), and Werne (2007, 49–51, 139–40). The charges against George Bartlett were probably true. Another quartermaster, F. W. Tansill, brought charges against him for "fraud upon the Treasury" after George Bartlett transported his private goods at government expense and tried to get Tansill to sign off on it (Tansill to Emory, Washington, March 1, 1853, box 6, f. 49, EP).
20. Draft, Emory to [James A. Pearce], n.p., ca. January 1852, box 5, f. 35, EP.
21. Werne, 2007, 57.
22. Goetzmann, 1991, 173–75; Werne, 2007, 54–57. The Disturnell map has been reproduced frequently, as in Rittenhouse (1965), where it appears in facsimile.
23. Timmons, 1990, 126.
24. Faulk, 1967, 57–65. Bartlett built part of his argument on his understanding of the boundary between Chihuahua and New Mexico under the Mexican Republic. One historian concludes he was right (Goetzmann, 1991, 191). See too Werne (2007, 165–66).
25. Faulk, 1967, 70–86; Goetzmann, 1991, 176–79.
26. Emory received his August 30, 1852, appointment with a covering letter from J. J. Crittenden to Emory, Dept. of the Interior, Washington, September 3, 1852, box 6, f. 43, EP. The quotation is from Edward L. Hardcastle's letter to Emory (Washington, September 10, 1852, box 6,

f. 43, EP), which included a full account of the proceedings in Washington. Goetzmann (1991, 192–93) discusses Emory's varying positions.

27. GCG to his sister Sarah, El Paso, November 27 and December 11, 1851. A letter from Emory indicates that he sent dispatches to Graham on the day he arrived (Emory to Sec. of the Interior, A. H. H. Stuart, Camp near El Paso, December 8, 1851, box 4, f. 34, EP).
28. For a more extensive summary of Emory's problems that spring, see Norris, Milligan, and Faulk (1998, 108–25).
29. Draft, Emory to Lt. O. H. Tillinghast, Magoffin's Ranch near El Paso, December 2, 1851, box 4, f. 34, EP.
30. Sumner to Emory, Head Quarters 9th Dept. near Albuquerque, NM, April 5, 1852, box 5, f. 38, EP. For Sumner's inability to furnish troops to protect the surveyors, see D. Miles to Emory, Fort Fillmore, March 21, 1852 (box 5, f. 37, EP).
31. Emory did not learn of García Conde's death until January 20, 1852, when he received a letter from José Salazar Ilarregui, writing from El Paso, January 19, 1852, and informing Emory that García Conde had died in Arizpe on December 20 (box 5, f. 34, EP). Salazar was off by a day: García Conde died on December 19.
32. Draft, Emory to Sec. of the Interior, A. H. H. Stuart, Camp near El Paso, December 8, 1851, and copy, Bartlett to Henry Jacobs, Santa Rita del Cobre, New Mexico, August 25, 1851, box 4, f. 34, EP.
33. Draft, Emory to Bartlett, Frontera near El Paso, January 1, 1852; Emory to A. H. H. Stuart, Sec. of the Interior, Frontera, Texas, 8 miles north of El Paso, January 8, 1852, both in box 5, f. 35, EP.
34. David R. Diffruduffer, U.S. consul, El Paso, to Emory, January 7, 1851, and draft, Emory to Diffruduffer, January 9, 1852, both in box 5, f. 35, EP.
35. George F. Bartlett to Emory, Magoffinsville, January [ca. February 6], 1852, reporting receipt of a letter from his brother, and draft, Emory to George F. Bartlett, February 7, 1852, Frontera near El Paso. Emory chides Bartlett for misdating his letter (box 5, f. 36, EP).
36. P. Crenion to Emory, Frontera, December 8, 1851, box 4, f. 34, EP.
37. Draft, Emory to [James A. Pearce], n.p., ca. January 1852, box 5, f. 35, EP.
38. By 1852 the old presidio had been dismantled by local residents in search of adobe bricks (Hendricks and Timmons, 1998, 73).
39. Draft, Emory to John J. Abert, Frontera near El Paso, January 1, 1852, box 5, f. 35, EP.
40. At one point, for example, Emory tried to solve the credit problem by making a draw, through the quartermaster, for $3,000 on the Department of the Interior, without permission but with apologies (draft, Emory to A. H. H. Stuart, February 8, 1852, box 5, f. 36, EP).
41. GCG to his sister Sarah, Frontera, Texas, April 9, 1852. Emory's letters to Washington that season resembled those he had written earlier from San Diego. He lacked everything, includ-

ing authority, and the work would have to stop. See, for example, drafts from Emory to D. C. Goddard, Dept. of the Interior, and to A. H. H. Stuart, Sec. of the Interior, February 17, 1852, both in box 35, f. 36, EP.

42. GCG to his sister Sarah, El Paso, November 27 and December 11, 1851. After becoming Emory's clerk, GCG initially signed his monthly reports to Emory as "clerk to the chief astronomer" or "clerk, etc." (see reports of February 1, February 29, and March 31, 1852, box 5, f. 36, 37, EP), but on May 1 and June 1 he signed his reports for April and May as "Asst. & Clerk" (reports of May 1 and June 1, box 5, f. 39, 40, EP). Clint continued to hold the title of assistant surveyor (GCG to his mother, Camp Opposite Presidio del Norte, Mexico, August 13, 1852).
43. GCG to Emory, Frontera near El Paso, February 1, February 29, March 31, and May 1, 1852; GCG to Emory, San Elizario, June 1, 1852, all in box 5, f. 36, 37, 39, and 40, EP.
44. That Monday was November 24, but Emory reported that he arrived on November 25 (Draft, Emory to J. J. Abert, near El Paso, December 1, 1851, and draft, Emory to Sec. of the Interior, A. H. H. Stuart, Camp near El Paso, December 8, 1851, both in box 4, f. 34). Perhaps Emory counted the twenty-fifth as the official date of arrival since that was when his "train" arrived.
45. A rapidly moving cold front from the north that causes the temperature to drop dramatically in a short time. An autumn phenomenon in Texas, it also called a "blue norther."
46. GCG has rendered the word to look like *Simpia*, but he must have meant to write *Limpia*. Emory apparently took his party through a recently opened wagon road through the Davis Mountains and along Limpia Creek, named for its clean water.
47. GCG traveled north of El Paso to Frontera, where Graham was stationed.
48. Also known as Franklin, on the site of today's El Paso. The St. Louis merchant Benjamin Franklin Coons purchased the rancho from Juan María Ponce de León after the U.S.-Mexican War, but Coons' financial ruin in late 1851 led Ponce de León to reclaim it. In 1852 a post office called El Paso was established there (Martin Donell Kohout, "Coons' Rancho," *NHBT* 2:312–13).
49. T. Frank White of Frontera.
50. The book appeared in 1854, in two volumes: Bartlett's classic, *Personal Narrative*.
51. The murders to which GCG refers include the shooting of a local citizen by a teamster as the commission made its way from Indianola to San Antonio, the killing of one teamster by another while the party was in San Antonio, and the shooting of a wagon master on the road to El Paso by Capt. S. D. Dobbins, a member of Col. John McClellan's party. Dobbins then killed himself. After the commission arrived in El Paso, another round of violence commenced, precipitated by a group of recently discharged teamsters. This time the casualties included a barkeep, a civilian, and Edward Clarke, the son of Bartlett's political sponsor, Senator John Clarke of Rhode Island (Goetzmann, 1991, 170–73).
52. Sometime that autumn GCG became Emory's secretary, or clerk.

53. This letter is in the Smithsonian.
54. Mail to El Paso could also be sent by way of San Antonio.
55. Emory contemplated going after Bartlett around January 1, once he got teams mobilized to continue the surveying (Emory to Sec. of the Interior, A. H. H. Stuart, Camp near El Paso, December 8, 1850, and Emory to Stuart, December 18, 1851, both in box 4, f. 34, EP). By December 19, when GCG wrote this letter, Bartlett had nearly completed convalescence in Ures, Mexico. By December 29 Bartlett had recovered sufficiently to set out for Guaymas and Mazatlán, finally arriving in Acapulco at the end of January. From there he took the steamer *Oregon* to San Diego, as GCG surmised he would, and spent the spring touring California (Goetzmann, 1991, 180–81).
56. Peter Rainsford Brady. See chap. 5, note 34.
57. James R. Sweet, who owned and operated a mercantile establishment with his brother-in-law John James in San Antonio from 1850 until the outbreak of the Civil War (Bernice Strong, "John James," *NHBT* 3:905–6).
58. An annual event sponsored by the Jackson Democratic Association at Jackson Hall in Washington on January 8, to celebrate Jackson's victory at the Battle of New Orleans. Advertisements appeared in local papers, as in the *(Washington) Daily Globe*, January 7, 1852, and the *Baltimore Sun*, January 8, 1853.
59. GCG and Emory were watching the moon pass in front of a planet or a star.
60. As clerk, GCG's pay now rose to $1,000 a year. Emory's was $2,400, and the pay of a computer, a position that GCG had earlier occupied, was $720 per year. Pay levels are found in "a true copy" made by G. Clinton Gardner, Clerk, of O. H. Tillinghast to Emory, Paso del Norte, Texas, January 1, 1852, estimating expenses for 1852 (box 5, f. 35, EP).
61. As GCG suggests, Bartlett was on his way to San Diego by ship, which on this date was becalmed off the coast of Mazatlán (Bartlett, 1854, 1:484–85). As to the second prediction, GCG was wrong. Bartlett did return.
62. The distance was closer to eight miles.
63. In 1848 T. Frank White, the newly appointed prefect of the El Paso district on the American side of the Rio Grande, established Frontera eight miles north of El Paso. By 1851 White's political influence and his hopes for a military post at Frontera had declined to the point that he offered to sell Frontera to Bartlett for $3,000, or to sell him two acres for a dollar and let him rent the buildings. Bartlett chose the latter course (Timmons, 1990, 105, 108–9).
64. We have found no other description of this twelve-course dinner, which was held on January 2. It later became the occasion for charges against Emory leveled by Lt. O. H. Tillinghast after Emory arrested him.
65. GCG is probably referring to Sir Charles Lyell's *Principles of Geology; or, the Modern Changes of the Earth and Its Inhabitants Considered as Illustrative of Geology*. This highly influential book

was in its eighth edition by 1850 and laid the foundation for future studies of Earth's development as well as evolutionary biology.

66. On the mail routes see chap. 5, note 23.
67. See GCG to his father (telegram), New Orleans, September 29, 1851, and GCG to his father, New Orleans, October 1, 1851.
68. An "upset," referring to his stage coach's tipping over.
69. The *National Intelligencer*, published in Washington. GCG may be responding to a lengthy article, "The Mexican Boundary Commission," that appeared in the issue of September 27, 1851, which gave the survey of the California boundary short shrift. The *Intelligencer* remedied that on October 9, 1851, with "Mexican Boundary Commission—Western Section," which provided a detailed account, mentioning the work of Hardcastle, GCG, and Ingraham.
70. GCG was correct. Gray headed for the Gila but turned back in order to escort Col. James Collier's party of migrants to San Diego. In a letter to Secretary of the Interior Thomas Ewing from San Diego on February 20, 1850, as published in a Senate report, Gray claimed that he had run the entire line from the Pacific to the Colorado River (SED 34, pt. 2:4–5). Hardcastle was furious, and he and Emory refuted the claim (Hardcastle to Emory, Camp Stuart on New River, May 2, 1851, and copy, Emory to A. H. H. Stuart, Sec. of the Interior, June 24, 1851, both in box 4, f. 30 and 31, EP; Rebert, 2001, 104).
71. GCG makes clear in the next sentence that the copy came from a published version of one of Whipple's maps, apparently one of the three maps of the junction of the Gila and the Colorado rivers drawn by and published in SED 34. One of those maps is reproduced in Wheat (1957–63, 3:206). Wheat discusses Whipple's work on p. 214 and provides descriptions of the three maps on p. 290.
72. By "gone over there," GCG means to the mouth of the Gila, but Gray did not go by choice and this time he went from the east rather than from San Diego. In July 1851, after Gray helped halt work on the Bartlett–García Conde line, Bartlett sent Gray and Whipple to survey the Gila. After a substantial delay while they hunted for supplies with Bartlett, Gray and Whipple surveyed 350 miles of the river, but on December 24 they stopped working sixty miles short of the Colorado when supplies ran out. They continued to San Diego, reaching it on January 8, 1852, three days before GCG wrote this letter from El Paso. Bartlett joined up with Gray and Whipple in San Diego a month later and soon after received dispatches from Washington that announced Gray's firing (Faulk, 1967, 87–93, and Gray to A. H. H. Stuart, San Diego, January 10, 1852, in SED 119, 305–7).
73. Lt. Col. J. D. Graham described in his report the inadequacies of the men appointed to Bartlett's survey. According to Graham, for all their enthusiasm and integrity as gentlemen, they had obviously received their appointments on the basis of political connections rather than practical experience: "Each young gentleman, when he joined, seems to have hoped

to find in his companions that information derived from previous experience which would help him on, by teaching him how to apply his theoretical school-lessons to practical usefulness. Alas! the answer in most cases was, 'This is the very thing I was in hopes you would be enabled to do for me!'" (Graham's report to Col. J. J. Abert, Washington, June 15, 1852, in C. M. Conrad, *Report of the Secretary of War, Communicating, In Compliance with a Resolution of the Senate, the Report of Lieutenant Colonel Graham [June 18, 1852] on the Subject of the Boundary Line Between the United States and Mexico*, 32d Cong., 1st sess., Senate Executive Document 121 [1852], 25).

74. Emory already held two positions on the survey, chief astronomer, the job previously held by Lieutenant Colonel Graham, and principal surveyor, previously held by A. B. Gray.
75. The reference is to GCG's brother "Doc" Jacob, who had married a woman named Rose, or Rosa, as GCG called her. He hoped to receive an account of the wedding from his sister Sarah. See GCG to Sarah, Frontera, February 16, and GCG to Charles Gardner, Frontera, March 18, 1852. For Jacob and Rose see the introduction.
76. This letter is in the Smithsonian Institution.
77. Gwinn was senator from California; Gen. Joseph Lane then served as Oregon Territory's delegate to Congress.
78. This could be the William T. Smith who bought Coons' ranch in 1852 from the heirs of Juan María Ponce de León (Timmons, 1990, 112).
79. Marine Chandler left Washington on April 10 (Chandler to Emory, April 28, 1852, Galveston, box 5, f. 38, EP). Marine Tyler Wickham Chandler was the assistant U.S. boundary commissioner. Chandler had a master's degree from the University of Pennsylvania and experience as a surveyor and engineer for the City of Philadelphia (Rebert, 2001, 25). His father, Joseph R. Chandler, a Whig, represented Pennsylvania in the House of Representatives (Goetzmann, 1991, 185). Early on Emory described Marine Chandler as "a gentleman of intelligence and integrity" and planned to put him in charge of a survey party (draft, Emory to A. H. H. Stuart, Sec. of the Interior, Frontera near El Paso, January 22, 1852, box 5, f. 35, EP). Emory trusted him to write the chapter on meteorology for the final report (Emory, 1857–59, 1:246–48). Bartlett credited Chandler with carrying out the survey's meteorological observations (Bartlett, 1854, 2:541). GCG, perhaps touched by envy, found him a windbag.
80. Emory had set out on January 25 from El Paso and traveled thirty-six miles toward the Gila when, as GCG suspected would happen, he was overtaken by a message from Charles Radziminski that caused Emory to return. Radziminski had arrived in the El Paso area on the night of January 25 with dispatches from Washington. The next morning he wrote to Emory urging him to return to the commission's headquarters at Frontera and read the dispatches before proceeding. Radziminski knew that Bartlett intended to visit surveyors on the Gila River, continue to San Diego, and return to Washington or El Paso by way of Panama. Radziminski feared that Emory would not overtake Bartlett. Instead of returning to Frontera,

Emory ordered Radziminski to bring the dispatches to his camp. Radziminski did, and Emory decided to abandon his search for Bartlett. He did, however, send Edward Ingraham to the Gila with a message for Lieutenant Whipple (Emory feared that Whipple would continue to Washington with instruments that Emory needed in Frontera) and wanted Ingraham to keep a lookout for Bartlett, although Emory thought Ingraham was unlikely to encounter Bartlett (Emory, 1857–59, 1:11; Radziminski to Emory, Doña Anna [*sic*], NM, January 26, 1852; copy, Emory to Radziminski, Cimialauke [Samalayuca] Springs, January 27, 1852; draft, Emory to [Ingraham], Cimialauke [Samalayuca], Mexico, January 29, 1852 [also in SED 119, 473–74]; and draft, Emory to Whipple, Cimialauke [Samalayuca], Mexico, January 29, 1852, all in box 5, f. 35, EP). Radziminski, a Polish immigrant who held the position of principal assistant surveyor, later worked with Emory on the Gadsden Treaty line as boundary commission secretary (Kajencki, 1988, 212).

81. GCG, like other employees, began on this day to send written monthly reports to Emory. We have reproduced only this one, which may be found in box 5, f. 36, EP.
82. John Henry Clark, whom Lieutenant Graham had appointed on the recommendation of Spencer F. Baird of the Smithsonian and who was to collect zoological specimens. Clark did, but he also gained skills as a computer and astronomer and at this point was working with GCG (Rebert, 2001, 24–25). GCG came to regard Clark as a friend, and Baird kept in touch with Clark, who continued to send zoological specimens (Baird to Emory, Smithsonian Institution, Washington, May 14, 1852, box 5, f. 39, EP).
83. At White's Ranch, or Frontera.
84. GCG was not exaggerating to impress his father. Bartlett (1854, 2:384) corroborates his story. Between September 1851, when troops were removed from Coon's Ranch, and January 1854, when Fort Bliss was established, the area of present-day El Paso, Texas, had no military garrison (Timmons, 1990, 131). Indian raids on the commission and its property were so common that Emory was ready to declare that "actual war exists with the Indians," yet the employees of the commission were "almost without arms, and without a particle of ammunition" (draft, Emory to Col. [D. S. Miles], Fort Fillmore, February 27, 1852). In a letter from Fort Fillmore on February 27, 1852, Colonel Miles told Emory that he had no guns or ammunition to spare and a scanty force (box 5, f. 36, EP). As weeks went on, Miles grew more discouraged at the territory he had to defend, which stretched from Fort Webster at the copper mines in the west to San Elizario in the east. A discouraged Miles told Emory, "I wish I was out of the country" (Fort Fillmore, March 16, 1852, box 5, f. 37, EP).
85. Magoffin's ranching activities were threatened, as GCG notes, by ongoing Apache raids. In addition to cattle and farm animals, Magoffin claimed to have lost sixty mules, kept within 150 yards of his house, over seven months (Timmons, 1990, 113).
86. GCG was correct. Col. John B. Weller, who had opened a law practice in San Francisco after his removal from the survey, was elected by the California legislature on January 30, 1852, to

fill the Senate seat vacated by Frémont, who had served as senator from September 9, 1850, to March 4, 1851 (Ellison, 1950, 276).

87. GCG apparently distrusted Smith.
88. Brady had not been killed, as we learn from a later letter Clint wrote to his sister Sarah, San Elizario, May 30, 1852. In it GCG chastises Sarah for telling Brady's family he was dead.
89. Meaning, apparently, what we would call a subsistence allowance, or per diem. The order apparently came from the secretary of the interior. Emory acknowledged receiving it and notifying "the gentlemen concerned" (Emory to A. H. H. Stuart, February 8, 1852, box 5, f. 36, EP).
90. As Emory's clerk, GCG was budgeted to receive a "commutation" of $1,460 a year ("true copy" made by G. Clinton Gardner, Clerk, of O. H. Tillinghast to Emory, Paso del Norte, Texas, January 1, 1852, box 5, f. 35, EP).
91. Charles Gardner was apparently lukewarm about an army career for GCG. In 1853, when his son brought up the issue again, he responded directly: "Discard the idea of a Lt's com.n and adhere to your profesion" (C. K. Gardner to GCG, Capitol Hill, July 5, 1853, GCGJC). GCG's father did, however, have an ulterior motive. He wanted GCG to work for him as a surveyor in Oregon.
92. At 8:00 on the morning of the fifteenth, Indians attacked the commission's camp five miles north of Frontera and took eighteen of its "best animals" (draft, Emory to Lt. D. C. Green, Frontera near El Paso, February 15, 1852, box 5, f. 36, EP).
93. Emory did have several men saddle up and pursue the Indians to a place where their trail crossed the river, about eight miles north of Frontera. Shots were exchanged, and Indians left the cattle behind. He reported the episode to Capt. A. Buford of the dragoons, telling him, "Knowing your limited means of pursuit I feel some reluctance in giving you the information" (draft, Emory to Buford, Frontera near El Paso, February 15, 1852, box 5, f. 36, EP).
94. GCG refers to Article 11 of the Treaty of Guadalupe Hidalgo, part of which bound the United States to discourage incursions by "savage tribes" into Mexico: "Considering that a great part of the territories, which, by the present treaty, are to be comprehended for the future within the limits of the United States, is now occupied by savage tribes, who will hereafter be under the exclusive control of the Government of the United States, and whose incursions within the territory of Mexico would be prejudicial in the extreme, it is solemnly agreed that all such incursions shall be forcibly restrained by the Government of the United States whensoever this may be necessary; and that when they cannot be prevented, they shall be punished by the said Government, and satisfaction for the same shall be exacted all in the same way, and with equal diligence and energy, as if the same incursions were meditated or committed within its own territory, against its own citizens." Recognizing the difficulty of the task, the United States sought and obtained release from that obligation in the Gadsden Purchase Treaty of December 30, 1853.

95. GCG's brother Jacob ("Doc"), who married Rose, or Rosa.
96. Ellicott's Mills, Baltimore County, Maryland, on the east bank of the Patapsco River opposite Ellicott City, with flour and lumber mills at its heart. Today it is a historic district on the National Register. It was the home of the Patapsco Female Institute, which the Butterworths' "fair niece" Mary attended (GCG to his sister Alida, New Orleans, October 3, 1851, and Frontera, Texas, March 21, 1852).
97. GCG's next-door neighbor in Washington, D.C.
98. GCG's sister-in-law, Rose, married to Jacob.
99. A mistress or concubine. He seems to be telling his brother that the camp followers at Frontera were beneath his social level, so he visited the respectable young ladies of El Paso instead.
100. Ingraham and GCG had gone through a good deal together and apparently become close friends. Ingraham had left in late January for the Pima villages on the Gila. En route Ingraham dropped south to Janos and Corralitos to find a guide and pick up mules. In reporting to Emory on that mission, Ingraham concluded by saying, "Give my respects to Gardner and all the rest at the ranche [*sic*]" (Ingraham to Emory, Barrancas 4 miles south of Corrallitas [*sic*], February 11, 1852, box 5, f. 36, EP). Ingraham returned to Frontera on April 12, having learned that Bartlett had sailed from Guaymas to San Diego. Ingraham had reached the Pima villages on March 7 and learned there that Whipple had also gone to San Diego (draft, Emory to A. H. H. Stuart, Frontera, April 12, 1852, and Ingraham to Emory, Frontera near El Paso, April 14, both in box 5, f. 38, EP). Emory praised Ingraham for "traversing an extensive region occupied by hostile Indians, without losing a man or animal." Ingraham had won Emory's confidence to such a point that Emory recommended him as someone who might be entrusted with the dispersal of all the commission's funds (draft, Emory to A. H. H. Stuart, Frontera near El Paso, April 17, 1852, box 5, f. 38, EP).
101. There may have been no "ladies," but the letter that GCG wrote to his brother Charles, three days earlier, indicates there were women.
102. Charles A. Snowden had come to El Paso with Colonel Graham and suffered from bad health during the winter of Emory's arrival. Snowden wrote to Emory on January 23, 1852, telling him that the surgeon expected him to be able to return to work in three or four weeks and asking for an assignment (copy, Snowden to Emory, Frontera near El Paso, January 23, 1852, and Emory to Snowden, Frontera near El Paso, January 28, 1850, both inbox 5, f. 35, EP). Emory responded that he could not give him an assignment until he reported for duty.
103. Secretary of the Interior Stuart had ordered that pay vouchers be cashed in the field. Rather than pay what he regarded as a high fee for cashing his paycheck, GCG sent the voucher home to Washington. He refers again to Stuart's order in a letter to his mother from Frontera, Texas, on May 11, 1852.
104. A 50 percent discount meant that when the men cashed their pay voucher, they received

only half of its face value. Cashing checks or drafts drawn on the government at a reasonable discount was a common practice among merchants at this time, but the 50 percent GCG mentions was usurious. Emory lamented that the commission employees, for lack of pay, were forced to get cash from "shavers and brokers" (Emory, 1857–59,1:14). One of those, as GCG indicates, was Bartlett's brother, George, who was in charge of the commissary (Faulk, 1967, 55–56; Werne, 2007, 140).

105. GCG was not the first to accuse George Bartlett of speculating. See Goetzmann (1991, 171–72). In San Diego, Commissioner Weller's brother was similarly accused. See chap. 2.

106. After García Conde's death the Mexican government appointed the surveyor, José Salazar Ilarregui, as commissioner ad interim and at the same time had him maintain his position as surveyor (Werne, 2007, 123). In a private letter, written in English on March 11, 1852, Salazar told Emory that he had received the appointment just the day before (box 5, f. 37, EP). GCG believed that the three highest offices of the American commission should also be united in one individual, and that was Emory. See GCG to his father, Frontera, Texas, April 9, 1852. GCG was wrong, however, about the Mexican commission's "uniting" three positions. The Treaty of Guadalupe Hidalgo did not require an "astronomer," so Mexico never appointed a counterpart to Emory, who was the official astronomer for the United States. Emory's preferred method for marking the line was astronomy; the Mexican team preferred the survey method called triangulation. Nonetheless, the Mexicans established observatories and made astronomical calculations (Rebert, 2001, 28–29).

107. Magoffin, born in 1799, was about fifty-three. How far back the "old" friendship went is not clear, but Magoffin and Emory knew one another from the recent war. They likely met in Santa Fe when Emory marched into New Mexico and Magoffin was among those trying to persuade Mexican officials to surrender the province (Werne, 2007, 117; Timmons, 1999, 38).

108. We have not been able to identify the French cook.

109. Soldier, as GCG slips into French.

110. A trip that would take him that month into northern Chihuahua to Lake Guzmán.

111. Again, GCG refers to the recent wedding of his brother Jacob to Rosa.

112. Gray (and Whipple), who had been surveying on the Gila, reached San Diego on January 8. Bartlett arrived a month later, on February 9. He had gone by way of Guaymas, as GCG supposed, but then took a detour south to San Blas and Acapulco before boarding a vessel for San Diego (Bartlett, 1854, 1:505; he incorrectly notes Gray and Whipple's arrival as January 10).

113. As Emory described it in his published report, the state of the commission at this point was chaotic. Five hundred thousand dollars had been spent, only one-fifth of which he thought had been put to good use (Emory, 1857–59, 10–11).

114. Fort Fillmore, named for President Millard Fillmore and established in 1851 six miles south of Mesilla (Frazer, 1965, 99).

115. Emory had begun to move instruments to San Elizario to use it as a launching point to descend the Rio Grande. GCG had begun packing boxes as early as April 27 (GCG to Emory, April 27 and 28, 1852, box 5, f. 38, EP).
116. Located in northwestern Chihuahua and fed by the Casas Grandes River, Lake Guzmán is part of a closed drainage basin that extends into southwestern New Mexico. GCG alludes to a previous letter announcing his trip to Lake Guzmán. We do not have that letter.
117. The botanist Charles Wright, who went along on this trip, wrote to his brother that the group left on April 16, crossing the Samalayuca sand dunes south of El Paso and the small community of "Cimieluque," as Americans spelled Samalayuca. From there they headed westward to Lake Guzmán and Lake Santa María. They returned to Frontera about April 26. On April 27 GCG was packing boxes at Frontera (GCG to Emory, April 27 and 28, 1852, box 5, f. 38, EP). Wright indicates that Emory wished to know whether Lake Guzmán was fed by the Mimbres River or the Santa María River (it is fed by the Rio Casas Grandes). Charles Wright suggested that Emory led this reconnaissance, but that seems unlikely. Emory's correspondence during this interval is dated at Frontera. GCG appears to have been in charge (Hartmann, 1993, 185–86).
118. Corralitos on the Rio Casas Grandes below Janos. Bartlett, making his leisurely way back to El Paso from San Diego, passed through Corralitos in August 1852 and described it as a silver mining town about 160 miles from El Paso with a population of about four hundred (Bartlett, 1854, 2:342–44).
119. After it leaves the higher elevations of what is today the Gila Wilderness and moves toward present-day Deming, the Mimbres is a seasonal stream. If he was actually following the Mimbres, GCG was fortunate to find it running.
120. Lake Santa María is also in the Guzmán basin.
121. See note 119. GCG has confused the Rio Mimbres with the Rio Casas Grandes.
122. A late-eighteenth-century Spanish fort, built on the Mexican side of the Rio Grande at present-day Ojinaga, more than two hundred miles downstream from El Paso. Emory and GCG would soon head there.
123. George G. Garner was an assistant astronomer for the commission (Garner to Emory, January 31, 1852, Frontera near El Paso, box 5, f. 35, EP). Two widows named Garner lived in Washington at this time, a Mary Garner and a "Mrs. Garner," and two blacksmiths, Charles and William (Hunter, 1853, 39). George Garner's sister was at Patapsco with GCG's sister. It is unlikely that a blacksmith's daughter would have attended that private school, so George Garner's mother was probably one of the widows.
124. Greasers—a common Anglo-American pejorative for Mexicans in that era.
125. The family's Black Rock property consisted of two or more lots in the independent municipality of Black Rock, on the north edge of Buffalo, along the eastern edge of the Niagara

River. Buffalo annexed Black Rock in 1853. GCG's mother owned the lots. Their family friend and relative John Butterworth owned a contiguous lot. GCG maintained an interest in these lots for several years, but at this point his offer to "clear" the property may be a reference to paying back taxes. He ultimately paid the taxes on the lots and purchased them from his parents (GCG to his mother, Salem, Oregon, March 18, 1856, and GCG to his father, Colville Depot, February 1, 1860, GCGP).

126. Tailors, it would appear.
127. Pants with spring bottoms flared out at the bottom to cover the shoe, much like bell bottoms in the 1860s and 1960s–1970s.
128. Weller had been elected to the Senate from California.
129. Emory gave Lieutenant Michler, a topographical engineer, detailed instructions for setting up an observatory at Eagle Pass, where he could obtain supplies and military protection from nearby Fort Duncan. He was to survey the river to its mouth, with Arthur Schott as a surveyor (drafts, Emory to Michler, Near El Paso, December 11, 1851; Emory to Schott, Frontera near El Paso, December 13, 1851; Emory to A. H. H. Stuart, Near El Paso, December 18, 1851, all in box 4, f. 34, EP).
130. The tensions between Smith and Emory began early. Lt. William F. Smith, a topographical engineer, was sent to San Elizario to set up an observatory and oversee the work of a surveyor whom Smith deemed incompetent. Smith early on claimed illness, and Emory came to believe that Smith was reluctant to do his job (draft, Emory to Smith, Frontera, near El Paso, December 16, 1854, box 4, f. 34, EP; copy, Smith to Emory, January 20, 1852, and draft of Emory to Smith, Frontera, January 21, 1852, both in box 5, f. 35, EP). Emory was correct. Smith asked to be relieved of duty that he found "exceedingly disagreeable" (copy, Smith to Emory, San Elizario, January 22, 1852, box 5, f. 35, EP). Emory honored Smith's request, pointing out that "no service is useful unless willing, and the dissatisfaction of one soon communicates itself to others" (Emory to A. H. H. Stuart, Frontera near El Paso, January 24, 1852, box 5, f. 35, EP).
131. Emory's charges against Smith are outlined in a letter Emory sent to Smith from Frontera on March 21, 1852 (box 5, f. 37, EP).
132. Trouble between Tillinghast and Emory began in February, when Emory asked Duff Green to move Tillinghast's grazing camp and take charge of it. Tillinghast challenged Emory's orders, and Emory shot back by saying that Tillinghast was "discourteous and disrespectful to your commanding officer" and demanding obedience (copy, O. H. Tillinghast to Emory, February 18, 1852, and copy, Emory to Tillinghast, Frontera, February 20, 1852, both in box 5, f. 36, EP). On the twentieth GCG personally delivered Emory's orders to Tillinghast and "then went into the Qr. Masters office, after which I came to the door and heard him say to Lt. Green that he would disobey the orders and that if he, Lt. Green, went out there and

attempted to move his, Lt. Tillinghast's, camp he would arm the teamsters and prevent it by force" (two copies, GCG to Emory, Frontera near El Paso, February 21, 1852, box 5, f. 36, EP). On May 2 Emory reported to Secretary Stuart that Tillinghast was under arrest. Tillinghast countered with charges against Emory, one of which was that he was drunk at a dinner and spoke badly of the government on January 2, 1852. Emory solicited statements from others who were present: James Magoffin, Charles Parry, Dr. John Bigelow (the surgeon), Marine Chandler, and GCG. All denied that Emory was drunk. GCG said, "I would also state that I have been associated with Major Emory intimately for the last three years and have never seen him under the influence of liquor except on convivial occasions when all were more or less the same and the imputation attempted to be cast upon him of intemperate habits is wholly unjust" (draft of Emory to A. H. H. Stuart, and copy of enclosure, San Elizario, ca. June 5, 1852, box 5, f. 40, EP; draft, Fort Fillmore, NM, May 2, 1852, box 5, f. 39, EP).

Presidio of San Eleazario.

From J. R. Bartlett, *Personal Narrative* (1854).
Courtesy of the DeGolyer Library, Southern Methodist University, Dallas, Texas.

CHAPTER 7
"God Knows What Will Become of the Survey"
Down the Rio Grande to Ringgold Barracks
May 30–December 11, 1852

In May 1852, his work in the El Paso area done, William Emory dismissed men he no longer needed and prepared to move down the Rio Grande with a small party. Two of his teams had surveyed sections of the river below El Paso. Moritz von Hippel, a skilled civilian employee, had covered more than two hundred miles from El Paso to Presidio del Norte; Lt. Nathaniel Michler had followed the river from Eagle Pass to Laredo, measuring the distance at 135 miles.[1] Nonetheless, great stretches remained to be surveyed, and astronomical observations had yet to be made. Emory planned to push the work to completion by following the river to its mouth. Along the way he hoped to rendezvous with John Russell Bartlett, who still controlled the commission's purse strings.

As Emory's clerk, Clint Gardner spent weeks, from late April through early June, packing for the trip. He boxed up files, equipment, and instruments to move from headquarters at Frontera to San Elizario, which would be the jumping-off point for the expedition. At San Elizario he set aside some items to ship ahead and store at Fort Inge, a military post on the San Antonio–El Paso road, and other items to use for the trip down the Rio Grande.

After various delays the party set out from San Elizario on June 11. Clint left the El Paso area with no regrets: "This place has no charm for me," he told his sister. Emory may have shared his sentiments but for different reasons. That same month Emory described Bartlett's operation as "the most senseless, rotten, and corrupt organization it was ever my misfortune to behold in my long service under the government."[2]

Leaving San Elizario, Emory's surveyors followed the Rio Grande valley accompanied by a small military escort under Lt. Duff Green. They worried about Apaches, but it was weather rather than Indians that assaulted them. On June 23 heavy rains caused the river to rise, and it washed out their camp in the middle of the night. Clint described the debacle vividly.[3] Four days later, on June 27, they left the Rio Grande and made their way over what he characterized as level terrain, reaching Presidio del Norte on July 8. The town stood on the site of an old Spanish presidio on the Mexican side of the Rio Grande, where a major river, the Conchos, flows out of Mexico to join the Rio Grande.[4] Emory dismissed it as "a miserably built mud town." Its eight hundred inhabitants suffered from famine. Indians had run off most of their cattle, and little rain had fallen in the previous three years.[5] Emory made his camp across from Presidio del Norte, on the American side of the river, calling it Camp Sumner.

Emory kept his party at Camp Sumner for two hot summer months until the last day of August. He had work to do but his men had had enough. "Disgusted with long-deferred 'promises to pay,' [they] became very mutinous. . . . A whole party rebelled and refused to proceed further."[6] Emory blamed the touring commissioner Bartlett, whose absence still left Emory without authority to draw funds.

One of Emory's tasks in Presidio del Norte was to authenticate the Bartlett–García

Conde agreement. In May Emory had agreed to meet his Mexican counterpart, José Salazar Ilarregui, at Presidio del Norte on August 1, with or without Commissioner Bartlett.[7] Emory and Salazar planned to sign off on the section of the survey from El Paso to Presidio del Norte and to sign a map that delineated the initial point at 32°30'. Emory, however, like Surveyor Andrew Gray before him, believed that Bartlett had placed the starting point too far north of El Paso del Norte and did not want to validate Bartlett's decision. At the same time Emory had orders from Secretary of the Interior Stuart to authenticate the arrangement. When Salazar presented himself at Presidio del Norte in late July, ready to sign the map, Emory put him off with plausible excuses. Emory apparently hoped that he would receive new instructions that would allow him to repudiate Barlett's compromise.[8]

Salazar kept pressing Emory to sign the agreement, and when Salazar and Emory met again at Presidio del Norte on August 26 and 27, Emory could delay no longer. He did, however, find a way out of his quandary. He signed the map and made a notation that said: "The Boundary line as agreed upon by the Two Commissioners April 24th 1851."[9] Or, as Emory later explained his action, he merely affirmed that the initial point had been agreed to by the two commissioners "and nothing else." This caveat, he believed, "left the government free to act, and repudiate the agreement by the two commissioners."[10] Repudiate it the United States did, but not until a Democrat, Franklin Pierce, replaced the Whig administration in the spring of 1853.

Among Emory's other tasks at Presidio del Norte were to set up an observatory, take astronomical readings, and push the survey below Presidio del Norte to Laredo.[11] Moritz von Hippel had finished surveying from El Paso del Norte to Presidio del Norte, but a daunting stretch lay before them. Two years earlier the Rio Grande beyond Presidio del Norte had defied a surveying attempt by a topographical engineer, and Emory had learned that even mules could not pick their way along the river's deep gorges in that area.[12] A month before Emory's arrival von Hippel had sent a party to reconnoiter the river below Presidio del Norte, and its report added to the dismal picture.[13] As Emory came to understand the situation, "the bed of the river for a large portion of the distance is located in great chasm of vertical rock, is full of rapids & falls and has never been traversed by white men tho many attempts have been made." To make matters worse, this area was "also the stronghold where the Kiowas, the Comanches and Apaches assemble preparatory to their general des[c]ents in the northern states of Mexico."[14]

Difficulties aside, Emory was determined to send a party to survey this daunting terrain, which we know today as Big Bend National Park. That stretch of the river contains its deepest gorges. His initial choice to lead the party that would descend the river was von Hippel. On July 19 Emory ordered von Hippel to survey as far as the mouth of the Pecos, well beyond the river's big bend. Earlier, an optimistic Emory had ordered Lt. Nathaniel Michler to stop surveying below Laredo and to work his way upstream from Eagle Pass until he met a survey team coming down the river.[15] Now, in the same spirit of optimism, Emory told von Hippel that if he met Michler's party before he reached the Pecos, he should "join your work to his."[16]

Von Hippel, who had worked thus far on a shoestring, was not about to lead another ill-equipped party. He responded to Emory's order with a detailed request for equipment and manpower. "I have made the estimate with a strict regard to the interests of the government, & the safety of the party," he told Emory. His list included fifty soldiers or fifty "well armed citizens" to escort a supply train that would have to follow the surveyors on the flatlands high above the river. Emory must have regarded von Hippel's list of essentials as excessive, if not unrealistic. On July 22, the day after submitting his list of requirements, Von Hippel resigned and Emory curtly accepted his resignation.[17]

Emory apparently hoped to replace von Hippel with another civilian surveyor, Marine T. W. Chandler, but Chandler was on an errand and his date of return uncertain.[18] Emory then turned to his clerk. He put Clint in charge of the party on the day that von Hippel resigned. As Emory's clerk Clint certainly knew how daunting the task would be. Nonetheless, with the breezy nonchalance and bravado that he occasionally affected when writing to his family, he explained to his sister that "the men seem to fear the work from here down the river—so I expect the Major will send me in charge . . . which I will be delighted to do." Emory, however, had put Clint on notice that he might not send him at all. He needed Clint's services as a clerk lest "his papers will become scattered."[19]

Clint quickly took charge of von Hippel's camp and conducted an inventory. The party was to consist of seventeen men, but only eleven remained. Clint sent Emory a list of the names and occupations of the eleven and recommended that Emory fill the positions of chainmen and instrument carriers with "Americans" and give the more menial positions to "Mexicans." Von Hippel's "Camp on the Rio Grande," was within a day's journey upstream from Presidio del Norte, far enough that Clint had to spend the night there. He took provisions, and he looked to his own comfort. He needed "mess furniture," a table he used for his desk, and some small camp stools: "It will also be adding much to my comfort, for we are compelled to take it como un Turco [like a Turk] at present." There were more things Clint would require if he remained "in charge."[20]

Clint's crew began surveying the Rio Grande, but his hopes for glory were soon dashed. Marine Chandler returned, and on August 8 Emory put him in "charge of the party temporarily in charge of Asst. G. C. Gardner, who on being relieved will resume his former duties." Emory ordered Clint to turn over all the instruments and public property to Chandler, and he instructed Chandler to "survey from the point where Mr. Gardner closed his work, down the Rio Bravo [Rio Grande] until you reach the mouth of the Pecos, or until you meet the party working up the River under the direction of Lieut. Michler."[21]

Emory seems to have assuaged Clint's disappointment by convincing him that his skills as a clerk had made him indispensable. As Clint complained to his sister Alida, the day after Chandler received orders to replace him, "I often regret having made myself so useful."[22] It may also be that Emory lacked confidence in Clint. Did a man who required the comfort of camp stools have the grit to carry out an arduous assignment that, in Emory's view, must not fail? "It is an operation that cannot be done twice," Emory told Chandler. "I therefore fully rely upon you for the successful accomplishment of this section of the work."[23]

Success eluded Chandler, and Clint was right when he told his mother he was better off despite his disappointment at not being able to conduct a survey through "unknown regions."[24] He missed the dubious reward of leading a miserable expedition on which the surveyors lost boats, mules, supplies, and clothing, and nearly starved. Chandler's party, which set out on September 12, followed the river around the Chisos Mountains, which force the Rio Grande to make its now-famous big bend, but six weeks and forty miles into the survey, the stupefying sandstone gorges of Boquillas Canyon forced Chandler to abandon the project. He left the river, made a harrowing three-week trek through northern Mexico, crossed the Rio Grande at Eagle Pass, and arrived at nearby Fort Duncan on November 24, 1852. His men were lucky to be alive but had suffered, as Chandler reported, "the effects of want of shoes & other clothing. Most of them lost everything in the boats on the river & have been without blankets or a change of clothes since the 16th of September. Their feet covered only with raw cow skin moccasins or left bare."[25]

Meanwhile, some of his shoeless men, feeling ill equipped to make the overland trek, braved the rapids of the Rio Grande in a small India-rubber boat and had the amazing good fortune to make it to Fort Duncan before Chandler did.[26]

Entrance to Cañon Bofecillos, Rio Bravo del Norte.

From W. H. Emory, *Report on the United States and Mexican Boundary Survey* (1857–59). Courtesy of the DeGolyer Library, Southern Methodist University, Dallas, Texas.

Emory and Gardner were not at Fort Duncan to receive them. Emory's party had left Presidio del Norte on August 31, two weeks before Chandler started the survey, and made its way to Fort Duncan. Duff Green and his soldiers were with Chandler, so Emory's party traveled warily through Comanche country without an escort. Signs of Comanches left Clint spoiling for a fight, regretting that he was "not more actively employed than at office work" and still eager for his father to get him a commission in the army.[27] They reached Fort Duncan in three weeks, on September 23, and remained there through November 12. Twelve days later Chandler reached Fort Duncan.[28]

Ten of Clint's letters remain from his time at Fort Duncan and have precious little to say about his surroundings. Founded in 1849, Fort Duncan stood on the American side of the Rio Grande two miles upstream from the new town of Eagle Pass, founded in 1850. Clint had looked forward to his arrival there, thinking it "a great place," but he soon dismissed it as "poor living." As he bluntly described it to his sister Sarah, "this is the dullest place, perhaps, you ever did see."[29] He told her of local dances, and of Emory's

From W. H. Emory, *Report on the United States and Mexican Boundary Survey* (1857–59).
Courtesy of the DeGolyer Library, Southern Methodist University, Dallas, Texas.

dislike of dancing, but nothing about the place or its inhabitants. He could have found much else to report, but perhaps he thought it of no interest to his family in Washington. A small, raw, multiracial, and lawless frontier town, Eagle Pass had more than its share of outlaws, smugglers, Comanche and Lipan Apache rustlers, and bounty hunters, some of whom crossed into Mexico in search of runaway blacks or black Seminoles, and some of whom crossed into Eagle Pass from Mexico to retrieve runaway peons.[30]

On November 13 Emory was on the road again. As he continued downriver, word of the wretched condition of Chandler's survey party would have caught up with him, and Clint would have learned how close he came to suffering Chandler's fate. In his surviving correspondence, however, Clint does not mention Chandler again or his own good fortune at avoiding the big bend.

Emory had expected Bartlett to join him at Fort Duncan. Clint's letters make that clear. "Bartlett is I presume on his way to join us," he wrote on October 8. Two weeks later he wrote: "Mr. Bartlett has not yet arrived altho' expected every day."[31] Emory continued to believe that Bartlett was coming and waited for him well into November.[32] But Bartlett failed to appear at Fort Duncan. Instead, he sent word that he would meet Emory farther down the river at Camargo, across from Ringgold Barracks.

Emory needed Bartlett, who had control of the finances, in order to complete the final stage of the survey, the long stretch from Laredo to the mouth of the Rio Grande.[33] A dispatch from Washington, which had reached Emory on November 11, made it clear that Emory's promissory notes would not be honored and that he was "to take no decisive step" without the commissioner.[34] Completely out of cash and authority, Emory suspended work that same day and ordered the discharge "of every man in the employment of the commission except those absolutely necessary to protect the public property, and accompany me to Camargo."[35] About sixty-five men, not counting those traveling in Bartlett's entourage, were in the commission's employ.[36]

Clint was among the indispensable. He followed Emory to Ringgold Barracks across from Camargo, where Emory hoped to rendezvous with Commissioner Bartlett and obtain a portion of the $105,000 that he believed Bartlett still had at his disposal. At the least Emory wanted to pay off his men. Emory also needed to see Bartlett because the instructions he received from Washington on November 11 made it clear that only Bartlett had the authority to continue the survey or shut it down.[37]

For his part Bartlett had returned to the El Paso area on August 17, 1852, from what Emory described as junkets to "Chihuahua and Sonora, and the Geysers of California—places sufficiently distant from the line."[38] Clint shared his boss's view but with a sharper edge, observing that nothing better could be expected from "a contemptible yankee."[39] Rather than take the road from El Paso to San Antonio and join Emory at Presidio del Norte, which he initially agreed to do, the commissioner went on another holiday. Professing to fear the danger of Comanches on the San Antonio Road, he decided to take a more circuitous and dangerous route through northern Mexico that would bring him to the Rio Grande. Initially, he told Emory that he expected to return to the Rio Grande at Camargo, across the river from Ringgold Barracks, but along the way he apparently got word to Emory that he would meet him farther up the river at Fort Duncan, where Emory had waited fruitlessly before moving to Ringgold Barracks.[40]

Ringgold Barracks and Rio Grande City.

From W. H. Emory, *Report on the United States and Mexican Boundary Survey* (1857–59). Courtesy of the DeGolyer Library, Southern Methodist University, Dallas, Texas.

During the wait for Bartlett at Ringgold Barracks, Clint wrote at least three more letters to his family. In one Clint summarized Emory's predicament. Only Bartlett had authority to draw funds, and he had yet to make an appearance. As Clint understood it, "there are also persons all along the line who will willingly credit the Commission but there is no one to take the responsibility since the Secretary gives all the power to the Commissioner and he is continually on a pleasure trip thro' Mexico."[41]

The situation was dire. Bartlett had sent two checks to Emory that merchants would not accept since the Interior Department had refused to pay on earlier ones. The commission was not only broke but the local merchants seemed unlikely to extend further credit. Emory complained bitterly to Bartlett, "I have now been one year on this work without seeing a dollar from you, and have been obliged to sustain it by a system of credits promises & threats wholly unknown to our government. Considering the munificent appropriations made by Congress, I cannot think the Survey has received anything like its just proportion of the funds." Or, as Emory lamented in a private letter to an official in the Department of the Interior, "I stand a chance of seeing the inside of a Texas jail and of seeing this work, now progressing so well with every prospect of being finished by the fourth of March, stopped and thrown again into confusion."[42] He did not see the inside of a Texas jail, but his other words were prophetic.

The thirty-one letters in this chapter begin with Clint's packing at San Elizario on May 30 and end on December 11 at Ringgold Barracks, where, as Clint told his sister that day, the party was "still in that state of uncertainty without knowing when or where our next move will be."

GCG to his sister Sarah, San Elizario, Texas, May 30, June 2, and June 5, 1852

Dear Sarah,

Having a few moments leisure I will occupy the time by writing you a few lines—Since I left Frontera I have been very busy in repacking our instruments at this place and invoicing the same to Leona [Fort Inge][43] with the exception of those we require on the river and even sundays have been busy days so I will have to watch my time to make out a letter or two home. The last mail from San Antonio brought no letters from home for me but I am delighted to say I received two from Sister Emma which have been the first for two years and gave me the news of a new niece of mine for the first time, it is a wonder you all never thought of writing it to me.[44]

I suppose as this reaches you Emma and her family will be in Washington spending her time with you all while Mr. Mouton visits the Baltimore convention.[45] I wish it were my luck to be with you all but I fear a year more will hardly cover the time ere I see Washington. I came near being on you all of a sudden but as my luck would have it something turned up to prevent [it] and the Major said he wished he [could] cut me in half and put one part in Washington and the other out here with him. He won't let me go now more than three days from the Office since I returned from Lake Guzeman for he said while I was out there his papers went to the old nick. I sometimes regret I accepted the place as Clerk and should not have done it, if I would have been obliged to drop my place as Assistant which I now hold and rank next to the Officers of the Army, and it increased my pay two hundred a year and if the commutation was reallowed it would be worth something.

How ever everything is for the best and I am in hopes Father will succeed in getting the commutation reallowed for I know Col. Weller will assist him if he asks it.[46]

June 2nd. Mr. [Marine] Chandler has arrived[47] but I am sorry to say with no letters for me. He told me the letters were to be sent to New Orleans to him but they never arrived such is my luck and he also told me that one of my letters home telling the report of Peter Brady's death created unnecessary sorrow. I think it was to you and if I remember right I told you it was but a report and not to repeat it and since then I corrected it so should you think it proper to communicate my letters to friends just see that you do it correctly.[48]

We had quite a pleasant dance the other night and I had an opportunity of seeing the beauties of the place but I cannot say much for them yet I think they are the best I have seen on the Rio Grande in this vicinity.

June 5th. We are going to have a farewell dance to morrow night after which I hope we bid farewell to this country. I regretted leaving San Diego but this place has no charms for me to be sure I hope I leave friends behind yet the country is what I allude to. It won't do for me to tell you any secrets for I see you all let out everything I write so I must keep dark. Tell father Mr. Chandler was much pleased with the attention he received from him and his visit at our house yet I think when he was announced at 8 a.m. there must have been some commotion for I am sure you got in [a] pigeon fit and if Alida has made her debut she must have created an awfull muss. Tell her Mr. C. says she has a splendid soprano or some other —no voice and his remark was "should I have turned my back on your sister which of course I could not do I should of taken her for Madame somebody,["] I forgot who, for he gasses so much it would take a week to think of the different remarks.

[Edward] Ingraham is now with us[49] but will go to Leona soon after we leave for Presidio del Norte so direct my letters to San Antonio hereafter so as they can be sent to me by the Post Master wherever I am. Give my love to all.

Your affectionate brother,

Clint

GCG to his father, San Elizario, Texas, June 5 and 7, 1852

Dear Father,

Mr. Chandler arrived here on the evening of the 2nd and I was sorry to find he had no letters not even an acknowledgement of the receipt of my pay account I asked him if it had been paid by the Dept. he said he did not remember but was under the impression it had been. My clothes and things which I requested I presume are in San Antonio yet I have never heard of any thing but the boots and shoes which Alida or Sarah wrote me Uncle John had in Charleston. The reason I think they are in San Antonio is that Lewis & Co. told Mr. Chandler he had some boxes for Major Emory or some of his party and the Major has not ordered any thing, so they must be for me, altho' I asked you to send mine to the care of J. R. Sweet & Co. and I can't see how Lewis & Co. comes to have them however I will write to Sweet & Co. and see if they can't be got by him and keep until my arrival in San Antonio. We are now about starting for Presidio del Norte and the clothes or boots I will not want until I get in that vicinity.

I wish you would write me and direct your letters to San Antonio to be sent to the Care of Major Emory wherever he may be which the Post Master at San Antonio will know as we will notify him of it so our letters can be received.

What ever became of that voucher of Ingraham's for the difference of pay due him? you told me mine had been paid yet his was not and I wish you would notify me of it if you collect it for I will have to return it to him. We calculate being at Fort Inge, Leona, some time about the 1st of September if nothing should happen, from which place we will go to Eagle Pass or Fort Duncan.[50]

I see Col. Weller has brought up the Commissioner and I suppose it will *result* in his dismissal if so I hope our government will follow the example of Mexico and unite the three, Commissioner Astronomer & Surveyor, in one.[51] I wish you would also keep them reminded of the necessity of allowing the commutation again on the ground that it is the only inducement for the Army Officers to remain with the Commission, their whole salary being consumed in purchasing the necessaries of life or that portion of it which is allowed by the Commission; of course they can not be expected to live on the provisions of the Commissariat alone, for the Commissary has not the kind that an Officer at the head of such a work should set on his table. He on account of the position which he holds must necessarily entertain more or less, which is rather a sinking business without the per diem.

June 7th. I suppose you must have received those letters I wrote making application for a Commission in the Army and hope you will do every thing in your power, altho' I don't suppose there will be a chance without the increase of the Army, yet if you think there is, I wish you would try it. We have had a great deal of trouble out here resulting from a premeditated attempt to put Major E's authority down, not by the citizens as one would suppose, but by the Officers of the Army and I am glad that he has caught one with a good hold and if the charges don't break him, I don't know what will. I don't know of ever having seen a more contemptible set in my life than were out here.[52]

I wish you would cut out from the newspapers all strips relating to the Commission and save them for me if you can get hold of the Home Journal of Providence you will find all the letters from Bartlett's crowd.[53] Tell the Doctor [Clint's brother, Jacob] I have never seen any Intelligencers that he so liberally offered to send me and when I get down in Texas I wish you would have those papers sent to me.

I will write every opportunity while on the road and wish you all would do the same for it makes no difference how many letters accumulate in San Antonio for if they are not sent to me I will get them when I go there myself.

Give my love to all,

Your affectionate son,

Clint

GCG to his sister Alida, San Elizario, June 10, 1852

I have just time Alida dear to acknowledge the receipt of letter of the 11th April [18]52.

Give my love to *All* I fear you won't hear from me for two or three months after this we leave here tomorrow.

Write often to San Antonio and if [I] ever do arrive I anticipate a treat in the way of letters, &c.

Your affect

brother,

Clint

GCG to his sister Sarah, Camp [?]eminst[54] the Cañon on the Rio Grande, June 18 and 20, 1852[55]

Dear Sarah

We arrived here day before yesterday with our party on our way to Presidio del Norte and as the Major is desirous to determine this point we remain here for some ten days and quizas [*quizás*—perhaps] longer as it has been cloudy and raining ever since we left San Elizario.

I have just finished putting up the observatory and now while waiting for dinner will endeavor to fill up the vacant space which would naturally occur in my correspondence home should I not write by the train which takes part of our party to Leona. The date of this will be as you see but a week or so later than my letters sent by the San Antonio mail yet, you will not receive it perhaps for a month after those as Ingraham will take in and mail it at Leona.

After leaving here we will not stop again until we reach Presidio del Norte which place should the weather favour us and nothing happen, we will leave on the second day of August for Leona and thence to the mouth of the river. The work is now progressing finely and should the appropriation be received by the Major we will all soon be in Washington and then I anticipate a gay time with but little to do.

When I arrive at the mouth of the river I shall then begin to think we are coming near home and will in all probability go over the gulf to see Emma[56] if time will allow, I believe I was going to say, when I was called away to dinner.

Now since I have eaten dinner I feel rather sleepy and good for nothing as you know you feel yourself so I will have to quit and should the train not come in tonight I will resume my letter tomorrow.

June 20th The train has not yet arrived and I fear they have been detained as they were to start the 15th yet however there is no hurry and I suppose they will take their time Ingraham goes down in charge of the Instruments &c. and there are one or two going home Mr. [Charles] Wright[57] the botanist and Charles Snowden, together with Lieuts. [O. H.] Tillin[g]hast & [William F.] Smith who the Major relieved from the work, neither of whom I have a high opinion of. Snowden resigned on account of his health and as he goes direct to Washington I will write a letter to Charlie by him and see if it is received as he says he has never received a letter from me.[58]

Write often and give my love to *all* you must not expect long letters from the field for we have nothing to write about except the monotony of the road.

Give my respects to the neighbors and all inquiring friends.

Your affectionate brother

G. Clinton Gardner

GCG to his father, Camp at the Canon on the Rio Grande, June 20, 1852

Dear Father,

Since I last wrote we have advanced this far on our road down the river this being the first stopping place and an opportunity occuring I will write you our anticipated movements.

We left San Eliazario on the 11th of this month and arrived here the 16th about 11 o'clock A.M. over rather a sandy road which was much improved by the rains we had daily.

Should the sky remain clear two or three nights we will again be on the move however we anticipate moving in order to be in time to leave Presidio del Norte about the 5th of August after which we will put in at the mouth of the Pecos river and from there to Leona at which point we are to establish a depot.

Ingraham goes down in the train in a few days with all the Instruments and other property not in use to remain until we get there.

From there we will go in again at Eagle Pass and along down the river to its mouth. The survey has been completed as far as Presidio del Norte from the initial point on the River and I expect a great deal has been done at Eagle Pass by the Surveying Party under Lt. Michler the last time we heard from there it had been completed as far down as Loredo [Laredo] which was some time ago.

The Astronomical points that have been determined are as follows Fort Fillmore, Frontera, San Elizario, this point, and Eagle Pass since we arrived here and I think if you sum up the amount of work done by Major Emory you will find it equals if not exceeds the amount done before our arrival when they had some $200,000. to do it with and he has not had a cent. We have now left a party at San Elizario doing nothing for the reason there is no money to set them at work and the extent of unfinished work on the parallel of Latitude west of the Rio Grande is not known by any one except Mr. Bartlett who is off in California or some other place finishing his book on a trip across the continent.[59]

These are the state of things here and I should think it ought to be known in Washington before this as they have been told so often.

Mr. [Moritz von] Hippel[60] who is now at Presidio del Norte has had to stop work on account of not having money to give his men sufficient to buy shoes and clothes which you must know are absolutely necessary, to be sure during the revolutionary war men marched without those things but then liberty was at stake here it is nothing but drag the chain all day with "stick" "stuck" which is quite different and over a country not like the states.

I wish you would get Uncle John to give the Major a puff in some of the papers in relation to the amount of work done yet don't publish any part of my letter for I don't approve of reporting the progress of the work, for that belongs to the Major himself yet I will of course mention my prospects of getting home soon.

I hope you have not forgot to make application for my appointment in the Army in which I hope you will succeed with the assistance of Col. W[eller] and Dr. Gwinn and tell Uncle John to use his influence for me also.

Write me often and give me the news about the Dept. &c. for every mail the Major asks me what you say of the movements in Washington and for the last three or four mails I have not received a single line from you.

Tell mother my health is in a perfect state and give my love to all. I should like to receive a letter from Aunty as a keep sake for her letters are scarce. Your affectionate

Son

G. Clint. Gardner.

GCG to his sister Alida, Camp on the Ciballo River [Cibolo Creek], July 6, 1852[61]

Dear Alida,

Here we are within two days march of Presidio del Norte having had quite a pleasant time since my last with the exception of a flood which happened a few days after my last home from the Cañon on the Rio Grande.

It was late at night when I was awoke by the fall of my tent the ridge pole of which knocked me on the head as much as to say it was time to get up but for what I did [not] know so I endeavored to fix it without dressing myself, so as to sleep the rain off which was falling in torrents but could not succeed and was compelled to abandon it and go in the large tent which was used for the office. I had hardly got to sleep when the Major was calling for me and I found upon getting up we were in water up to our ankles and the river was rising so I immediately took hold of my desk which contained all the public letters &c and started for the nearest hill but soon found myself up to my waist in a strong current and every step I took went deeper and deeper in mud and water so I called for assistance and we soon got the desk to the hill and all the other property with the exception of my things which were in my tent. I went across frequently and fished for my things but found my trunk and rifle [and] my shoes. Dressing case and camp stools were afloat and were picked up in an awful condition but, however I lost nothing but yet everything spoilt. My books all got wet and the next morning the whole hills were covered with clothes papers books &c to dry. The Major lost nothing and got very few things wet. I believe I was really the only sufferer and if I had of [*sic*] taken the warning of the tent pole I should have been up in time to save all without any trouble for the roar of the waters thro' the cañon were heard some time before the flood but no one knew what it was until the river commenced running in and when it was too late for in a few moments every thing was afloat. These are the enjoyments of camp lifle [life] with all the fuss and fright of the men I enjoyed the fun of seeing ever body running thro' the water. Then after I secured all I could I watched others and the first thing that attracted my attention was one of the men swimming down the stream, as I thought on a log, I hailed him and he answered "bound for the nearest landing with Lt. Green's trunk."[62] after which I did not dare to cross again as I could not swim so I watched the men come ashore some with nothing but shoes on[,] others with a shirt on[,] some with a vest[,] some with a coat. Yet there was hardly any one dressed for they all took what was left and found.

I am sorry for my dressing case but I have glued it up and it is quite respectable for an ante diluvian affair[63] together with my other things which have been all well dried and are only a little stained with the Rio Grande mud.

We left the Canon on the 27th of June and have been on the march ever since and have had quite a pleasant time with only a few showers of rain which were quite cooling. Our road has been quite good until the last two days when we have been compelled to make it w[h]ere nature did not do it for us with the exception of that it has been over the open plains and tomorrow is the last day we will have to work the road. We have been very fortunate in finding water along at our camps and the road is one of natures best never having been travelled before.

This is a beautiful stream when it runs I expect, but like all the rivers out here goes dry once a year and often stays dry this however is running in one or two places and is the most delightful water I have tasted for some time or rather the only water that has no taste, it is very clear something that is never found in this country. The valley is quite large and there seems to have been a ranch or some thing else here for I see two or three houses a little distance up it I suppose the Indians have driven them off who ever they were.

We have not yet seen any Indians yet, but I should not be supprised to see them come on us at any time however we are in a good condition to meet them.

I wish you would write to me often and should you see John McCalla or John Lyles[64] tell them I have written several times to them and would write now but I have not time while on a march for you see from this it is rather hurried. When you write to Eunice & Mathilde[65] remember me to both for I think they must forget I am living however I forgot Emma has written it seems by the two last letters I received so I [word illegible] in yet previous to that I was thinking about. Give my love to all the pretty girls and when you write to Orange Co. give my love to all my cousins.[66]

Ask Mother to keep Father in mind of my request constantly for I fear he will not exercise his influence and also tell him to write me the news political &c. What Chandler was about in Washington for he speaks often of Fathers attentions.

Tell aunty not to forget my request for a letter from her tell [her] she must send it in duplicate so as I will certainly receive one copy.

Remember me to all the neighbors and give me an account of all the young ladies.

Tell Miss Jane Sprigs [that] George Garner is still at San Elizario and Miss Emlie Houston that Lt. [Duff] Green is still with us and hopes to go home with us which I hope will be soon.

Give my love to all and every body included individually and collectively Aunt Sucky tambien [también—also].

Your affectionate brother,

G. Clint Gardner

My letter stamps were washed a way in the flood so lend me five cents to pay the postage and don't consider it an insult because I did not prepay it, as Chandler says the ladies do now in the States but let me know the rules & regulations first. G.C.G.

GCG to his father, Camp opposite Presidio del Norte, July 10, 1852

Dear Father,

Not knowing that the Express would remain with us this long I wrote to Alida while on the route thinking it would leave the next day.

We arrived here day before yesterday and have now fixed our camps and are building our observatories.[67]

We are luxuriating on watermelons, green corn, grape pies, &c., &c., notwithstanding tho' the want of funds.[68] This is a new field for credit. They have not found us out yet.

I wish you would write me often and give me all the news that is afloat around the Departments.

Don't forget my wishes concerning a commission.

Give my love to all

Your affectionate

Son,

G. Clint. Gardner

GCG to his sister Sarah, Camp Sumner opposite Presidio del Norte,[69] July 23, 1852

Dear Sarah,

We have arrived here so far on our way down the river and will I hope soon be again [on] our march. I received by the last mail from San Antonio a letter from Alida which you will please acknowledge for me as my last was to her. I write this to you for fear you might be inclined to complain however it makes no difference as I presume my letters are read to the whole family—to which there is no objection provided they don't contain any love affairs which Alida seems desirous of bringing about by her remarks of the young ladies in general and one in particular—to which I would reply if I were writing to her. Give my love to Miss Mamie [Mary] Butterworth, so you can tell her and ask her to give me the benefit of her intimate acquaintance with the one referred to.

Since my arrival here there has been a little disturbance in camp caused by a fear of Indians or something else, which in all probability will result in my taking the field. The chief of one of the parties has resigned, and the men seem to fear the work from here down the river—so I expect the Major will send me in charge of a party to carry on the survey, which I will be over delighted to do—as it will increase my pay and be quite a handsome piece of work if accomplished. The Major tho' seems to think if I leave him, his papers will become scattered and only fear it will lead him to make some other arrangement.

I wish you would tell Mother to thank Cousin Ann McLean for the *housewife*, but as yet I have not received it however I have no doubt it will be just the thing required.[70]

I also wrote some time ago concerning the Black rock property, which I hope has been received, for if the arrangements can be made I will be very glad to do it.[71] At present there are no funds but as soon as they can be obtained I will have it all arranged as I mentioned in my previous letter. It seems strange to me I have not heard from Father for several months not even the acknowledgment of my pay account delivered by Mr. Chandler altho' I have asked him to write me and give me the news afloat at the Departments I cannot but suppose he has written since you all told me that you wrote to me in California when I did not receive a single letter from home but got them from my friends and now of course since I receive leters from different members of the family I must believe Father has written me also and therefore will not ask you to remind him of it. Should you hear him say anything about my *numerous* applications for a Commission, I wish you would mention it in your letters and also tell Alida the same for even if you write the same that Father does write or has written, there is no harm done.

When you write to Emma [Mouton] give my love to them all and also Mathilde [Mouton], who you will oblige me by telling that I should like very much to hear from that little white head nephew of mine as I suppose him to be since you say he is like his Dad.[72] Remember me also to Mr. & Mrs. Sam Butterworth and congratulate cousin Sib for me on having found a friend, lover and husband, together with a pleasant home, I hope.

Give my love to all inquiring friends—in particular to the young ladies. Give my love to *All* at home and tell Aunty I have not yet received that letter from her.

Your affectionate brother,

G. Clinton Gardner

GCG to his sister Alida, Camp on the Rio Grande above Presidio del Norte,[73] July 29, 1852

Dear Alida,

Before finishing my preparations to go down the river I will commence this perhaps the last letter home for some time. I have been placed in charge of the surveying party here in consequence of the resignation of its chief [von Hippel] and have orders to proceed down the river but I am at present in a quien sabe [*¿quién sabe?*—who knows?] case as the Major seems determined not to let me leave him if any one else can be found to take charge of it and as Mr. Chandler is expected here in a few days he says he will send him and my only hopes are that he will refuse it as rather a rough place for such a nice personage as himself.

[letter apparently unfinished]

GCG to Emory, Surveying Camp Opposite Presidio del Norte, August 1, 1852[74]

Major,

I have the honor to report that I was engaged in my duties as Clerk up to the 26th of last month when in obedience of your order of that date I moved up to the Camp of the Surveying party of M. Von Hippel, and, took charge of the same on the day following.

I commenced the survey of the Rio Grande on the 30th and have now completed the field notes to the head of the island near my camp. I found one small island above this point belonging to the United States.

I am very Respectfully Your obt. Servt.

G. Clinton Gardner

Assist. Surveyor in Charge of Party

GCG to his sister Alida, Camp, August 9, 1852

Dear Alida

My last I believe was to Sarah so I address this to you as there is an express to leave tomorrow to intercept the mail for San Antonio.

I was in hopes this express would take my last letter home for some time, as I have been placed in charge of the surveying party on this portion of the work but as my services are so very valuable to the Major he has relieved me from the same and I am now to go with him. I often regret having made myself so useful, but however there is nothing lost for this is only one opportunity and nothing very desirous, except the credit of having carried a survey over an unknown country.

I was delighted to receive by Lightner's train from San Antonio two pair of shoes and one pair of boots, which have not exactly been covetted, but wished for, often, by others. I would make a few comments on them but I feel thankful for them to such a degree that [I] must refrain, and knowing Charlie to be near with such a small foot, I have nothing to say. He no doubt told Mr. Wheatley that my foot was much larger than his so they are abundantly large and peged [pegged] which in this dry country stands but little chance.

The Housewife has not yet come to hand but I expect will in a few days if it was sent by mail, you must return my sincere thanks [for] it any how, for if it is not received I will be still thankful to mother and the doner [donor] for their trouble.[75] It is now with great pleasure I look to our soon departure from this place, after which we will be in close correspondence as it were, comparatively speaking. I hope also to get in among civilization once more, and be able to see a little society of a different kind from this on the Mexican frontier. Since we have been here Col. [Cayetano] Justiniani, the Chief of customs for the State of Chihuahna has given us two baylés [*bailes*, or dances], to show us what the people of this place can do, telling us at the same time not to judge of the interior from this. He seems to be quite a fine old gentleman, very fond of seeing others dance, and often joins in himself.[76] Since our arrival here we have been eating musk-melons and watermelons constantly so you can imagine our appearance in these warm days seated around two or three watermelons. I have eaten for the last two or three weeks from one to two a day myself, if it were at home what would mother say? for I well remember when she would not allow one to come on the table. This is the bountiful season here every one is doing well but as soon as winter begins to show itself, then comes poverty. I have been told in the winter the inhabitants are almost starving; they never save anything for the winter—not even sufficient corn for tortillas.

You must excuse, Dear Alida my brevity but I feel quite tired as I have been plotting all day the work which I have done and expect [to] be hard at work again in the morning, as my successor, Mr. Chandler, seems desirous of taking charge.

Give my love to *All* and ask father to write me some of the news.

Your affectionate brother,

G. Clint. Gardner

GCG to his father, n.p., August 10, 1852

Dear Father,

I begin to think that you are never going to write me again for the more I ask you the less it seems you are inclined to write. I received the boots & shoes for which I return to Mother and yourself my sincere thanks and if you will give me a few items of news from the Dept.[77] perhaps I can return something more acceptable. I believe I have asked you several times if the pay accounts I sent home by Mr. Chandler were paid but as yet in vain.

And about the things I desired you to use your influence for me in I have heard nothing and would beg that you tell Alida or Sarah so as they can write me on the subject.

I wish if possible you could arrange that property at Black rock for me for I have no doubt it will be a good investment and you seem to care but little about it.

Give my love to all and do write soon,

Your affectionate Son,

G. Clint. Gardner

Aug. 10th / 1852. Hurrah for Pierce & King.[78]

GCG to his mother, Camp Opposite Presidio del Norte, Mexico, August 13, 1852

Dear Mother,

Perhaps I can imagine you all seated in the front parlor about this time enjoying the cool breeze from the river, after a very warm day, making casual remarks about Congress or perhaps the coming election of President—and should your thoughts for a moment wander this far perhaps it would be I, you would try to trace the steps of, on some old map if any happened to be at hand, wondering at what point I was to sleep for the night and if your finger passed over this place, I think I see it stick, for it is here I rest for a few weeks longer.

I feel quite disappointed, with out cause for the same, for it was but a week ago I was delighted with the idea of conducting the survey thro' unknown regions—altho' the Major told me often that it was only temporarily that I took charge of the party, yet something told me I would remain there; however, here I am again after a week's work back to my old duties, with only the name of Asst Surveyor. Well, I often think to myself—is it not better? for here I enjoy all the comforts such a barren desolate country is capable of giving, with not a third of the labor which necessarily attends the Chief of a party, and with the same pay and emoluments. Yes, better pay—for here I am entitled to the same quantity of provisions &c. as the Officers, and in the field, as the Chief of a party, only a ration is allowed, and besides the small allowances, the dangers of different kinds are great. So I think I am content—or at any rate I may as well think so, for I have to be. It is once in a while perhaps I show my displeasure—and then the Major assures me my position is the highest, and that when I take charge of a party I fall instead of rising—but somehow or other I can't begin to think so, and if an other opportunity should occur I will certainly come down for it; for I have the advantage of age in the service, if not in years, for I now can say I am the oldest Assistant on the work (civil I refer to).

We all expect to move in a few weeks—and then home, once more, will be in our bow [?]—yet far off. I dislike very much the idea of returning by the same road to Leona; and have been trying to pursuade the Major to go *thro' Mexico* to Eagle Pass, instead of travelling on our side of the line. And if he does go he says I go only because he knows I won't fall in love, and if I do, I told him I would return and see about that after wards. I long to see more of the world for the idea of travelling over the same road twice is rather dull when one can help it.

Since we have been here I have been to two dances and had the pleasure of seeing the beauties of the place which, comparatively speaking, will pass; yet I now wish to see the *elit[e]* of Mexico—which the frontier don't show; I want to see if the upper ten hold bayles [*bailes*] like "quaker meeting," which they do here, for all the young ladies collect together in one end of the room and leave the gents at the other, so there is no chance for flirtations, with out one flirts with all.

I was all fixed, some days ago, to go to Chihuahua, but as usual was disappointed and as some would say "just my luck" however, every thing is for the best—and will show itself soon—but I am only impatient.[79]

Well I must close as I begin to be sleepy so I bid you good night.

Give my love to *All* and remember me to the neighbours.

Your affectionate son,

G. Clint. Gardner

Please send me some postage stamps and direct to Eagle Pass Fort Duncan Texas.

GCG to his sister Sarah, Camp on Las Moras Creek,[80] September 21, 1852

Dear Sarah

We having halted here to day on our march to Eagle Pass I thought I would acknowledge the receipt of the letters I have received since I left Presidio del Norte. The Major sends an express to San Antonio to morrow with his mail so you see I avail myself of the opportunity to keep up the correspondence.

I have to acknowledge the receipt of yours of the 6th of June, 25th July, 9th August; the last two I received on the 18th of this month near the first crossing of Devil's river[81] by the mail on its way to El Paso. They had a fight with the Indians and were compelled to return some distance which makes them far behind their time. The first I received about the 20th of August together with the house wife [*Housewife*] which was after the one of Alida's of later date announcing that the house wife had been sent and I began to fear it was lost. I have also received a letter from Emma who seems to be in high glee and she also announces the advent of a niece in Infantry order,[82] which I happened to read before yours so you see you were a little too sure however I am delighted that you are going to keep up a correspondence weekly which I will do also when I get to Eagle Pass if there is a possibility of sending letters that often. I am glad at last to hear from my pay account sent by Mr. Chandler for the letter from Father of earlier date was never received nor the gold pen so I presume they are lost.

My clothes you say will have been ready the first of August and if you have not sent them yet I wish Mother to send them to San Antonio Texas care of J. R. Sweet

& Co. and I will address him a letter concerning them. I should also like some fine white shirts &c. however I will make a list and send to Mother. The boots &c. I have acknowledged some time ago the receipt of.

It is I believe the opinion that Gen. [Franklin] Pierce will be elected as there is some talk of Old Dan's [Daniel Webster] opposing General [Winfield] Scott which I think will insure the election on our side or rather father's side. I am neutral myself, for I can't see any advantage to be gained altho' I am of age I don't intend to vote.[83]

You must excuse the brevity of this which I know you will do when you know I have to write all the Major's Doc[uments] for the same mail.

Give my love to All in haste, Your affectionate

brother,

G. Clinton Gardner

I have just seen the death or rather Capt. Prince of the Army did of the death of Frank's daughter Cecelia R. Gardner the birth of whom you announced in your letter.[84] I regret it exceedingly and pity Mathilde & Frank but "God's will be done".

Your brother,

Clint

GCG to his sister Alida, n.p. (Las Moras Creek), September 21, 1852

Dear Alida,

I almost tired of penning it but I will also acknowledge the receipt of your letters of the 18th July & August 1st which were received with Sarah's by the mail on the road. I can assure you all that it was the most pleasant thing imaginable to receive in this wild country so many interesting letters why I can read myself to sleep every night for weeks with out the least difficulty now.

We have had quite a pleasant trip from Presidio del Norte over a new and unexplored route to the Leon[a] Spring on the San Antonio road and from there to this place, making about 20 miles per day.[85] The Indians have not been seen by any of our party but their tracks are in abundance. The Major has us up as Officers of the night and the men as sentinels as we are travelling without escort and for some time past I have been honored with the command of the advance guard composed of four bold dragoons, alias an expressman, tailor, carpenter and blacksmith mounted on chargers of the Don Qui[x]ote order. For the last two days I have been compelled to give up my command being unable to mount my horse having got poisoned on the Devil's river so I ride in the Ambulance but hope tomorrow, to take the saddle again.

I am delighted to hear of the success of my friend and school mate John H. Lyle you must congratulate him for me.[86] Remember me to the Young Ladies and all inquiring friends.

Tell Aunty if she is as much of a florist as ever I will send her some of the seed of a purple thistle which groes [*sic*] in this country. And ask the Doctor what has become of those news papers he was going to send me so regular for about this time they will be very interesting.[87]

I would be glad dear Alida to write more but really I have not time so adios.

Your affectionate brother,

G. Clint. Gardner

GCG to his father, n.p., n.d. (Las Moras Creek, September 21, 1852)[88]

Dear Father,

I send herewith enclosed a pay account for three hundred dollars and am glad at last to hear of the safe arrival of my other one which was sent by Mr. Chandler. The letter in which you acknowledged the receipt of it was never received so you see I knew nothing of it until the other day and if I don't keep account of these things I may certify too much. You have never as yet told me of the account which I signed blank or rather the sheet of paper which was for a difference of pay due caused by a mistake of that rascal Charlie Weller's paying me at the rate of 33? dollars per month. The one for the difference of 50 or 60 made in the pay roll Sarah says you have received [it] but she says nothing of the other. As soon as you receive the amount of Ingraham's account let me know for I must give it to him after you get it.

If there is any possibility of getting some of the Senator's to talk for the per diem allowance you would much oblige several and in fact the dropping of that allowance has only as it were placed a premium on the discharge of the Officers from this work for that is all they wished to remain on it for and now it is taken away their pay remains the same and they have more work to do than they would have by being with their Corps or Regiment and they are consequently glad to quit.

I am also of the same opinion that it would not be advisable to leave my profession to take a temporary commission without it would insure a permanent one.

My clothes I wish you would send to San Antonio to the care of J. R. Sweet & Co., and I will notify them of my place of residence at the time.

Mother wrote some time ago about the Black rock property and I answered it but have never mentioned it since if it is a good investment and the money is not necessary for the wants of the family I should like to have it invested in that way but I always feel bound under the circumstances to contribute to the wants of those who have so long contributed to mine knowing that in the end I will be remunerated by a portion of what is left and will have the credit of doing what I can if nothing else.[89]

Write me often and acknowledge the receipt of this so in case one letter fails I will receive the next.

Give my love to all and believe me as ever your

affectionate son,

G. Clinton Gardner

I desire to keep one quarter ahead for contingent expenses and when I return I want to visit Emma if agreeable to your wishes, as well as see something of the country on my way home, and it is for that reason I want my clothes before getting home.

GCG to his father, Camp near Fort Duncan, Texas, September 27, 1852

Dear Father

I just [have] time to say before the mail leaves. I sent a pay account by the last mail for $300. which you will oblige me by acknowledging the receipt of in *one or two* of your letters, so I may know if it has been received.

We arrived here on the 23rd and I expect will leave soon to go on down the river send your letters to San Antonio to be forwarded wherever we may be.

I have not had time to get a letter from the major in recommendation of my commission will do so the first opportunity if necessary.

My love to All in haste,

Your Affectionate Son,

G. Clint. Gardner

GCG to his sister Alida, Camp near Eagle Pass, Texas, September 30, 1852

I take great pleasure, Dear Alida, in acknowledging the receipt of your letters of the 29th of August and 5th of September which I received yesterday by our express from San Antonio and have commenced this letter now so as I may not be hurried when the mail leaves as I was the other day in my last which was to father. We are now living quite comfortable comparatively speaking but really I must say I was quite disappointed in finding this place so destitute of mechanics &c. which are much needed. I am afraid I will have to trouble Mother again with numerous requests for clothes &c. and in fact I know I will for some shirts, for I think she can get them made better and cheaper than they can be purchased any where; however this is not entertaining you much, which I believe is the object of this letter, so I will drop the subject and try to give you an idea of the pleasures of a roaming life where one can breath[e] the pure air of the plains.

When we left Presidio del Norte which was on the last of August we travelled over a country without a road until we reached the San Antonio road to El Paso. The country was quite well supplied with grass and water when we crossed and our animals improved from the time we started altho' now they are not in the best order but while at Presidio they got nothing to eat the grass being very poor and you must know animals out here generally don't get any thing else. It seems strange to you no doubt when I tell you we ride a horse all day and at night picket him out until morning when the journey is commenced again if some of the horses in [the] states were used in that way they would not stand it very long. Father asks why I have not mentioned the names of the Officers of our escort the reason is the best of all for we have had none; Duff Green,[90] whom you must remember, since the death of Col. Craig[91] is the only officer and he is on the river with the distinguished Marine T.W.C. [Marine Tyler Wickham Chandler][92]—and our escort consisted of "Old Fisher" and [an?] old soldier who was too old to march over the rough country so we brought him along with us and as I mentioned in my last from Las Moras I was the Capt. of the advanced guard and we all done military duty as Officers of the Day with the Teamsters as privates and when I wrote that letter from Las Moras I was on duty as Capt. of the guard for the night and had to set up half of the night when one of the other young men relieved me. Altho' we have had no occasion for soldiers we might have had for we would see signs of the Comanches every day and as I was always in advance, I had a good opportunity of inspecting them and must say I profisied [prophesized] a fight once but was disappointed; the next morning however we met the mail on its way up who had been detained by quite a nice little fight which accounted for all the signs we saw crossing and recrossing the road. They fought for a whole day on defensive and got off without any serious injury only one man being wounded in the arm while it is supposed they killed three Indians. I regret very much I am not more actively employed than at office work but it is all for the best for if Father don't succeed in getting me a commission I am in hopes I will be retained in the Office at Washington at the conclusion of the field work.

Ingraham is now here he having just come over with the express from Fort Inge where he has been for two months in charge of the Depot but will return today to bring over the property after which I think he will be sent to the Brazos[93] via New Orleans and if so I am only sorry I could not have sent word before so as he could have brought out some things for me however they will come by express as well. He has been quite sick with the billious fever and has not yet gained his former appearance of health.

In reading your letter I find you have forgotten the young lady your previous letters spoke of but before reading many lines you seem to remember well but without saying so; however I had almost forgotten who you ment [meant] when I first commenced the sentence for it is not much time I have spent in thinking of the beauties of the states. I think when I return it will be time enough and "sufficient for the day is the evil thereof."[94]

I anticipate a long letter from Charlie on his return from the North. I wish you would tell him not to disappoint the bright expectations of youth by keeping silent on the subject for I should like to hear from him very much.

You must give my love to all the pretty girls and continue your letters with all the gossip of the same. I am sorry I cannot make my letters as interesting as yours are to me.

Give my love to All and excuse the blots &c. for I am writing with a drawing pen having broken my gold pen.

Your affectionate brother,

G. Clinton Gardner

GCG to his father, Camp near Fort Duncan, Texas, October 2, 1852

Dear Father,

I sent by the last mail my pay account for *three hundred dollars* and enclose herewith another for *two hundred*, which I desire you to draw and dispose of as follows: I am indebted to Ingraham to the amount of one hundred and five dollars, together with the amount of his pay account for the difference of his pay, if you have been able to draw it, so I wish you would place the amount of my indebtedness to him in Corcoran & Riggs' bank subject to Edward Ingraham's order and send [it] to him at New Orleans, to the care of Judge Strawbridge, a certificate of deposit and also inform him what house in New Orleans he can draw on in Corcoran & Riggs' name.

Ingraham was here the other day and told me he had received a letter from Frank Holley stating, I think, that he, Frank Holley, had not received the forty dollars, or rather that he had not received any thing from you, which I think is a mistake, he must have intended to say that he had not received any money from you on Ingraham's account, for if I remember right in one of your letters to me you stated that *you had returned him the forty dollars* which I borrowed of him and if so I only hope you have his acknowledgment of the same.[95]

Ingraham will leave here I expect in about a week or perhaps two weeks and I desire you will if possible have it fixed so as he can obtain the money upon his arrival in New Orleans for it is money he borrowed for me when I was there. I wish you would also acknowledge the receipt of the pay accounts as soon as possible.

If you will give this matter prompt attention you will add much to the comfort of others as well as oblige greatly your affectionate son,

G. Clinton Gardner

GCG to his sister Sarah, Camp, October 3, 1852

Dear Sarah,

Having a few moments again to write before the mail leaves I will reply to yours of the 18th of August which I received in company with Alida's letters.

I have forgotten how that memorable day was spent by me but as it created no excitement, contrary to what you supposed, it is of little consequence but I feel much obliged to Aunt Suckey for her intentions and will take the "will for the deed" and without the Lord calls the good old soul to rest soon I have no doubt I will have the pleasure of visiting my birth place [Washington] once again where I will see her give my best love to her and her worthy son Step.

I am sorry you were interrupted in your letter to me but upon hearing of your occupation I only regret I was not present for I must confess I have entirely forgotten both of the young ladies and would like very much to renew my acquaintance. Should you see Johnny Ingle I wish you would remind him of numerous letters which I have written to him without receiving one[96]—Stop I have just been obliged to run after your letter which a gust of wind has blown away and now it is so wet with the rain that I can't handle it. I am delighted to see the rain for it has been so extremely hot for several days back I could not do any thing and I am in hopes this will cool the air a little.

Since I wrote to Alida I have been to a dance here which is quite a rural affair they are in the open air in front of some log house the place lighted up by lanterns the girls are not very beautiful, but all dance well. I endeavoured to teach one the polka and succeeded quite well, for they dance well for a ground dance and it reminds one of Mr. Payne['s] opinion of the polka.

I remember Austin Smith[97] very well the next time you see him give him my best respects but tell him I differ in opinion with him for I think "*gold*" is more readily made than "*fame*" and with the latter the former is easily acquired and if not acquired is not much loss for I go on the Doctors plan "the world owes me a living."

I regret that Father had a headache but I think the gin sling must have cured him. I should like to join him in one and hope it will be before many years pass. We expect now to stay here some time as the Indians have looked very hard at one of the parties on the river but I believe they managed to stare them back as hard by moving off a little[98] and the Major for the want of an escort has sent his men or those of his party to join them and we will not move until they return.[99] I have applied in the mean time to make a reconnaissance on the other side of the river up towards the Mouth of the Pecos and perhaps altho' rather doubtful may be sent.

Write often. Give my love to all and believe me as ever your affectionate brother,

G. Clinton Gardner

GCG to his father, Camp near Fort Duncan, Texas, October 8 and 10, 1852

Dear Father,

I have but a few moments to write by this mail being very much occupied in getting up my records of the Office, so as, if possible, to take command of a party that will be sent in a week or so up the river on the other side for the purpose of communicating with the Surveying parties on the River. Since I have been here I have spent quite a pleasant time. I have visited Col. & Mrs. Moris,[100] and Mr. & Mrs. Colquhorn the latter are from Virginia, perhaps you know them, the[y] are quite ho[s]pitable persons and should the we[a]ther favour us, we will have them to dine with us to morrow.

With the exception of our correspondence I think we are as far out of the world as at El Paso, for nothing can be had here without sending to San Antonio for it, however it is not far and we get a weekly mail.

Mr. Bartlett is I presume on his way to join us at last but God knows when he will reach here. I see by the papers that they are abusing the administration more than him and I have also seen a letter to the Editors of the Union, dated in July, from El Paso[101] which I think is the most base lie I ever read, with the exception of the expenditure of the funds on the Boundary without accomplishing any work by Mr. Bartlett, but he says the work is wrong from the Initial Point to the Presidio de San Elizario which is false, the observations and the Survey have been finished in the most scientific manner and the person who wrote that letter must have known it—he also states that the Mexican Commission have done so and so with twelve, which is false, for we left the Mexican Commissioner, Astronomer and Surveyor at Presidio del Norte about the date of that letter and they had not accomplished one tenth of the work which has been done since Major Emory took charge of the work on the 8th Dec. 1851.[102]

You will do me a great favour if you will inquire the name of the gentleman who signs himself "Melnotte" and at the same time inform the Editors that it is false as I have stated. I don't mean to enter into any details about it for my opinion is it was written by Mr. Vandincourt [Vaudricourt] who as I learn was discharged from this Commission for insolence and contrary to all the rules of etiquette was employed in the Mexican Commission because he is a fine draughtsman.[103]

I enclosed to you by the last mail a pay account for $200 and in a mail from Las Moras one for $300 which I am anxious to hear from don't fail to acknowledge them. My cloth clothes I requested to be sent to San Antonio. I suppose you have done so and when the next wagon goes in I am in hopes of getting them.

Give my love to all and make numerous excuses to Sarah and Alida for my not writing to them but really I am very tired of writing and disgusted with a steal [*sic*] pen having broke the last gold one I had.

I would give something great to receive a letter from Aunty for it would really be a treat never having seen one of her epistles in my life. I would keep it as a relic for time out of mind.

Write soon to your affectionate

Son,

G. Clinton Gardner

Oct. 10th—5 P.M. As the mail closes I would just remark I have taken dinner roast pig, beef stake [*sic*] sweet potatoes Rice &c. &c. with a des[s]ert of pound cake made by a most agreeable lady and sent to the Major ain't bad to take for a camp dinner, what is the opinion of the family? G.C.G.

GCG to his father, n.p., n.d. (ca. October 15, 1852)[104]

Dear Father

I received the letter stamps day before yesterday and the housewife [*The Housewife*] I acknowleged some time ago for it was received at Presidio del Norte. I wrote last night a letter to send by Lt. [Nathaneil] Michler of the Engineers in case he should go to Washington it seems the Department has protested Maj. E[mory]'s drafts which seems very strange after giving him the authority they did to make them, but God only knows what they mean in Washington for it seems to me at that Department they do nothing but cause trouble.[105] I expect the next thing will be my accounts will be repudiated. However since Lt. [O. H.] Tillinghast has arrived of course he can make it look as awful as possible[106] and if the Secretary listens to him he will soon stuff him for if he don't gain ground soon he may not have a chance. I wish you had have told me his whole conversation for [I] expect there were only a few lies in it. If he or the others who have been sent home talk about the Survey I wish you would let me know what they say for I am desirous to know how this game is played there.

You seem to think your letters has [*sic*] to remain in San Antonio until the 1st of the month for what reason I can't see for we have a weekly mail here. I have written you about my clothes and desired them to be sent to the Care of J. R. Sweet & Co. San Antonio who will send them to this place.

I have also sent home two vouchers in all amounting to $500. which I should like to hear from however it is not time yet, but there is no harm in reminding you of it for I know you are forgetful and I have no doubt you have forgot all about my Commission you must push that strong and I think with my service on this frontier, I can get it.

Write to me often and tell all the news from the Dept.

Your Affectionate Son,

G. Clinton Gardner

GCG to his sister Alida, Camp near Fort Duncan, Texas, October 16, 1852

Dear Alida

I take great pleasure in acknowledging the receipt of yours of the 19th of last month which was received by the last mail. I suppose you have not received my letter in which I acknowledge the receipt of the house wife [*The Housewife*] it was dated just after the receipt of the boots, &c. you all seem to think it is lost but I find that it has done good service. We are now living in a delightful state of uncertainty not knowing exactly what will turn. Lt. [Nathaniel] Michler left here this morning for San Antonio and perhaps for Washington. I gave him a letter to Father. You will find him a very fine young man quite handsome but married.

I received some time ago a letter from John McCalla and he says that none of the family know him for he sees them every sunday at Church and you all never speak. So it is not his fault that he don't speak for it is the ladies who are to recognize first according to my rule.

I fear there is little chance of my being home in time for the "Inauguration ball"[107] without I get that Commission which Father is applying for then perhaps I may come home before going elsewhere. Ingraham left here two or three days ago for New Orleans and to go from there to the Mouth of this river so I suppose in a little time he will be home once more well I think it is time for he has not been there for four years now.[108]

What has become of all the girls that used to live on the hill are they married yet it seems that Miss Jane Sprigs was not the flame of [George] Garner but of Mac [Malcolm] Seaton who I expect will be down here in a few days with the Commissioner. I believe they are on the road.[109]

We are enjoying ourselves as well as can be expected with dance every little while there was one last night but I did not attend having a great deal of work to do before Lt. M. left here. The Major thinks it is not proper to dance so much he objects to them of course like all old married men.

You speak of Mr. & Mrs. Allen well I should like to know who they are for I don't know them and so "Charlotte and *your friend Mamie* [Mary Butterworth]" are at Newburg[h]. I thought the latter was at school and the former was to be married. I think you wrote me that some time ago.

I wish you would write in full every week for we get a weekly mail now which you all don't seem to understand. I will bring this to a close by placing a few lines to Aunty on the opposite page with your permission so adios

Your brother,

G. Clinton Gardner

GCG to his aunt, Camp near Fort Duncan, Texas, October 16, 1852

Dear Aunty,

It would afford me great pleasure to receive a short epistle from you for I know you can give me all the news about Old Joe Nurse &c. who all go [to] the four and a half street church by the by who is the pastor now?[110] I have almost forgot how the old place looks with the exception of that old corner in the gallery where Old Aunt Peggy used to take her seat.[111] I remember that well. I often call her to mind and always with a gusto for I think it was amusing to see her travelling along on sunday with her head [cocked to?] one side. What has become of the Miss Underwoods and all those of the Coyle[112] family don't fail to remember me kindly to all the neighbours and write soon to your affectionate Nephew,

G. Clinton Gardner

GCG to his sister Sarah, Fort Duncan, Eagle Pass, Texas, October 22, 1852

Dear Sarah

The mail arrived yesterday but I am sorry to say that it brings me nothing so I suppose you all are waiting for Nov. 1st as Father remarked that his letter would remain in San Antonio until that time.

This is the dullest place, perhaps, you ever did see we have no news at all but depend entirely on our friends at home and once in [a] while we see sufficient in the papers to talk of as for instance at present we see by the news papers that Cuba has again caused a little excitement and we are in hopes, or rather I am, that the Government will now take charge of that little hacienda and send some troops there for that purpose which will in all probability give me what I have applied for, a Commission.[113] Ingraham tells me he has also done the same so my only wishes are that we may succeed, what do you think are the chances? My opinion is that if Father will push the matter on the ground that I have served four years on the Frontier that it will succeed but I see that while I am out here I have to contend against two obsticles [*sic*] one is Father's objections and the other the difficulty of getting the Commission however the former is I hope overcome and come over the latter.

Mr. Bartlett has not yet arrived altho' expected every day and we are detained here on account of cloudy nights which seem to rise with [the] moon at the very time they are not wanted—I am very anxious to get off for I look to high living when we get down on the coast but however one is never satisfied when I was on the road I thought this a great place, well it was when I first arrived but now I begin to think it poor living, so I expect it will be the same down below, so I will content myself with what I have and anticipate nothing for fear of disappoint ment, one that has been disappointed *once* will naturally avoid it the second time, so I am content.

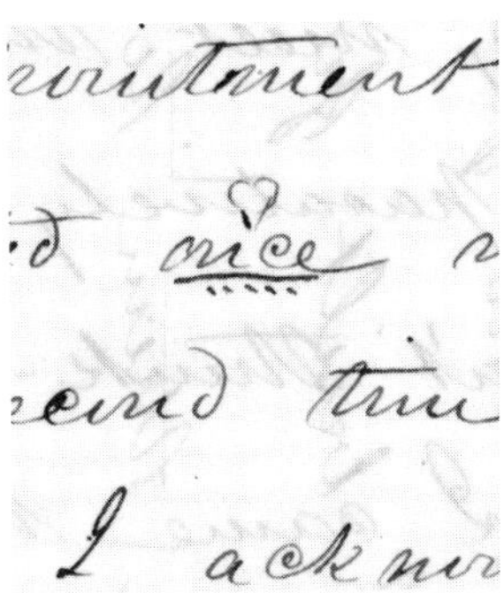

I acknowledged the receipt of Alida's & Father's letter last week and also the post stamps which were quite unexpected as I had sent to San Antonio an[d] received quite a number. It seems that my last letters from Presidio del Norte had never arrived the cause of which I can not imagine as there has been no failures of the mail as I have heard.

We have had several dances but at present we are so busy looking at the stars above that the earthly ones have no attraction being far eclipsed and when one thinks of those in the States they still remain in the dark. The young men out here seem to think when they get to the states they will fall in love with the first they meet thinking them beautiful well comparatively speaking they are but I don't think that was my opinion when I came home last however perhaps Alida may contradict that since she underscores "*my friend*" with great care in all her letters.

You must give my love to all the young ladies and write to me of their gos[s]ip and I am sure of having all the news then for I know they never let any thing pass unobserved.

Remember me to all the neighbours and tell Maj. Barker that I waited some four years ago some time for him to give me some ideas to put on that large star banner which I built for Cash & Butler and now he can give them to some of [the] other rising generation for Pearce & King as I believe it is still in the wood house if Old Aunt Sucky hasn't split it up for kindling wood or taken the cotton to mend her with.

Note that in this letter Gardner has written both vertically and horizontally, apparently because he was short of paper.

Give my love to *All* and tell Aunty I am waiting for an answer to my communication addressed to her last week for I desire to have her opinion of one who has not seen civilization for four years and has never troubled the sexton of a church to sweep up the dust from his shoes for the same length of time. By the by I found today Dr. [Clement M.] Butler's sermon on the death of H. Clay but had not time to read it being so much interested in the Cuba business.[114]

Tell Mother I wish she would keep the Doctor [brother Jacob] in mind of his promise to send me those newspapers but however it is generally the case if one offers to [do] any thing for you, you may be sure it will never be done but I think Mother will remind him of it or get Sister Rosa [Jacob's wife] to do it, it will be done.

You must excuse this scrawl but I hope the letter is interesting enough to take your attention from the writing. Oh by the by this only calls your attention to it being sure it was not observed before so I will close bid[d]ing you good night.

Your affectionate brother,

G. Clinton Gardner

GCG to his sister Alida, Fort Duncan, Texas, October 31, 1852[115]

Dear Alida,

I have to acknowledge the receipt of yours of a very late date via Santa Fée and also Sarah's of the 3rd Oct. which came by the last mail it seems the weekly correspondence is fast drop[p]ing off for several mails have arrived with no letter for me and the Major has not received one from Mrs. Emory for some mails back.

We are now in a doubtful state no money no credit and the Commissioners gone to Comargo [Camargo] or some other out of the way place—God knows what will become of the survey the Major has more patience than I would have under the circumstances he still hopes on but I fear Bartlett has the up[p]er hand or at least his is backed by the [Interior] Dept:—I gave Mr. Wilbur[116] a letter of introduction to Father he you will find quite a clever gentleman but I fear he will not do well at the Dept. for he is too quick. I hope father will aid him in all he asks. Mr. [Nathaniel] Michler I suppose you have seen ere this he promised me he would call on the family—Ingraham is in New Orleans by this time and I fear you will not succeed well as Sarah says you are going to set your cap for him altho' not doubting your good looks I don't think you have any dinero [*dinero,* or money] to back it which is the substantial attraction now a days—He left me his horse a most splendid animal to take care of and I am annoyed to death about him, some rascal has given him a blow near the eye and I fear he will lose it, however he has all my attention if it will be of any service to him. It is a great pleasure to ride him for his head and tail are as high as possible yesterday evening he came near putting me on the ground by taking [me] by surprise.[117]

I wish you would tell Father I send him by Mr. Wilbur a specimen of silver ore taken from the San Pedro mines about 25 miles NE of Coralitas [Corralitos] in Chihuahua by Ingraham and also an other with a letter to Johnny Lyle for Bery

Hallowel [?],[118] you must tell Johnny I have written several letters to him but receive no answers and the same to John McCalla. I send also a letter once every week but am sorry to say do not receive from there every week. I don't mean to complain for I know altho' Sarah says it is very dull in Washington that the society there is better than the correspondence of one in the wilds of America. Write often & believe me in hast your affectionate brother.

G. Clint Gardner

GCG to his father, Camp near Fort Duncan on the road to Laredo, November 12, 1852[119]

My dear Father,

I received yesterday your letter of Oct. 13th and Sarah's of Oct. 17th and *the bcases[?] of despatches*. Geo. F. Bas[tle?] also arrived after his baggage which came the day before. We are in a quien sabé [*¿quién sabe?* or uncertain] state at present but as we were on our way to Laredo we will not turn back it being just as easy to return home by that route as any other. The work could have been finished in a short time and perhaps would have been if there had have [*sic*] been any authority given to act upon but without authority nothing can be done no one is certain fool enough to keep some hundred men employed without authority for so doing altho' they could be keep [kept] if one would take upon himself the authority.

I regret exceedingly that my pay accounts have been stopped for I fear Ingraham will be in want at New Orleans but however this is a thing that can't be helped. As for myself I don't think I will return home for some time at any rate as I don't intend to quit the work if I can avoid it for there is no object in my coming to Washington without employment and it has been for these reasons I urged my application for a commission which I hope you have not dropped.

As yet the major has not decided what to do and how we are to get home. I am of the opinion that I will remain in charge of the Inst. [instruments] rather than go begging home. I have seen enough of the Cities without money. I do not desire to see any more of them yet. I do not fear for a minute I could be placed in a position of actual want. Do not suspend the correspondence for I am not off yet.

Genl. Smith[120] was here the other day and I had the pleasure of dining with him at Col. Morris's quarters rather a shabby suit buckskin pants and a borrowed coat to visit a lady but however in this country such things are not spoken of often. I must close as we make a march to day of some twenty miles.

Your affectionate son

G. Clint. Gardner

Give my love to *all.*

GCG to his sister Alida, Ringgold Barracks, Texas, November 28, 1852[121]

Dear Alida,

Here we are this far on our way down the river and as the Commissioner has not arrived here yet we will be detained here some time—I am sorry to here [hear] from your last that you have given up writing to me with expectation of my coming home I fear that there is no such good luck as that or at any rate I hope it will not happen until there are some funds for I dislike to travel with out sufficient to see every thing as I go along.

I did intend going out to see Emma [Mouton] but if we go home flat broke there is no chance for I certainly will not start out for enjoyment on borrowed capital so you can tell her in your next letter for as she never honors me with a letter of late I don't trouble myself much and as for Frank,[122] he is without pens or some thing else for he never writes and Mathilde has too many cares on her mind so now my correspondence has dwindled down to nothing since you all have stopped writing home and therefore in consequence of the same this will be my last for some time until I hear from you all again. I have now arrived in a civilized country w[h]ere steamboats[123] come and go and I can easily say "no le asi"[124] write if you like but if you don't I won't.

This post is garrisoned by two companies of the seventh Inft. [infantry] and when I arrived here the first question asked me was when I heard from my brother [Frank] and I had to reply that I had not seen his writing for three years they told me he had been in the habit of writing the news concerning the regiment but that now he never wrote—I suppose his family requires all his attention if so thank God I ain't married and I am sure never will be under those circumstances.

We are now living in fine quarters which Major [G. R.] Paul[125] has been kind enough to give us and I anticipate spending quite a pleasant month Lt. [Edward Augustus] Palfrey[126] is here also he and the Major are the only officers at the Post do you remember the latter? he was in Washington when Alfred was on furlough from the Point and was one of his intimate friends but has not heard from him since he left the Point.[127] You can tell them of it when you write there again.

You must remember me to all the young ladies but don't flatter yourself with the idea of seeing me soon for I certainly now have not the anticipation of seeing Genl. [Franklin] Pierce inaugurated[128] and perhaps will not be in Washington until father gets me my commission in the Army so if you have any desire to see me you had better push him to get it soon.

We have left the numerous [Boundary] Commission all along the road and Major E[mory], Mr. [John Henry] Clark, Mr. [James H.] Houston and myself are the only ones that have arrived at the head of navigation—I have of late been doing all kinds of duty Secretary, Assistant, QrMaster, Comsy: &c. so my positions now are quite numerous but I only rise to fall further.

I feel now the want of my coat & pants for I have nothing decent to put on but I suppose they are waiting my arrival in New Orleans and at present I am unable to give Mr. Taylor [in New Orleans] any directions as I would write further. I hope in making them there was no allowance made for growth for I have been compelled to make a gues[s] at what those boats[boots] cost me and give them to [John Henry] Clark they being so large I could not walk in them. You all must have thought my *understanding* great.

Ingraham has gone to New Orleans and I am afraid will bless me for not having known sufficient about my pay accts but that can't be helped poor fellow I know now pity him [*sic*].

Give my love to *all* the family and believe as ever you affection brother

G. Clinton Gardner

A notation by Sarah, written upside down on the last page of this letter

Dear Father—I enclose you the letters received *to-day* from Clint & Emma. We all wonder very much that we do not hear from *you*. I suppose you celebrated yesterday at Tammany Hall.[129] We shall expect you home next week. How did you spend New

Years day? Did you call upon Mrs. Motley? Who is her husband? All join me in much love. Let us hear from you soon.

Your affec daughter

Sarah

GCG to his father, Ringgold Barracks, Texas, November 28, 1852

Dear Father

We have just heard the result of the election not very correct I don't suppose but they say Pierce is elected by a majority of four to one[130] so I congratulate you on the success of your party and hope you have not let your friendship for [Winfield] Scott influence you in taking a part in the different meetings &c. I anticipate now with a democratic congress and president of holding thro' your influence the Commission which I have so constantly urged you seem to think that it is best for me to remain in civil life but the reasons I have for a change comes more and more stronger for should the sudden stoppage of this work take effect I would ask what I could do in Washington come and live off of my relatives and friends I think [that] is the only answer as for taking a part in the political world it is a thing I never intend doing my only ambition being to get in some permanent place where I will have a chance of travelling over some of our extensive country about which little is known at present. It is also my opinion that ere long there will be again a war for how we can get over taking Cuba I can't see but however should that not bring it on there is such a feeling on this frontier not on this side of the line but in Mexico that will eventually result in drawing our people into action. The State of Chihuahua is now verging on a revolution and if it does come to it it will result in the fall of the Mexican republic which in all probability will effect this portion of our country.[131]

I regret to see the work at this advanced state drawn to a close without being finished. I hope it will not be done for it will cost the government more by twice to resume it than it will to finish it now and as for the appropriation that would make but little difference for there are none of the working men that are not willing to remain on the work and wait for their pay. There are also persons all along the line who will willingly credit the Commission but there is no one to take the responsibility since the Secretary gives all the power to the Commissioner and he is continually on a pleasure trip thro' Mexico.

I see that [Thomas] Rusk has said in his speach that the Com[mission]er was bribed but I hardly think that yet I don't conceive it is too great a crime for him.[132] The initial point as established is certainly wrong but how it can be rem[e]died I can't see for the object of placing the line to the south is to gain the road across the continent which we cannot get even by the treaty line and as for the loss of land it is of no value so the only way I see is for the government to take the road by war.[133]

I wish you would write me by every mail for I think there is no chance of my coming home yet and send your letter via Brazos Santiago so they will come up the Rio Grande. Give my love to all and tell Mother I think my coat and pants will be rather old when I get them but I hope they will not spoil.

With Much love I remain

Your affectionate Son

G. Clinton Gardner

John H. Ingle to GCG, Alexandria, Virginia, December 3, 1852

Dear Clint,

Your letter, or rather note was recd. last night from home, together with the specimen of silver ore. The former I was glad to receive and the latter I delivered as you requested.[134]

When I told him that you had sent it, manufacturing in your behalf a very pretty speech, he expressed himself to be very much delighted at your remembrance of him, and said that you had sat next to where I now am established, and that you "were a great comfort to him." I can't exactly tell what he meant by the word "comfort" *but perhaps you can. He desired to be remembered to you when I wrote, and sent his thanks &c.*

I have written you more than two letters certainly, and you count such little notes as the last letters, but would be quite offended at one of that sort from me. So my dear fellow you must not quarrel, really as though I confess to a part of the charge, yet you have not much right to say anything.

You may have heard of a certain Lieut. Cash, who used to hang around our ranch, pretty considerably well, he succeeded in carrying off Molly last Wednesday—that is, at twelve o'clock on that day they were married. They left for Philadelphia in [the] evening train where they will remain till Christmas and then come back to Wash[ington] to live.

What do you think of the election? Ever in your life see such a defeat as we *poor whigs have suffered? routed entirely, put to flight—horse, foot, and dragoons. So far as you are from the seat of war, I suppose you don't hear much of politics. But the whigs in Washington look as if they were going to be hung, they certainly will have their heads cut off. The loco's look as if there wasn't room for them to walk, and are as proud as peacocks.*

There is a talk of your party being recalled, is it a fact? or will Major E[mory] follow the illustrious Jackson and take the "Responsibility" upon himself.

I should like very much to see you. I have some curiosity to see how you look after four year[s] campaigning. Couldn't you send me your portrait: take it yourself and forward it by the next Steamer.

I was much obliged to you for your advice, which proved of much service to me. All your family are well I believe and all your friends.

Joe Adams sent his love to you which I was to deliver the first time I wrote.

I remain,

Your Sincere friend,

J. H. Ingle

GCG to his sister Sarah, Ringgold Barracks, Texas, December 11, 1852

Dear Sarah

I have but a few moments to write and remind you all that I am still in the land of the living enjoying fine health and spirits. We are still in that state of uncertainty without knowing when or where our next move will be, but as long as we remain here I anticipate comfort; finding all the officers very agreeable indeed and disposed to do every thing to make themselves agreeable. Major Paul, of Frank's regiment, is in command and Lieut. Palfrey, who if you remember was Alfred's intimate friend at West Point, is the only Officer with him. We have an arrival all most every day which brings some agreeable personages. Day before yesterday Col. [William Wing] Loring of the Rifle regt. arrived with Capt. [William R.] Palmer his Adjt. on a tour of inspection he being the commander of the frontier posts. I have no doubt you remember Capt. P—for when I was introduced to him he named over the whole family and reminded me of the times when I went to dancing school with my "twin brother Charley" but I soon told him the latter had been left far behind he being still quite small but "ould" more so than I. However the Capt. said that *he* was not old yet and when he met me last I was quite a shaver.[135]

The Steamboat leaves in a few moments so I will close this uninteresting scrawl to be resumed some future day. I can now take the liberty of writing short letters as I never receive any.

Give my love to all
Your affectionate
brother,
Clint.
Direct your letters via Brazos Santiago. G. C. G

---

1. Draft, Emory to A. H. H. Stuart, Secretary of the Interior, San Elizario, June 10, 1852, and Michler to Emory, Eagle Pass, June 28, 1852, both in box 5, f. 40, EP.
2. Draft, Emory to Texas congressman Volney Howard, June 1, 1850, box 5, f. 40, EP. Norris, Milligan, and Faulk (1998, 128) reproduce part of this quotation but with an error.
3. GCG to his sister Alida, Camp on the Ciballo River while on route to Presidio del Norte, July 6, 1852. Duff Green noted the date of this episode (Tyler, 1968, 50).
4. At today's Ojinaga. Although Mexicans and Americans alike think of the Conchos as flowing into the Rio Grande, "historically, geographically and hydrologically" the Rio Grande flows into the Conchos, and it is the Conchos that flows into the Gulf (Morgenthaler, 2007, 9).
5. Emory, 1857–59, 1:85.
6. Ibid., 1:12.
7. Norris, Milligan, and Faulk, 1998, 125; Emory to A. H. H. Stuart, Frontera near El Paso, May 15, 1852, box 5, f. 39, EP.
8. In a letter to Volney Howard, Emory explains that he "put off Salazar until the 28th of August hoping to hear something or get some shadow of authority for reopening the business at that time. But Mr. Bartlett never made his appearance" (draft, Camp near Fort Duncan, November 7, 1852, box 6, f. 45, EP).
9. Emory repeated this language in several communications as he sought to explain his position; see, for example, Emory to Capt. George Thom, Ringgold Barracks, July 7, 1853, box 6, f. 53, EP. Werne offers the clearest account of this episode (2007, 128). See too Hewitt (1991, 569). In August, Salazar wrote several letters to Emory asking for a clear answer and a meeting (box 5, f. 42, EP). Goetzmann (1991, 183–84) cites the minutes of the meetings between Emory and Salazar, August 26–27, 1852, that he found in Emory's papers. This document was missing or misfiled when David Weber searched for it in January 2008.

10. Emory, 1856–57, 1:17.
11. Draft, Emory to A. H. H. Stuart, San Elizario, June 10, 1852, box 5, f. 40, EP.
12. The topographical engineer, Lt. Col. Joseph E. Johnston, had been around the edges of the big bend in 1850, but his party did not penetrate its canyons (Tyler, 1975, 75–80). In June 1850 William F. Smith had headed down the Rio Grande from El Paso in boats that sank; he stopped at Presidio del Norte (Tyler, 1968, 46). As early as mid-May 1852 Emory had learned that beyond Presidio del Norte the trail was so bad that mules could not navigate it (Von Hippel to Emory, Camp near Presidio del Norte, May 15, 1852, box 5, f. 39, EP). Emory also had a map by Emelio Langberg that revealed the difficulty of the terrain (Tyler, 1975, 82).
13. The group, led by Thomas Thompson, left on June 11 and returned on the sixteenth (copy, Thomas Thompson to von Hippel, Presidio del Norte, June 16, 1852, box 5, f. 40, EP).
14. Draft, Emory to A. H. H. Stuart, Camp opposite Presidio del Norte, July 23, 1853, box 5, f. 41, EP.
15. Michler to Emory, Eagle Pass, June 28, 1852, responding to Emory's orders of April 19, and letters of May 25 and June 2, all in box 5, f. 40, EP.
16. Draft, Emory to von Hippel, Camp Sumner Opposite Presidio del Norte, July 19, 1852, box 5, f. 41, EP.
17. Von Hippel to Emory, Camp near Presidio del Norte, July 21 and 22, 1852, and Emory to von Hippel, July 23, 1852, all in box 5, f. 41, EP.
18. See chap. 6, note 79 for more information about Chandler.
19. GCG to his sister Sarah, Camp Sumner opposite Presidio del Norte, July 23, 1852. See too GCG to his sister Alida, Camp on the Rio Grande above Presidio del Norte, July 29, 1852, and GCG to his sister Alida, Camp, August 9, 1852.
20. Two letters from GCG to Emory, Camp on the Rio Grande, July 27, 1852, box 5, f. 41, EP.
21. Draft, Emory to Chandler, August 8, 1852, box 5, f. 42, EP. Parry was to accompany Chandler, at least part of the way, until Parry had acquired all the botanical information he desired.
22. GCG to his sister Alida, Camp, August 9, 1852.
23. Draft, Emory to Chandler, August 8, 1852, box 5, f. 42, EP.
24. GCG to his mother, Camp Opposite Presidio del Norte, Mexico, August 13, 1852.
25. Chandler to Emory, Fort Duncan, November 25, 1852, box 6, f. 45, EP. Chandler set out on September 12 and abandoned the project around November 1, at Boquillas Canyon. Tyler (1975, 80–97) and Werne (2007, 118–22) offer overviews. Duff Green's account is indispensable; Tyler (1968, 43–60) reproduces Green's map of the detour through northern Mexico (facing p. 49). Green led a dozen soldiers to supply Chandler. Chandler's understated report, dated December 1, 1852, from Fort Duncan, is in Emory (1857–59, 1:80–85).
26. F. W. Tansill wrote to Emory from Eagle Pass on November 14, 1852 (box 6, f. 45, EP) to report the arrival of the India-rubber boat and news that Chandler had halted the survey.

27. GCG to his sister Alida, Camp near Eagle Pass, Texas, September 30, 1852.
28. Emory's last letter from Fort Duncan bears the date November 12. On the thirteenth he wrote letters on the road to Laredo (box 6, f. 45, EP).
29. GCG to his sister Sarah, Fort Duncan, Eagle Pass, Texas, October 22, 1852.
30. Kerrigan (1998, 275–302) offers many complex, contradictory views of Eagle Pass.
31. GCG to his father, Fort Duncan, October 8, 1852; GCG to his sister Sarah, Fort Duncan, Eagle Pass, Texas, October 22, 1852.
32. Draft, Emory to A. H. H. Stuart, Camp near Fort Duncan, November 12, 1852, box 6, f. 45, EP.
33. Draft, Emory to Col. J. J. Abert, Camp near Fort Duncan, November 1, 1852, box 6, f. 45, EP.
34. On November 5, as Emory prepared to leave Fort Duncan, he received word from George Bartlett, the commissioner's brother, that he had dispatches from Washington. Emory waited until the eleventh for George Bartlett to bring the mail from San Antonio, and it contained bad news (draft, Emory to A. H. H. Stuart, Ringgold Barracks opposite Camargo, November 30, 1852, box 6, f. 45, EP).
35. Draft, Emory to F. W. Tansill (the quartermaster), Camp near Fort Duncan, November 11, 1852, and draft, Emory to A. H. H. Stuart, Camp near Fort Duncan, November 12, 1852, both in box 6, f. 45, EP.
36. List of amounts due to officers and employees of the U.S. boundary commission, F. W. Tansill, Eagle Pass, Texas, November 20, 1852, box 6, f. 45, EP. The commissioner's entourage included people like Charles Radziminski, who was not on Tansill's list.
37. Drafts, Emory to A. H. H. Stuart, Camp near Fort Duncan, November 12, 1852, box 6, f. 45, and Emory to J. J. Abert, Ringgold Barracks, December 1, 1852, box 6, f. 46, both in EP.
38. Emory, 1857–59, 1:12.
39. GCG to his father, Frontera, Texas, March 27, 1852.
40. Bartlett to Emory, El Paso del Norte, Texas, October 5, 1852, box 6, f. 44, EP. We have only this letter from Bartlett to Emory. Copies of Bartlett's other letters to Emory are not preserved in Bartlett's papers, which are at Brown University and also available on microfilm. Emory acknowledged receipt of an earlier letter that Bartlett wrote from El Paso and told Bartlett that he had "finished all my work to Presidio del Norte and started a survey party down the river to meet the one coming from Eagle Pass up." Emory's letter did not mention a plan to meet Bartlett or Bartlett's intention to meet him (Emory to Bartlett, On the road near the first crossing of the San Pedro River, September 18, 1852, Bartlett Papers, and box 6 f. 43, EP). Werne (2007, 129–30) makes a strong case that Bartlett's concern about Indians was a pretext for a pleasure trip.
41. GCG to his father, Ringgold Barracks, Texas, November 28, 1852.
42. Drafts, Emory to Bartlett, October 30, 1852, and Emory to George Whiting, chief clerk, Dept. of the Interior, both at Camp near Fort Duncan, both in box 6, f. 44, EP.

43. Built on the Leona River a mile south of present-day Uvalde, Fort Inge was established in 1849 as Camp Leona but was renamed Fort Inge later in the year; GCG used the two names interchangeably.
44. His sister Emma Gardner Mouton was married to the governor of Louisiana. The newborn niece was apparently named Marie. See the introduction.
45. Alexandre Mouton apparently planned to attend the Democratic national convention that met in Baltimore, June 1–5, 1852.
46. Former commissioner Weller was now a U.S. senator from California.
47. The well-connected Marine Chandler, whose father, Joseph R. Chandler, was a Whig member of the House of Representatives from Pennsylvania, had been in Washington on an errand for the commission that included visits to GCG's and Emory's homes (Chandler to Emory, April 28, 1852, box 5, f., 38, EP; GCG to his sister Sarah, San Elizario, Texas, May 30, June 2, and June 5, 1852).
48. For Brady see chap. 5, note 34.
49. Ingraham had been sent to the Gila in late January and returned to Frontera on April 12. See chap. 6. He would be sent to Fort Inge at Leona to take charge of the "depot," or storage facility, that Emory intended to establish there.
50. Fort Duncan, founded in 1849, on the U.S. side of the Rio Grande near present-day Eagle Pass.
51. Here GCG returns to a wish he had expressed earlier to his father. See GCG to his father, Frontera, Texas, March 27, 1852, and chap. 6, note 106.
52. A reference particularly to lieutenants William F. Smith and O. H. Tillinghast. See GCG to his father, Frontera, Texas, May 16 and May 18, 1852.
53. The *Providence Journal*, the newspaper in Bartlett's home town. Goetzmann (1991, 189) notes an undated clipping in the Bartlett Papers at Brown University.
54. The initial letter of this puzzling word is not clear, but the remaining letters, "eminst," are crystal clear.
55. This letter is in the Smithsonian Institution.
56. His sister Emma Mouton, who lived in Louisiana.
57. Hartmann (1993, 171–87) chronicles Wright's activities. Emory sent Wright back to Washington with botanical specimens for some of the leading scientists of the day—John Torrey at Columbia, Professor Spencer Baird at the Smithsonian, and Dr. George Engelmann in St. Louis—and the "skull of a Comanche Indian killed at the Escondido springs, in the fall of 1850" to deliver to Col. J. J. Abert, the head of the Corps of Topographical Corps (copy, Emory to Abert, San Elizario, June 1, 1852; Emory to Wright, San Elizario, June 8, 1852, box 5, f. 40, EP). Wright was to travel from San Elizario as far as Leona in a party headed by Ingraham but asked Emory for permission to collect botanical specimens along the way because "Mr.

Ingraham takes no pains to conceal from us, botanists [Drs. Charles Parry and John Bigelow], his contempt for our pursuits" (Wright to Emory, Camp near San Elizario, June 7, 1852, in box 5, f. 40, EP).

58. Snowden, whom GCG knew in Washington, had been in ill health since January, asked to go on leave in May, and resigned on July 8 (Charles Radziminski to Emory, Fort Fillmore, May 21, 1852, and copy, Emory to Snowden, July 11, 1852, box 5, f. 41, EP).
59. Emory left Radziminski in charge of a small group at El Paso, including George Garner at San Elizario (Emory to Garner, San Elizario, June 16, 1852, in box 5, f. 40, EP).
60. Moritz von Hippel, a German-born draftsman hired by Bartlett who stayed on to work with Emory as an assistant surveyor (Rebert, 2001, 25). For von Hippel's predicament, which was characteristic of other surveying parties as well, see Norris, Milligan, and Faulk (1998, 123–24; 128) and Werne (2007, 118–19). As GCG makes clear, Emory was aware that von Hippel had to stop work. His men lacked basics, such as shoes and clothing, and had neither money nor a source of supplies (Von Hippel to Emory, Presidio del Norte, June 4, 1852, box 5, f. 40, EP).
61. GCG wrote this letter, which is in the Smithsonian Institution, from Cibolo Creek, which flows southward into the Rio Grande at present-day Presidio, Texas. GCG describes the creek later in this letter. Emory apparently planned to follow the creek to Presidio.
62. Lt. Duff Green, head of the expedition's military escort. In his report Green noted that the flood occurred on June 23, endangering the mules. His trunk was rescued "only after all the contents were wet and filled with mud, [and] most of my Books and papers were destroyed" (Tyler, 1968, 50). Emory (1857–59, 1:92) also described this event, dating it June 25 and noting that GCG could not swim.
63. Perhaps he means postdiluvial.
64. In a September 21, 1852, letter to his sister Alida from an unnamed location, GCG refers to a John H. Lyle. We have not identified him.
65. Mathilde was the stepdaughter of GCG's sister Emma; Eunice was probably another member of Emma's Louisiana family.
66. GCG's cousins on the McLean side of the family lived in or near Newburgh and Little Britain, Orange County, New York.
67. On July 9 GCG filed his monthly report late with Emory; it said that he had been doing "my duties as Asst. & clerk up to the time of our departure from San Elezario [*sic*] and have also made notes of the route from the San Antonio road to this point" (GCG to Emory, Camp opposite Presidio del Norte, July 9, 1852, box 5, f. 41, EP).
68. Arriving at Presidio del Norte, Emory noted that he found "watermelons ripe and the corn in tassel" (1857–59, 85).
69. The camp was named for Edwin Voss Sumner, commander of the Military Department of New Mexico.

70. This could be a book by Laetitia Montague, *The Housewife. Being a Most Useful Assistant in All Domestic Concerns, Whether in a Town or Country Situation* (London: J. Dixwell, 1785?). This book contained recipes and hints for food preparation but also contained "Cures for most of the Ailments, Accidents, and Indispositions with which the human body is chiefly smitted." With this book one could "restrain Hunger, keener Thirst, and Direful pain," according to the title page. *The Housewife* reached GCG about August 20 (GCG to his sister Sarah, Camp on Las Moras Creek, September 21, 1852).
71. The property in Buffalo, New York, which is discussed in chap. 6, note 125.
72. Perhaps one of Emma and Alexandre Mouton's two boys, Charles A. or Paul, whom the 1850 census put at ages four and two, respectively (see the introduction). Alexandre Mouton apparently had white hair, the source of GCG's joke.
73. GCG is at von Hippel's former camp, upstream from Presidio del Norte.
74. This letter is in the Emory Papers, box 5, f. 42.
75. See note 70 for information about *The Housewife*. The donor was his cousin Ann McLean.
76. Justiniani, born in Jalapa in 1800, had a long military and political career in Chihuahua. A brief biography appears in *Diccionario Porrúa: Historia, biografía y geografía de México* (1971, 1:1138–39).
77. By this GCG seems to mean the Department of the Interior.
78. Franklin Pierce and William R. King, whom the Democrats put forth as their candidates for president and vice president in June.
79. Emory sent couriers into Mexico to search for Bartlett on several occasions. Emory apparently planned to send GCG to Chihuahua at this point, then changed his mind.
80. Just seventeen miles long, Las Moras Creek flows out of Las Moras Mountain near present-day Brackettville and runs southwest into the Rio Grande ("Las Moras Creek," *NHBT*, 4:92). The road to Fort Inge and San Antonio crossed the creek.
81. A southwest-flowing river that emptied into the Rio Grande (today, into the waters behind Amistad Dam), Devils River was popular with east-west travelers ("Devils River," *NHBT*, 2:612).
82. Emma Mouton's new daughter, apparently named Marie. GCG had learned of her birth in the spring (GCG to his sister Sarah, San Elizario, Texas, May 30, June 2, and June 5, 1852).
83. Pierce defeated Scott in a landslide. GCG's father was a lifelong Democrat and often received political appointments, so GCG's expression of political neutrality is curious.
84. Captain Prince was apparently William Edgar Prince, who served in Texas at this time and whose diary for 1850–61 is at the Beinecke Library at Yale (Heitman, 1965, 1:807–8).
85. *Unexplored* seems too strong a word, but it was a new route. Emory noted that on this occasion he "opened" a new road "from the Presidio del Norte to the Leon springs" (1857–59, 1:88).
86. We have not been able to identify John H. Lyle.

87. "The Doctor" was GCG's brother Jacob.
88. GCG wrote this undated letter from Las Moras Creek on September 21, the same date that he penned letters to Sarah and Alida to be sent by express to San Antonio. He notes in a letter of September 27 to his father that he sent the $300 "by the last mail" and mentions specifically in another letter to his father that he sent the $300 from Las Moras Creek (GCG to his father, Camp near Fort Duncan, Texas, September 17, 1852, and October 8 and 10, 1852).
89. We note GCG's interest in the Black Rock property, in Buffalo, New York, in chap. 6, note 125. Generously, he put the needs of his family first, offering to send them money rather than invest it in real estate.
90. Lt. Duff C. Green, born in Washington, D.C., in 1828, was just three years older than GCG and a West Point graduate. Tyler (1968, 47) contains a brief sketch. GCG makes clear that his family knew Green, who appears to have been close to Ingraham and GCG. In a brief note to Emory, Green expressed pleasure at Ingraham's return from the Pima villages and noted: "I shall be very much pleased to see Ingraham. Please present my compliments to him & Gardner" (Ingraham to Emory, Fort Fillmore, NM, April 17, 1852, box 5, f. 38, EP).
91. Col. Lewis S. Craig was murdered by two deserters in June 1852, as he led a small military escort from San Diego to the Gila with Commissioner Bartlett. For sources and the context see Faulk (1967, 97–98) and Werne (2007, 105).
92. GCG may have been sarcastic in referring to Chandler as "distinguished," since he once described him to his family as a "bag of gas" (GCG to his father, Frontera, Texas, February 6, 1852). See too GCG to his sister Sarah, San Elizario, Texas, May 30, 1852.
93. The port of Brazos Santiago on Brazos Island near the mouth of the Rio Grande.
94. From a biblical passage, Matthew 6:34: "Take not thought, therefore, for the morrow; for the morrow will take thought for itself. Sufficient for the day is the evil thereof" (*The New Testament . . . The Common English Version*, 2nd rev. ed. [New York: American Bible Union, 1866]).
95. Money GCG borrowed from Holley in Montgomery, Alabama (GCG to his father, n.p., March 1852). GCG's father had not repaid the debt, but GCG did (GCG to F. M. Holley, Washington City, March 9, 1853, which is reproduced in this chapter).
96. GCG's next-door neighbor. See chap. 2, note 45.
97. A. Austin Smith was a lawyer who lived in Washington (Hunter, 1853, 93).
98. GCG perhaps refers to an incident on September 23 when Chandler and Green encountered a group of Comanches crossing the Rio Grande en route to Mexico. Green gave them a cow, which he thought pleased them, but later learned that they had killed four Americans on the San Antonio Road (Tyler, 1968, 54–55).
99. Emory had no military escort so apparently sent some of his men to aid Chandler's party. Perhaps they did not reach them. In their reports neither Chandler nor Green mentioned their arrival.

100. Heitman (1965) shows no Morris with the rank of colonel at this time. This Morris may have been an honorific colonel or a member of the Texas militia. See, for example, Thomas Cutrer, "John Dabney Morris," *NHBT*, 4:844.
101. A number of institutions have issues of the *(Washington, D.C.) Union* for 1852, either in the original or on microfilm, but we have been unable to locate this letter to the editor.
102. José Salazar Ilarregui, like Emory, lacked money and an escort, and his survey had stalled below El Paso (Hewitt, 1991, 571–72; Werne, 2007, 126–27).
103. Augustus de Vaudricourt, an artist and draftsman whom Bartlett appointed to his team. Vaudricourt resigned, disgusted with Bartlett's disorganization, and found employment with the Mexican team, as GCG says (Werne, 2007, 58). Perhaps, however, Vaudricourt was not the "Melnotte," for he subsequently worked for Emory on the survey of Gadsden line in 1854. See too Taft (1953, 277–78).
104. In the GCG correspondence at DeGolyer library, the original letter is filed after one dated January 14, 1853, but it is clearly out of order. This undated letter was written shortly before Nathaniel Michler left for Washington on October 16. In it GCG told his father that he has written a letter to send by Michler in case he goes to Washington. A letter GCG wrote to his sister Alida on October 16, 1852, while he was still in camp near Fort Duncan, indicates Michler left for Washington that morning.
105. To keep the work going Emory had written drafts totaling $5,269 on the Department of the Interior. He had received authorization to do so. He learned on October 14 from San Antonio merchants Lewis and Grosebeck that the drafts had been protested in Washington on September 11. Furious, Emory dispatched Lieutenant Michler to Washington to deliver a letter to the secretary of the interior explaining that "we have not a cent on hand and until it is settled nothing can be done" (drafts, Emory to A. H. H. Stuart and Emory to Michler, Camp near Fort Duncan, Texas, October 15, 1852, both in box 6, f. 44, EP).
106. Emory had quarreled with Tillinghast. See chap. 6, note 132.
107. The traditional inaugural ball was not held in 1853. Franklin Pierce was mourning the recent death of his son and asked that it be cancelled.
108. By November 3 Ingraham was in New Orleans, where the Canal Bank also declined to cash Bartlett's check to Emory for $8,000 that Bartlett had written in El Paso on September 18 and that Emory had endorsed to the cashier of the Canal Bank (Ingraham to Emory, New Orleans, November 3, 1852, box 6, f. 45; draft, Emory to A. H. H. Stuart, Washington, February 3, 1853, box 6, f. 48, both in EP).
109. Malcolm Seaton was part of Commissioner Bartlett's party (Bartlett, 1854, 2:120). On returning to El Paso from San Diego with Bartlett, Seaton spent the month of September assisting Radziminski (M. Seaton to Emory, San Elizario, Texas, September 30, 1852, in box 6, f. 43, EP). Then, if GCG was right, Seaton left El Paso with Bartlett to join Emory.

110. The First Presbyterian Church was at 4½ Street, between C Street and Louisiana Avenue (Hunter, 1853, 19, 120). Old Joe and Nurse might have been black servants employed, or once employed, by the Gardners.

111. This is the only mention of Old Aunt Peggy in these letters. She too might have been a household servant, or former household servant, of the Gardners'.

112. A John Underwood, a claim agent, lived in Washington at this time, and at least five Coyles maintained residences there (Hunter, 1853, 102, 24). This is the only time an Underwood or a Coyle is mentioned in these letters.

113. Like many southerners, GCG hoped the United States would annex Cuba to prevent Britain from obtaining it and freeing its slaves. The Taylor-Fillmore administration had acted with restraint toward Cuba, but the Democrats, whose candidate for the presidency, Franklin Pierce, triumphed in the autumn of 1852, favored the acquisition of Cuba, and some were willing to acquire it by force if they could not do so by purchase. They imagined Cuba as a future slave state (Langley, 1968, 24–42).

114. Clement M. Butler was the rector of the Trinity Episcopal Church.

115. This letter is in the Smithsonian Institution.

116. A. P. Wilbur, an assistant in the U.S. commission whom Emory had ordered to report "without delay" to the secretary of the interior. Wilbur had apparently been part of Bartlett's entourage, but Bartlett fired him "in consequence of false and malicious statements made by him in [a] letter from him, which was read by Colonel Weller before the U.S. Senate, tending to injure my character and that of other officers in the commission." Bartlett sent Wilbur to Emory, who put him to work before sending Wilbur on to Washington with a letter to the secretary of the interior and to answer Bartlett's charge against him. Little wonder that GCG found Wilbur "clever" (Bartlett to Emory, El Paso, Texas, October 4, 1852; draft, Emory to Wilbur and Wilbur to Emory, Fort Duncan, October 27, 1852, both in box 6, f. 44, EP).

117. Ingraham's horse was neglected after GCG left him in the care of one Captain Granger. See E. E. to GCG, Fort Inge, Texas, February 16, 1853, and E. A. Palfrey to GCG, Ringgold Barracks, Texas, April 23, 1853.

118. We have not been able to identify a Barry Hallowell. The U.S. Census shows two Hallowells living in Alexandria, Virginia, both teachers.

119. This letter is in the Smithsonian Institution.

120. Persifor F. Smith, who had distinguished himself in the U.S.-Mexican War and served afterward as military governor of Mexico City, was head of the army's Pacific Division when GCG arrived in California in 1849 (*DAB*, 17:33; Berthold, 1937, 60–61).

121. Ringgold Barracks (later Fort Ringgold), about 225 miles from Eagle Pass and about one hundred miles from Brownsville, was established in 1848. It was adjacent to newly founded Rio Grande City and across the Rio Grande from Camargo. This letter is in the Smithsonian Institution.

122. GCG's brother Frank, who married Emma's stepdaughter Mathilde.
123. Steamboats brought passengers and cargo up the Rio Grande to this point, connecting Rio Grande City to New Orleans and other ports on the Gulf.
124. Perhaps *no le hace*, by which he means to say, "I won't do it."
125. Capt. Gabriel René Paul, who graduated from West Point in 1829, was the post commander. He was a captain in the Seventh Infantry in 1852 but had been brevetted as a major in 1847 for gallant and meritorious conduct at the Battle of Chapultepec (Heitman, 1965, 1:766; Paul to Emory, Headquarters, Ringgold Bks. Texas, December 3, 1852, box 6, f. 46, EP).
126. Palfrey, from Louisiana, graduated from the U.S. Military Academy at West Point on September 1, 1846 (Heitman, 1965, 1:766).
127. We cannot identify Alfred, who does not show up in any other letter.
128. That would occur on March 4, 1853.
129. Headquarters of the Democratic Party in New York City.
130. Pierce was elected on November 2. GCG has received an inflated report: Pierce won with 50 percent of the popular vote to Winfield Scott's 44 percent Pierce's lead in the electoral college vote was more commanding, 237 to 42, but the college did not meet until December.
131. Resistance to Governor José Cordero was growing that fall. On December 23 he was overthrown by Gen. Ángel Trías. What GCG characterized as a revolution was merely a barracks coup, and he overstated the consequences for Mexico (Almada, 1955, 238–39).
132. Texas senator Thomas Jefferson Rusk was the leading critic of Bartlett's compromise, arguing that it violated the Treaty of Guadalupe Hidalgo. Rusk's proviso that no money be spent on the survey until Bartlett's compromise was reversed was added to an appropriations bill in late August.
133. GCG shared a widespread view that the best route to the West Coast was south of the Gila in the area of Tucson, a part of northern Sonora the United States did not acquire from Mexico until 1853. Moving the boundary south to El Paso would not, in itself, open the way to the Pacific.
134. The silver ore went to Barry Hallowell (GCG to his sister Alida, Fort Duncan, Texas, October 31, 1852).
135. Captain Palmer, born in New Jersey, became a second lieutenant in the Corps of Topographical Engineers on July 7, 1838, when GCG was seven years old (Heitman, 1965, 1:768).

BROWNSVILLE — TEXAS.

From W. H. Emory, *Report on the United States and Mexican Boundary Survey* (1857–59).
Courtesy of the DeGolyer Library, Southern Methodist University, Dallas, Texas.

CHAPTER 8
Back to the Rio Grande, Then Home Again
January 2–November 18, 1853

On December 11, 1852, Clint told his sister Sarah: "We are still in that state of uncertainty without knowing when or where our next move will be." Nine days later the uncertainty ended with the arrival of Commissioner Bartlett and the discovery that he had no money to continue the work.

Bartlett had taken his time getting back to the Rio Grande. He left El Paso on October 8 after making lengthy preparations for his trip. Two and a half months later, on December 20, he reached Camargo and crossed over to Ringgold Barracks that evening.[1] By then Bartlett had learned that Congress had frozen his funds. Democrats in Congress, unhappy that the Bartlett–García Conde agreement seemed to cut off the only potential southern railroad route to the Pacific, encumbered the money appropriated for the survey until the New Mexico question could be resolved. Emory had probably received the first inklings of this development in late September.[2]

Bartlett had heard directly from Secretary of the Interior Alexander Stuart that Congress had appropriated $120,000 for the fiscal year but had added the proviso that no part of the appropriation could be used "until it shall be made satisfactorily to appear to the President of the United States that the southern boundary of New Mexico is not established by the commissioner and surveyor of the United States further north of the town called 'Paso' than the same is laid down in Disturnell's map."[3]

Stuart, who supported Bartlett's compromise with García Conde and, like other Whigs, hoped to see the Whig-appointed Bartlett succeed, assured him that Congress had not officially closed the boundary survey. Stuart did acknowledge that a shortage of funds would have the same effect. If Bartlett had remaining funds in his possession, however, Stuart urged him to continue surveying the lower Rio Grande since that was not a point of contention. Bartlett lacked the means to transport the members of his commission back to Washington, much less continue the survey. Without money the situation had become "impossible," as he told Emory.[4]

Following Emory's advice, Bartlett disbanded the commission, sold what he needed to pay off the men, and secured their transportation home. He stored the instruments for what seemed to him the likely event that funding would come through eventually and the survey would resume.[5] The men went home.

Clint accompanied Emory on a vessel that took them from Ringgold Barracks to Corpus Christi, which they reached on or before January 1, 1853.[6] From there Emory went ahead, sailing to New York by way of Indianola and New Orleans. He left Clint at Corpus Christi in charge of some of the commission's public property, with instructions to turn it over to Bartlett's designee.

January 11, 1853, found Clint in New Orleans, ordered by Emory to return to Washington but wondering if he would be sent back to the Rio Grande. By then he had learned that Congress had had a change of heart. On December 16 Congress authorized the use of funds to finish the survey of the Rio Grande below El Paso, but the news had not reached the commission until it was on its way home.[7] Later Bartlett would write: "The interruption of the survey delayed its completion about eight months, at an additional expense of fifty thousand dollars."[8] That was one of the few points on which he and Emory agreed. The real waste, Emory charged, came from Bartlett, who, by Emory's calculations, misspent $500,000 on needless employees and side trips.[9]

The fiasco gave Clint a chance to return home. He remained in Washington until late April when he left to resume the survey with Emory. Clint's extant letters tell us nothing about his months at home, but we know that he continued to work as Emory's clerk.[10]

By early May, Clint was back in New Orleans, having sailed there from New York with a stop in Havana, whose delights he recounted in a letter to Sarah.[11] The trip from New Orleans to the mouth of the Rio Grande on a small steamship, the *Yacht*, was less pleasant. The "small dirty and dangerous tub of a sea boat," as Clint described it, came close to crashing onto Brazos Island one frightening night.[12] Gardner and Emory believed the crew incompetent and the ship not seaworthy. Perhaps they were right. Later that year, Clint reported, the *Yacht* wrecked off of Brazos Island.

On May 16 Emory and Gardner, relieved to be alive, came ashore at Brazos Santiago, a busy port on Brazos Island where steamers like the *Yacht* transferred their cargo to shallow draft vessels that could negotiate the bars at the mouth of the Rio Grande

(Brazos Santiago disappeared in a hurricane in 1867 and was never rebuilt). Emory got his equipment and supplies onto the mainland and quickly put four crews to work to survey different sections of the river from Laredo to its mouth. Clint took his place alongside Lt. Nathaniel Michler, Arthur Schott, and Charles Radziminski as one of the four section leaders and, at $300 a month, one of the highest-paid civilian employees of the commission.[13]

Mouth of the Rio Bravo del Norte.

From W. H. Emory, *Report on the United States and Mexican Boundary Survey* (1857–59). Courtesy of the DeGolyer Library, Southern Methodist University, Dallas, Texas.

Clint's particular task was to build an observatory at the mouth of the Rio Grande to house a "Zenith Telescope and a large transit instrument."[14] Readings taken at this point, coordinated with those taken from a vessel sent by the superintendent of the coast survey, were essential to fixing the beginning point of the boundary on the Gulf of Mexico. Article 5 of the Treaty of Guadalupe Hidalgo had specified that the boundary line "shall commence in the Gulf of Mexico, three leagues from land, opposite the mouth of the Rio Grande, otherwise called the Rio Bravo del Norte, or of its deepest branch, if it should have more than one branch emptying directly into the sea."

After assisting Emory at Fort Ringgold in May and June, Clint began work on the observatory in early July and had it operational by the end of the month. Some of his letters to Emory describe the problems of building the observatory on sand and operating it on cloudy nights.

In early August long-delayed letters from his father reached Clint at the mouth of the Rio Grande. They would change the course of his career. His office-seeking father had received a position from the new Democratic administration of Franklin Pierce: surveyor general of Oregon. A political appointee like Weller and Bartlett, under whom Clint had worked, Charles Gardner needed experienced surveyors to work for him, and he strongly urged his son to join him. His father's importuning put Clint in a "quandary."[15] His letters suggest that he wished to work with his father, but he could not leave his post at the observatory until November at the earliest. Emory planned to return to the States and leave his former clerk to record lunations through November.[16] Clint believed this presented him with an opportunity to advance his career, and he also knew that Emory had no other assistant who could do the work. He decided to stay at his post, but he planned to join his father in Oregon early in the new year—and did. Charles Kitchel Gardner approved of his son's decision. Clint stayed at the observatory through the fall, taking lunations with the assistance of John Henry Clark and supervising his small crew.[17]

Emory had spent the month of August at the observatory taking his own readings, then left for Washington late in the month. Along the way he contracted yellow fever but survived the ordeal. He was lucky. New Orleans had been in the grip of an epidemic since June, and it had spread throughout the summer. By September yellow fever had reached Matamoros and Brownsville, seventeen miles up the Rio Grande from Clint's camp, where it lingered into the autumn. It sickened some members of the commission, including Radziminski and Malcolm Seaton, and killed one young lieutenant.[18] "The mortality at Fort Brown, Brownsville & Matamoras [Matamoros] has been and yet continues to be frightfull," Radziminski told Emory in late November.[19] Isolated at their camp at the river's mouth, Clint and his crew were spared. "Don't either of you come here. It is very sickly," Radziminski wrote from Brownsville to Gardner and his companion, the assistant astronomer James Hamilton Prioleau.[20]

As the surveying parties under Schott, Radziminski, and Michler finished their work that autumn, they followed Emory's instructions to leave their instruments and other equipment with Gardner. He, in turn, boxed up the instruments and shipped them back to Washington and sold other items at auction. By late November he had finished his work and left for New York either that month or in early December. He had planned to go by way of New Orleans to visit his sister Emma Mouton at her plantation, and perhaps he did. By December the yellow fever epidemic had begun to wane. It seems likely that he returned home through New York, where he would have picked up instruments he had shipped ahead and taken them to Washington.[21]

Fifty-four letters follow. Most are brief business letters. They begin with Clint heading back to Washington through Corpus Christi and New Orleans in January 1853. They continue through his return to Texas in May and end in November as he closed down his observatory and prepared to return to Washington.

Emory to GCG, Corpus Christi, Texas, January 2, 1853[22]

Sir,

You will take charge of the duplicates of the notes of the survey this day handed to you and after turning over all the public property now in your possession to whoever the U.S. Commissioner may elect to receive it you will proceed without delay to report to me in person at Washington City.

unsigned

GCG to his father, Corpus Christi, Texas, January 2, 1853[23]

Dear Father

I have arrived here this far on my way home and as the mail leaves to morrow hasten to give you all a few items of my prospects, &c. The Major leaves to morrow morning to go by land to Indianola and take steamer from there and I am to remain here to get rid of the property which has accumulated on my hands in the shape of mules wagons &c. in other words I have turned deputy Qr. Master or wagon master in General quien sabé [*¿quién sabe?*—who knows?] which.[24]

I expect to get off in about a week or ten days in a fishing boat to Indianola and there take the steamer for New Orleans. I regret I will not be able to visit Emma but my orders are to "proceed without delay" so I have to come. James Houston the nephew of your friend on the hill is my companion what a delightful time we will have.[25] I met Mr. [Forbes] Brit[t]on late of the 7th Inft. who offers me any thing from a shirt to a hay stack the latter I think I will accept as my mules have nothing to eat and out of respect to him I won't d—n the country for he is a very fine person and goes it strong for Texas.[26] I met several Officers of the Army all of whom treat me well as the brother of a brother Officer I begin to think Frank [Gardner] is quite a fine fellow altho' he never writes to me.

Give my love to all and soon I will meet with the fond embrace of a returned uncle brother son, &c., &c.

Your Affectionate

Son,

G . Clint. Gardner

Excuse haste.

GCG to his father, New Orleans, January 14, 1853

Dear Father,

I write you a few lines to keep you apprised of my whereabouts. I arrived here on the 11th and as yet do not know which way I am to go my orders say to Washington without delay, but I am awaiting the movements of Major E[mory] perhaps I will be sent back on the Rio Grande without coming to Washington, altho' possible yet hardly probable.

I will write and let you know if I return to the field of operations if not I will inform you in person to the contrary. I received my clothes and feel much indebted to you and Mother for your kindness having them made for me. They fit beautifully and [I] am making good use of them. Give my love to All and excuse the brevity of this.

Your affectionate son,

G. Clint. Gardner.

P.S. Ingraham has returned to the Brazos Santiago.[27] G.C.G.[28]

James Green to GCG, New York, January 18, 1853

Dear Sir,

I return your check for $84 on Bank of Washington in payment of my account against U.S. Boundary Commission—it wanting signature.

Yours,

Very respectfully,

James S. Green

Asst. U.S. Bound[ary] Commission

E. E. to GCG at Washington, Fort Inge, Texas, February 16, 1853[29]

My Dear Sir,

Agreeably [?] to your request I take this occasion to acknowledge the receipt of one horse sent by you in charge of my servant to our late camp El Salado Texas. The horse reached me safely & in good condition & at this time is looking well. A few days after we parted I returned to Ringgold Bks again where I met Capt. Smith U.S.A.[30] *who told me that he had an order for the horse from the owner a Wm. Ingram [Edward Ingraham] then in Brownsville who was desirous of having him sent down. I have not as yet had any opportunity of sending the horse to Capt. Smith but I am now under orders for Bellville or Redmond's Ranch & on my arrival I shall avail myself of the first opportunity to deliver the horse to Capt. Smith at Ringgold believing this will be in accordance with your wishes. I am daily expecting you back again since Congress has removed all impediments & directed the survey to be concluded. I hope to have the pleasure of soon meeting you & the major at my headqs at Bellville.*

There is nothing of interest among us here since you left some few Indian depredations & murders only

Remember me particularly to our mutual friend Major Emory & Believe me yours sincerely

[name?]

PS When you return I wish you would have the goodness if not too much trouble to bring me a thousand of Maynand's primers for the Sharp's Rifle I got from yourself. I find it a most splendid arm and in my opinion exceeds in utility all others I have ever seen.

E. E.

PS. 2d If there is anything I can do for you in this direction write me at Bellville, Texas & your wants will be promptly attended to.

Your sevte

E. E.

GCG to Francis M. Holley, Washington City, March 9, 1853

Francis M. Holley,

I send herewith enclosed a draft for *one hundred and seventy five dollars* made payable to your order which is the amount of the indebtedness of Ingraham and myself to you. The items are as follows:

viz. Amount loaned to both of us at Montgomery, Alabama in September, 1852	\$100.
Amount of voucher placed in Mr. Ingraham's hands for collection, which was lost	\$ 75
	\$175.

You will please acknowledge the receipt of this immediately as Mr. Ingraham is now here and we are desirous of having it settled.

Very Respectfully Yours &c.

G. Clinton Gardner

GCG to his father, New York, April 22, 1853

Dear Father

I arrived this morning safe with every thing and find my papers in my coat pocket which should have been sent in before I left, so I enclose them to you to deliver at the Dept. and say nothing about it for it was a neglect of duty on my part. I do not wish the Major to know I left in such haste.

I send also a check for the balance due me from the bank—I request that you will bid all good bye for me and dispose of the following funds for me as soon as you receive this all my bills I have left you to pay—they are all in your name except the one at Mr. Wheatley's.

Give Mother	\$5.00
Aunty	5.00
Rosa	5.00
Sarah	5.00
Alida	5.00
Amount brought up	\$25.00
Aunt Suckey	2.00
Stephney	1.00
Eliza	1.00

\$29.00 to purchase any little thing they may want as a present from Clint.

You may think I am extravigent [*sic*] but hope not as I think I should give all as s[t]ated and certainly not less.

I have seen Charlie & the Thompsons. I will write again before I leave in haste
Your affect. son,

G. Clint. Gardner

Direct to me at the Veranda, New Orleans. G.C.G.

Please send me a statement of our account to New Orleans.

E. A. Palfrey to GCG, Ringgold Barracks, Texas, April 23, 1853

Dear Gardner,

Your very welcome epistle of the 20th ultimo came safely to hand a few hours ago, so you will perceive that it was a little more than one month on the way.

I am happy to hear that you are enjoying yourself so much and wish that I was with you. We too have been having our little spree on the morning of the first we surrounded "Gomorrah" *and took the whole filibuster host consisting of some hundred men. We sifted out the Ringleaders handed them over to the tender mercies of the U.S. Marshal and they are now in prison to await their trial before the Federal Court which sits in June—We have thus purged our Sweet Village and it is now quite respectable to what it was some time ago.*

Remember me to [James] Houston. I am glad that he is so well and that his taste is so immeasurably improved. Tell him that I discharged that little obligation of his and be kind enough to hand him the enclosed receipt for the amount. It is a little dirty perhaps but has been in my pocket for one scout which accounts for it fully.

I am happy to hear my dear fellow that you are soon to be with us once again the time that the U.S.B.C's were at this post was decidedly the most pleasant I have passed here. Now we have three companies here—one of Rifles commanded by my classmate Selford whom you may remember having met on the Laredo road in French boots and a Standing collar.

Has Houston rung in any where in Washington this winter advise him by all means to quit the practice.

Remember me to Maj. Emory if you please. The instruments are all right I suppose as they have not been touched since he left. Say to him that if he *does not come out here that no one can have them unless at his order or that of the Secretary of the Interior as I promised him.*

Remember me to your kind parents, and also to your Sister. We heard from your brother, Frank, the other day he is well. Remember me to [John Henry] Clark[,][James H.] Houston and to all enquiring friends.

Yours Sincerely,

E. A. Palfrey

I am much obliged for your kind offer to take charge of any thing to me. I know now of nothing that they may want to send me from home. But do call, and they will always be happy to see you.

Mr. Ingraham's horse left by you in charge of Capt. Granger of the Rifle Regiment,[31] *arrived here the other day in very bad condition—broken down. Capt. Smith of our Regiment had an order for him from Mr. I[ngraham] but now refuses very properly to take him as he directs me to say, and also that he shall send him back by the first opportunity to Capt. Granger as he cannot receive him in his present condition.*

Yours —

E. A. P.

GCG to his sister Sarah, New Orleans, May 6, 1853

Dear Sarah

Mr. Durive [Frederick Durrive] having just told me that a friend of his was to leave for Washington this evening, I thought I would write you a few lines, and enclose a rose, which altho' not handsome, came from the most fairy like place I ever was in. that is the garden to the Bishop's Palace at Havana. We arrived in that sea port about 11 o'clock on the last day of last month, & being anxious to see all I could I visited the city, which was not a very great curiosity to me as I have seen other spanish cities, but the grounds in the vicinity were beautiful. We got a volante, or to make myself better understood a machine "commo esta" [*como está*—like this], with a big nigar on the smallest horse imaginable and drove to the Bishops palace some three or four miles. It is a large house, well guarded by soldiers and surrounded by the most handsome grounds through which runs a most beautiful st[r]eam, the commencement of which is a cascade of some twenty feet over an artificial cave about ten feet in diameter, in which I stood & imagined myself in some of [the] delightful places described in some fairy tale.

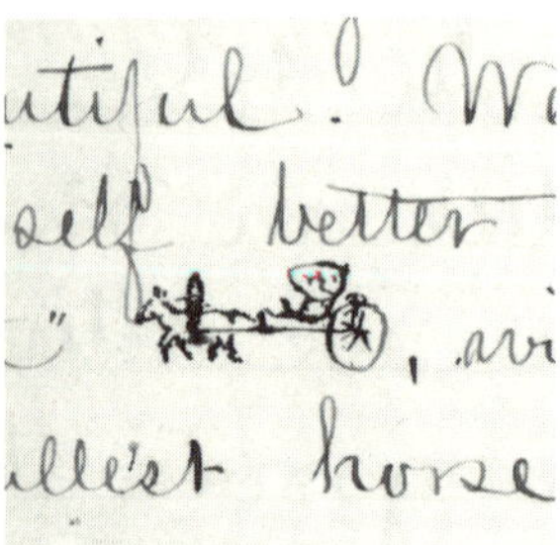
utiful! Wi
self better
", avi
llest horse

The rides in the vicinity of the city are beautiful and as we returned we met all the young ladies taking their evening drive and some most beautiful ones at that. Time was all I wished that I might make myself acquainted with some of the fairies that inhabit those roads. We went in the evening to the theatre and heard the opera, but we did not go there so much to hear that as to see the ladies in the dress circle after which we retired to one of the delightful groves to partake of some ices which they serve up so take my visit to Havana all together [and] it seem[s] more like one of those dreams described by the Opium eaters.

I have just left Uncle John who is writing a letter to Father I believe to send by mail. The Major has not yet arrived so I can't tell whether I will make Emma a visit or not. Do not fail to write letters to me and direct at this place. Give my love to all in haste yours truly,

G. Clint. Gardner

GCG to his sister Alida, Brazos Santiago, May 17, 1853

Dear Alida

I arrived here yesterday after rather an unpleasant trip from New Orleans on the "*favourite steamship Yatch* [*Yacht*]" a small dirty and dangerous tub of a sea boat.[32] We came off the bar on the morning of the 15th, but owing to the heavy sea on the bar and the want of water, we were compelled to lay out, and towards morning we found our-selves within a short distance of shore in the midst of the breakers, all hands being a sleep and the steam gone down, one minute more and we would have been wrecked on the beach, perhaps all our lives lost, delightful situation was it not, for one not prepared to leave this world of pleasure.[33]

I have been tempted to write an article for some of the papers to show the neglect of those in command of this steamer, as well as to warn the public in general of the danger in taking passage on such a boat, which has an undeserved credit of being a fine sea boat but to say the most of her, she is nothing more than a small river boat and now not fit for that, her age having rendered her incompitent [*sic*] to perform any trip without great danger to her passengers. However we have escaped from the brink of a watery grave being urged on by the strong hand of Messrs. Harris & Morgan while in their possession,[34] and once more tread the firm ground, altho' in [a] rather desolate place, which I hope to leave soon.

The Major will I expect go up to Brownsville[35] in the stage he being anxious to go on so as to be in time to have every thing ready by the time we arrive and we will go up the Rio Grande in the steamer *Grampus*,[36] with the chronometers and other instruments too delicate to take by land. You can tell Sarah that she may tell Mrs. Emory that the Major is enjoying good health and also Winder [Emory] altho' the latter was a little unwell on the steamer but now is quite a *tar*, so now with good luck we have nothing to fear.[37] The Major says he feels as if he had just been saved from some terable death & not unlike all of us feels very great full.

I anticipate receiving several letters upon my arrival at Ringgold Barracks from home and was a little disappointed at not getting one more before I left New Orleans You must write me all the news concerning the young ladies and remember me to them all. I often think of you all in the neighbourhood running from house to house and having a dance in [the] evenings and wish I could be with you all for a while again, but "variety is the spice of life" and I feel quite at home once more in the climate of Texas with the gentle sea brease blowing all day and this portion of it is new to me.

Tell Sarah she must remember me to Miss Bertha Macklin and write the city gos[s]ip all of which I take great interest in. Oh! I believe I never told you of a look, recognition and almost a bow from a young lady in Broadway, New York. Who it was I can't tell, but I think it was Mamie [Mary] Butterworth, she looked so handsome and I did make up my mind to go and call on her to see if it was her but sorry to say I forgot all about it, and the only reason it occurred to me now was as I came from dinner some one asked me if I did not dream last night about that steamer, when my dream was any thing else but that for, I thought I was seated alongside of one of the most beautiful young ladies I ever looked at in my life and was most desperately in love. It being mutual I was about to propose, when I asked the advice of a friend who recommended me to do so it being done I discovered a rival who became desperate and fired upon me and as I was about returning the fire I awoke to see nothing except my friend [John Henry] Clark who I was sleeping with on the floor of the Qr. Master's store house, but feeling quite relieved after the only night's rest since I left New Orleans.

I see my paper is drawing me to a close so I think I had better finish by giving my love to *all* and just leaving room to sign the name of your affectionate brother,

G. Clint. Gardner

GCG to his sister Sarah, Ringgold Barracks, Texas, June 18, 1853

Dear Sarah,

The mail came in today bringing your letter of May 29th, Alida's of May 24th, and Emma's of May 17th, all of which have been read with great pleasure. I was expecting to hear something of the anticipated war in some way or other perhaps, by receiving a commission to join some of the regiments. I hope father has not failed to apply for me for now is the time to look for such things. Santa Anna is collecting troops in all directions[38] and we learn by the newspapers that President Pearce has given Genl. [John] Garland instructions to occupy the disputed valley near El Paso, if so it is a declaration of War and we will get flog[g]ed in that quarter rather bad at first.[39]

I would be delighted to get a letter from Father given [*sic*] me the items of news about the Depts., for we are all ignorant of what is going on. I begin to feel a little pugnatious [pugnacious] now and if the war, which must eventually come, does commence before I leave the country I will be tempted to join any party that turns up.

We got all our news from the states, from a Delta,[40] a day or two before the mail got up and when the mail [arrived] it did not bring any thing later. I must make my letter short, for the mail leaves to morrow morning at day light so answers have to be written in a hurry.

I see you still remember that I said I was coming home in November, but after receiving your letter I told the Major that Mrs. E. expected him home in November and that you mentioned that of course I would be with him, to which he replied that he was going to leave me at the mouth of the river in charge of the observatory there, so of course, I prefer staying, altho' sorry to disappoint you.

You must remember me to all the young ladies of my acquaintance and tell Alida to give me some news of her friends and schoolmates. [Lt. Edward Augustus] Palfrey and myself have concluded to get married if suitable ladies can be found with incomes of six or eight thousand each. Should you hear of any such persons about just let us know and if there is any chance for ordinary looking men we will come on.

Give my love to Rosa and Effie and when I return I expect to find the latter quite a young lady. Remember me to all the neighbours and tell Mother not to let father loose a single chance to apply for my Commission for I see we must have a fight now.

Give my love to all and by the next mail I will write to Alida.

Your affectionate brother,

G. Clint. Gardner

GCG to his father, n.p., June 16, 1853

Dear Father,

I wrote to you & Mother yesterday but will repeat my request that you will get the mortgage and keep it for me.

You have never sent me word whether you sent in my papers to the Dept. or anything concerning them, I do not know if you received them, please tell Sarah or Alida concerning them and the other affairs I have about so as I may hear from you if you have not time to write yourself for I feel much interested about them.

You seem to forget the very items of news I most like to hear and therefore I expect you have forgotten the contents of all my previous letters which is an excuse for repeating my request so often made that is to get me in any additional regiment that may be raised.

I should like also to hear of your prospects as to getting a fat Office.[41]

Write to me once in a while and give me the news about the Depts. the prospects of a War for Cuba and what is to be done about this initial point & also concerning the railroads to the Pacific.

Believe me truly

Your affectionate son,

G. Clint. Gardner

GCG to Emory, Brownsville, Texas, June 29 and 30, 1853[42]

Dear Major

I arrived here on the morning of the 26[t]h with every thing in good order and have now every thing on board the *Grampus* to sail to morrow morning. I have purchased the lumber &c. necessary to erect the observatory but find great difficulty in obtaining the services of a carpenter. the only one who is willing to leave here is Smith. the man who wanted to go up the river with us, and he is on this court as one of the jury, however I think he will be released as the court is to adjourn to Galveston. There is one other who I could get but he wants his $3. per day and roast beef which I think rather extravigent [*sic*].

I have been in search of a house for you ever since I have been here but as yet am unable to find one[.] Capt. Ring tells me he thinks he can get one near the bank of the river, which I have asked him to do in case I do not succeed with one on Washington square that I have been after.

I take from here all the provisions left by Mr. [Charles] Radziminski but find that more have been sent him than his requisition calls for. Of some of the articles there is not sufficient so I have purchased some from the Comisy. Dept. and have now sufficient to last me one month after arriving at the mouth, and more, of a few articles of the ration.

I find the box which you supposed to be brandy contains a level for the Zenith Inst. I think, and supposing you will not want it, I take it with me.

June 30th

I have not been able to get you such a house as I think you want but I find there are a great many in town for rent, which you can get at any time but they are all on some of the main streets in the centre of the town and some distance from the river.

I have all the Insts [instruments] from the Qr. Master's store house on board of the *Grampus* and we expect to get off in a little while.

Remember me to Lt. Palfrey and the other officers also [John Henry] Clark and Winder [Emory].

I remain with respect

Yours &c.

G. Clinton Gardner

Col. Charles K. Gardner to GCG, General Land Office, Washington, June 30, 1853[43]

Dear Clinton:

I have found this morning, at the Dept. of the Interior—that in the Cabinet yesterday, the President had ordered my appointment as "Surveyor General of Oregon", which includes the territory of Washington. I have got this appt. on your account as well as my own. I can share with you the proceeds—at least guarantee to you $1200 salary, besides the *fees you get from surveys &c. I have agreed to it, as more desirable than any other appt. in this City, solely on the reliance I have in your assistance, for the knowledge I had of surveying at school, has been obliterated, long since.*

I wish you to inform me the earliest day you can join me at N. Orleans, on our way to Oregon by Nicaragua. Write immediately when you receive this. I shall write you again more particulars in a few days. I do not know what obligations you may be under, to finish any survey, or duty you may be engaged in, under Major Emory, but he will appreciate my wants and anxiety to have you immediately—by your appeal to him in my name. You might follow me, but I am most anxious to have you with me.

You can write to the N. Orleans *telegraph, if you appoint an early day. Affectionately, C.K. Gardner.*

GCG to his father, Mouth of the Rio Grande, July 3, 1853

Dear Father,

I have arrived here at last but I fear you all did not get my letter directing you to address my letters to the Brazos so I will have to wait for them to go to Ringgold Barracks and return. I am going over to the Brazos to morrow to spend the 4th and I am in hopes the steamer will get in while I am there if so I will then see if you all got my letters in time.

I am now very much in want of Lee's formulaes and tables which I asked you to get for me and I presume you have forgotten all about it. I would be very happy to have a copy if you can possible get one I would now willingly pay any price to have it. Please send me a copy as soon as you can.[44]

Have you ever got that mortgage yet from Uncle Wash if not please send after it for I think I would not let it lay there. I would like very much to know how much taxes are due on that property and all about it if you will give me all the information on the subject you can I would be very much obliged to you. The reason I am so anxious about it is because if anything should happen to you. I would be left in the lurch with nothing at all and it won't take you more than a half an hour to fix it all straight for me.[45]

Give my love to all and *please* answer this soon & oblige

Your affectionate son

G. Clinton Gardner

GCG to Emory, Mouth of the Rio Grande, July 3, 1853[46]

Dear Major

We arrived here on the 30th of June about dark and encamped for the night. After looking around for a suitable place to build the observatory I find that no other foundation than sand can be had, so I think had better place it on the hill. I have rented the house belonging to Mr. Clark it being the most convenient to the place selected.[47] I was under the impression that a better foundation could be had on the flat but after digging five and six feet thro' sand you came to water.

I have engaged four men and w[h]ile I think have to have one or two more when I get the observatory fairly commenced for the log at the Brazos is very long and heavy. There is but one log there that I think will answer for a stand. I will have to get one here for the Zenith inst. & one or two small ones for cross timbers in order to make the stands immoveable. By placing one or two cross timbers in the bottom of the pit for the log to stand on I don't think there will be any danger of its settling and a floor to the observatory will prevent the ground from being disturbed by walking.

I will have to work hard to get every thing fixed by the 14th which I believe is the first night for the next moon[.] The nights here seem to be quite cloudy I have not been able to see any stars except thro' clouds since I have been here.

I remain with respect

Yours &c.

G. Clinton Gardner

Col. Charles K. Gardner to GCG, Capitol Hill, Washington, July 5, 1853[48]

Dear Clinton,

We received, this morning, your letters of the 6th and 18th June—the last being a surprisingly short transit of 16 days. We get, however, letters from San Francisco, at N. York, in 18 days—the regular line of Steamer to California, via Panama, being well organized—by which I intend to go. I shall probably get a free passage. If you are not ready to join me, immediately, I shall be much disappointed: but, I cannot dispense with your assistance, immediately after my arrival in Oregon City. I shall want your aid in the first organization of the Office. Mr. John B. Preston, my predecessor, is considered in the General Land Office here, one of the best Surveyors General.

If you explain to Major Emory this necessity, from me, I am confident he will enable you to depart speedily. Charlie *will accompany me, who, you know, is raw in the business, and unacquainted with land surveying, except what he & you acquired at Nourse's school.*[49] *He stands well with Lt. [Maxwell] Woodhull,*[50] *who reported him to me as very bright and well informed. You must get all the pay you can, as I can get no advance, and no allowance for transportation to Oregon, and am, as yet, destitute of funds. So is Charlie I presume, and have also debts to arrange for—tho' my credit is good.*

I have to give two Bonds—as Surveyor General for $30,000, and as Receiver of public monies, for $50,000, required by the law of the last session.

If you can write an open letter to me, under cover to Fred. Durrive Esq., N. Orleans, stating when you can meet me there, *or perhaps come to New York (and pay a short visit to Washn. before you go to Oregon) and follow me in the next Steamer as you may arrange to do. Such a letter may be most expedient. I should prefer, still, to have you follow immediately, if you cannot join me. My calculation is to take the Steamer from N. York of the 5. August, which you might be able to join at N. Orleans, or Havana, but I would wait for the next regular trip, if I could be certain of your joining me in that Steamer.*

Your appointment as my Principal Deputy *in Oregon, is worth more than* five *young Lieutenant's commissions put together, and more than three* Captain's *commissions—all in time of peace. Your people out on the Rio Grande are entirely mistaken about the chances of a war—and about the instructions to B. [Brigadier] Gen. [John] Garland. The Govr has no such design—and Col. [James] Gadsden (who is here) will settle by negotiation with Santa Ana all that we want south of the Gila.*[51] *If it (a war) was ever so probable, there is* no law *for any additional regiment, but an effort at the last Session to reduce the Army. So that a commission in any levy would be but temporary, at all events. Discard the idea of a Lt's comn and adhere to your profession, which, in Oregon, in the post I can give you, besides its present value, will give you the opportunity of making a fortune to last your life.*

Do not allow anything to divert you from availing yourself of this fine opening!

The balance of your service in the Boundary comn. is of no value comparatively—tho' honorable enough. I should not be surprised if Frank *accepted my offer & left the Army!*[52]

Yrs. ever

C. K. Gardner

GCG to his sister Alida, Boca de Rio Grande, July 18, 1853

Dear Alida,

The steamer was due this morning from New Orleans but as yet has not arrived, and for fear she will leave the sooner for getting in late, I will commence this short epistle to remind you all that I am still in the land of the living, which you all seem to forget every now and then for your correspondence does not seem very regular. I hope you will not take this all to yourself, for I believe I have received more letters from you than Sarah yet I think there is room for improvement with both.

As for Father I fear he has given up ever again letting me see his hand writing, altho' I have written him several letters on business, which if of no importance to him, [is] of great to me; however perseverance conquers all things, so I will try it again and if I don't get any letters by the advent of the steamer I will write to him again.

What has become of John and Charlie.[53] I have not heard from either of them for some time. I should think Charlie might have written to me some time since as I left him several little things in New York to attend to for me.

We have now quite comfortable quarters here and altho' the thermometer stands above 90°—we live quite cool for the sea brease is almost always sufficient to keep off the mosquitos who are very annoying when the wind stops blowing. Our house is a two story frame with a long ten pin alley[54] attached which we have converted into a store room.

I have almost finished the erection of the observatory and am very desirous to have every thing fixed right before the Major gets here.[55] He is still at Ringgold Barracks with Genl. [Robert Blair] Campbell who from the last accounts I hear is sick, but however that won't detain the Major up there after he finishes his work.[56]

Sarah was right in supposing the Major will be in Washington by the last of November but I am fixed here until Jany. or February next, so the Major told me when I left Ringgold, for which I am not sorry as I am always willing to remain in the field as long as I have a party.

You must give my love to *all*, including your schoolmates.

Your Affectionate bro[ther],

G. Clinton Gardner

July 19th. Your letter of June 19th has just been received from Ringgold Barracks for which I am much obliged. G. C. G.

GCG to his sister Alida, Boca de Rio Grande, July 19, 1853

Dear Alida,

I have written a letter to you to send by this mail but after receiving your long and interesting one of June 19th, I will not let it go without expressing my great pleasure in being so agreeably disappointed at the receipt of it, for I was just coming to the conclusion, or rather had come to the conclusion, that I had been quite forgotten which I could not account for on your part, but fully expected it from Sarah, as she has her thoughts occupied by her little but interesting friend Mr. A—.[57] I am quite delighted also to hear that Father has brighter prospects now, for you all having remained so silent on that subject since I left home, I was really afraid he would succeed rather badly, and more so since I have seen by the papers that B. B. French had been appointed commissioner of public buildings, altho' I never was of the opinion he would get that place.[58] You do not mention which office it is that he is in hopes of getting. Uncle Clint I should think was depending on a broken reed if he waits for his beautiful and accomplished daughter to get his for him,[59] altho' I acknowledge that the fair sexes have an influence, not to be accounted for, in their application for a thing yet I should suppose she was rather unsophisticated and it would result like the many applications for office by the seekers expending all he had and having to retire rather disgusted.

I wish you would remind Father of my last letter to him asking for a book from the Topographical Bureau "Lee's Tables" and also keep him in mind of his good intentions concerning that mortgage for he is apt soon to forget and without you and Mother [to] remind him of it once in a while I fear he will not attend to it.

Tell Mother I am now quite comfortably fixed altho' many do laugh about the U.S.B.C. being quartered in a ten pin alley, but my opinion is it never was put to better use not meaning of course to say that it would not be put to better use, for me thinks I hear Aunty say "a ten pin alley, for the want of a better place, would do for a church," in which I agree with her, yet I won't acknowledge there would be more work done in it for we are all well occupied and even the molasses is following our example by commencing to work as I noticed today at dinner when I took my old fashioned des[s]ert bread and molasses.

I have just been thinking of getting a cow to add to our comfort but I fear she will not come up every night and I can't spare a man to look after her; then it is also very difficult to keep the milk in this warm climate however the clabor [clabber] is quite as good as the milk to me but I am afraid then the sugar will suffer for Aunt Suckey always said I eat it for the sugar.

You must congratulate the bride for me and tell her I glory in the spunk of one who will not be imposed on for I believe Grafton is quite a nice person and no one has a right to interfere without they can find out something traducing his character altho' I don't think he would have the same spunk, if one can judge from his actions at the commencement of the war.[60] By the by ask Father what is the prospects of another war if any. Or are we to let Cuba and the initial point of this Boundary stand without claiming and taking our rights.

I have noticed the Mexicans on this work who are stationed just opposite are rather cool, but perhaps it is from necessity, yet my opinion is they do not wish to keep up the sociability that has always existed, for when they first came down there I crossed and paid them a visit and they have not yet thought proper to return the compliment altho' several weeks have passed. But who cares, I would as leave fight them as any other of their countrymen at this pugnacious time.

You must remember me to all the neighbours and I will not trouble you to read any more of my nonsense. Give my love to *All* and return the same by writing often to your affectionate brother,

G. Clint. Gardner

P.S. I am in great want of that book please get Father to send it to me soon. G.C.G.

GCG to Emory, Boca de Rio Grande, July 27, 1853[61]

Dear Major

I have finished the observatory, and placed the transit on the meridian last evening with but little difficulty. I have now commenced to determine the metrimental [?] errors. The stands seem to remain perfectly firm and the small level with which I have tried them does not change its position, so I feel quite certain that the instruments will stand well.

It will be a great improvement to have the hill around the observatory covered with some clay, which I understand can be had a little distance up the river. The weeds around the house have been all worn off and as it is now the sand will blow in every direction so if you will let me have one or two wagons after your arrival in Brownsville I can soon have it fixed so as to prevent that, and will also make a walk from the house up to the observatory.

It was necessary to employ several laborers while building the observatory in order not to delay the mechanics all of whom I have now discharged except two and at the end of the month I will dispense with them and will then have but one cook and one attendant on the observatory, which I believe was your intention.

The coast survey has arrived at the Brazos and they have commenced their observations of the tide and weather.[62] Mr. Wurdeman who I believe is in charge of the meteorological party was over to see us and asked if we had an engineer's level, altho' he did not ask to borrow yet if he does I should like you to say whether I shall lend him one.[63]

Mr. [J. Hamilton] Prioleau wishes me to present his regards to you and together with myself to Mr. [John Henry] Clark & Winder [Emory].

With respect I remain your truly

G. Clinton Gardner

GCG to his sister Sarah, Boca de Rio Grande, July 30 and August 1, 1853

Dear Sarah,

It has been almost a month since we have received a mail so I anticipate receiving four upon the arrival of the next steamer, which we all look forward to with a great deal of pleasure. I have written one or two letters but am afraid they will not arrive at their destination before this one gets there as there has been no communication with this out of the way place of late except thro' the little vessels that leave once in a while. Our place of late has become quite a business place for within the last two or three days there has been [*sic*] three vessels here that discharged their freight a thing quite unusual since the war.

I was quite sorry to hear to day of the death of Thos. Jones one of the Asst. on the Survey who you perhaps remember about Washington the son of Genl. Walter Jones and the brother of Charles Lee Jones. He was drowned in the river just below Reynosa. The particulars of the case I have not heard except that they were returning from their work in the field in one of their boats under sail and it capsized. He attempted to swim ashore and went down the others held on the boat and were saved it seems from the account I had that the body had not been found yet. I expect it will be in a day or two. It has cast quite a gloom over us and I expect will distress his family in Washington a great deal. The last time I met him was as I came down here in the steamboat and he was in high spirits little supposing he was so near his end.[64]

The Major is expected here tomorrow, so I expect I will hear all from him as it has, of course, been reported to him.[65]

August 1st. The Steamer came in sight this morning and we are hourly expecting our letters from Brazos, and if I should be disappointed I will be tempted to burn this notwithstanding it has cost me a good deal of trouble to write it, for I have not been to bed since I commenced it & am now waiting to get my letters before going to sleep. I have been up all night observing and expect to be up again to night so if I cut this rather short, you can draw your conclusions as to the reason. I have this moment received Father's bond and will sign it and send it back by the return mail which takes this. And am delighted to see even in this way that Father has been provided for yet I hope soon to receive the letters intervening for I have not heard any thing about it which Father does not consider when he sends me the document without saying any thing of his prospects.

You must give my love to all and tell Mother I regret to hear of the Yellow fever in New Orleans and hope Uncle John will get off clear.

With much love for all, I remain

Your affectionate brother,

G. Clinton Gardner

GCG to his father, Mouth of Rio Grande, Near Brazos Santiago, August 1, 1853[66]

Dear Father

I enclose herewith your bond signed and witnessed, but do not understand your note. You say you have filed your bond for $30,000 and then in parentheses say that two are necessary, so I presume you intend this one to be the second[67]

The last paragraph of your note where you say that the Major will undoubtedly release me on the exposition of the necessity seems to imply that you wish me to go to New Orleans to sign the other yet I can hardly think that possible for it could be sent to me sooner at this place.

I am now here in charge of a party and the Major is not here[68] and will not be here until tomorrow or next day and I am confident it would be injurious both to myself, as well as the rapid conclusion of the work, to leave at this time, or I would send up immediately and ask leave of absence.

I am delighted to see you have been so well provided for and would like to know something of the manner in which you are going to do the duty and the extent of your powers. I have always had an idea of settling in Oregon and if you appoint your surveyors, I would make application if the pay will warrant my leaving my present position, which I consider more valuable now than it ever has been before, for I have now charge of an observatory.

I wrote in my last letter requesting you to send me a copy of Lee's tables and formulae which I hope you have done, yet you seem to think I am on my way to New Orleans which I do not understand without it is that you have ordered me home to go to Oregon with you.

I am in hopes there are some more letters for me which perhaps have been put in the Mail for Brownsville[,] if so I will get them in a day or two but perhaps not in time to answer by the return of this Steamer.

Give my love to all.

Your affectionate son

G. C. Gardner

Brownsville, August 2, 1853

Received, Fort Brown, Texas. Augt. 2d. 1853 of L. W. Emory, five sacks of oats, being for forage for six mules from Brownsville to Rancho Blanca and back again. J.H. Prioleau Asst. U.S.B.C. Brownsville—Augt. 2d. 1853.

GCG to his father, Boca del Rio Grande, Near Brazos Santiago, August 4, 5, and 9, 1853[69]

Dear Father,

Your letter of the 5th July 1853 has just come to hand and fully explains your note of the 14th which I answered by the last steamer. The Mail in which it came was sent from New Orleans in a sail vessal and has just arrived so you see I have not been aware of your possition until today and must confess I am in a great quandary as to what to do. The Major suggests a very good reason for my remaining several months and the case as it stands is this. I took charge of an Astronomical party to be stationed here and have commenced work by erecting an observatory and making transit observations and there is no one that the Major would place in charge of it except Mr. Clark who is assisting him and is to assist me when the Major leaves so to resign now would look rather ungrateful for the many kindnesses I have received at his hands so I have not urged the matter and am uncertain as to what I shall do.

The Major advises me to write to you and get you to write me after you get out there what instruments &c. will be necessary so as I can bring them out with me but altho' that may seem like good advise [*sic*], yet I do not feel like following it so I think I will have a talk with him & if I come, I will not return to Washington, but go on from New Orleans.

August 5th I have come to the conclusion that it will be better for me to wait three or four months before leaving the Commission for the following reasons. I at present do not know the duty of a Surveyor General nor his deputy and before coming to Oregon I should like to be prepared with the necessary instruments to make a survey such as may be required and at present do not know what kind of work I am to do and have not the funds to take me there and pay for such instruments as will be necessary so you see I am difficient [*sic*] in the most important item. Again, Genl. Campbell is up at Ringgold Barracks and has in his possession all my receipts for public monies which will have to be settled, for before I can draw what pay is due me and he is the only one who has money to pay with, the Major has not sufficient and it would be some time before I could get that settled and in the mean time I would be spending my funds by hurrying up there. Whereas by remaining here in charge of this observatory I will make myself more proficient in my profession and will have my name associated with the determination of this place which is now fairly under way and will be finished by the end of October and thereby become known in my profession, it will also give me a standing which I would not get otherwise. And at the end of that time I will have received letters from you informing me what is necessary to provide myself with and I can then join you with every thing that is necessary and without being stinted for want of funds. I could now borrow sufficient but I do not wish to do that for it will place a drawback upon me, which I could not get over for some time. After this work is finished I think I can also get a leave of one or two months which will pay a great deal towards taking me out which I could not do at this time. After the parties return to the states it will be, as it were, commencing again to begin the Office work and then I think no objection can be made to a leave of one or two months at the end of which time I will resign. It has been done before for others and I cant see why it will not be done for me.

So if you will write to me while on your way and after you get there and let me know the exact state of things I can provide myself accordingly and join you at any day you may name after the last of December. If I have to start without going to Washington, I will do so. I hope you understand and will think well of my conclusion for I assure you I have thought well of it ever since I received your letter and have come to this conclusion after taking the advise of all my friends who are on the ground and see exactly my possition.

The Majors advise is the same altho' he wishes me to state to you that he does not nor will not place any obstacle in way of my leaving in the next Steamer for New Orleans but thinks since I have commenced the determination of this point, I had better finish it and then join you with such instruments as will be necessary.

Tuesday, August 9th.

The Major does seem disposed to advise me to stay since you have written for me to come yet he still holds the opinion that I stated above and I am further convinced that is the correct one for I will not be able to join you as there is no conveyance from here before Thursday week by which this letter will go and the only advantage I could be to you would be in travelling out there with you, having been over the route twice I could releive [*sic*] you of any care and annoyance in travelling and now since there is no possibility of my being able to join you I think I may as well wait until the conclusion of this work which I have commenced. I therefore told Major E[mory] and Genl. Campbell that I would remain to take four lunation[s] and then return to Washington with the instruments in my charge. The Major will not remain here longer than the end of this month so you see I will have complete control and will have the credit of making all the observations after that time which will be of great advantage to me in a professional point of view.

I hope I will hear from you soon and it will give me great pleasure to hear that after this explanation that you agree with me entirely on the subject for I can't suppose for a moment that my absence from Oregon City for a few months could possibly make any great difference when by remaining I can accomplish so much here and in the end do more after I join you.

I hope you will write to me from Panama and while on your way up the coast or have Charlie write to me so as I may hear all about your trip. I will wait in anxiety to receive an answer to this.

Your affectionate son,

Clint.

GCG to his father, n.p., August 11, 1853[70]

My dear Father

Your letter of the 30th June came to hand yesterday so you see the last one of July 14th, arrived first and placed me in rather a quandary as to what you intended me to do.[71] My letter which will accompany this will tell you my decission on the subject and I have now taken charge of the observatory at this place and made my estimate for funds up to the end of the year, by which time I expect to be in Washington awaiting your orders. You wrote me you were to leave New York on the 5th of August, so it would be impossible for me to join you in New Orleans if I had decided to start in the first Steamer and as it will be better for me to go out there prepared to work in the field with all my instruments, I cannot suppose for a moment that you will differ with me in the opinions expressed in my accompanying letter.

I will by that time be able I hope to purchase and *pay for* all instruments necessary for my work in that country and after receiving one or two letters from you which I hope to do before that time, I can make my outfit complete.

The only advantage I can imagine it would be for me to join you immediately, would be to accompany you on your journey out there, as I have been over the road twice I could easily make the trip again without any inconvenience and be of great service in leading the way and showing you and Charlie across the Isthmus. I would warn you against the fruit of the country at this season of the year but fear you will not receive this before you are across the Isthmus, for no one can stand the climate who indulges in it, yet still I would advise you to live liberally for total abstinence is as bad as the other extreme and when you find one diet disagrees with you do not try it again for one can not experiment as easily there as in the states. The hot sun is also very injurious and very apt to give you the fever which is not easily gotten rid of.

Write to me often and tell me all the news, if, I receive any thing further from you by this coming steamer I will reply to it immediately.

Your affectionate son

G. Clint. Gardner

Col. Charles K. Gardner to GCG, Capitol Hill, Washington, August 13, 1853[72]

Dear Clinton,

I have just received, this evening, and am much obliged to you for your promptitude, the Bond with your signature—which enables me to perfect it with [Samuel] Butterworth's signature (for which he has given me his assurance). You have done this too, without having my previous letters to you: the first one announcing my appt. of the 30. June last—and two or three others, addressed to Ringgold Bar[racks] and one to San Antonio*—by way of an additional chance. And my very anxiety to get the information to you, seems to have paralysed [paralyzed] its expedition.*

You may well enquire, in answer to my short note of the 14th July, "of the manner in which I am going to do the duty, and the extent of my powers". I wanted you for my principal Deputy, and I spoke but slightly of your compensation, or fees, but I certainly can guarantee to you $2500 every year, and the probability of much more.

I do not know what instruction to give you, at this date, as you may have acted on some of my previous urgent letters, and you will know, better than I can instruct you, what course to take to effect the object desired. *I write this chiefly to apprize you, that I am making my preparations and expect to be ready, to embark on the* 5. August Steamer *from N. York by the Panama route. If this reaches you in the continued charge of* the

Observatory, *I suppose it will be out of your power to join me, unless you can sail direct from some Texas port to "Aspinwall" on the Isthmus.*[73]

Charlie will accompany me, and probably John[74] *(who writes from Greenbay that "I cannot go without you" as you must be my "right-hand-man"), he says, as he is not established in Wisconsin, he will go to Oregon, where, in fact, his knowledge of law & conveyancing may make his fortune.*

The powers of a Surveyor General you can learn from Major Emory or any of your elder associates or country-people. A copy of the last session's Laws, printed as "Statutes at large" which must have been sent to the Major, or to your Commissioner Gen. Campbell, will tell you—my increase of salary to 3500, &c &c.

But, I am in hopes that you have embarked, early in August, in some packet or Steamer to N. York—avoiding N. Orleans on account of its dreadful calamity[75]*—and will be ready to embark with me.*

Affectionately
C. K. Gardner
[to] G. Clinton Gardner Esq.

GCG to his father, n.p., August 1853[76]

Dear Father,

I have written you long letters explaining the reason why I did not resign and join you, which I have sent to Mr. [Frederick] Durrive in New Orleans but as I learn from Sarah's letter that you will not leave New York until the 20th of this month so I will repeat the main reason which is that I have not sufficient now due me to take me to Oregon City and purchase what instruments I will require to commence work and by the end of the year I hope to have enough without borrowing. I will also have to go to New York or Phila to make my purchases which will necessarily be of great expense to me, whereas if you leave on the 20th I can get letters from you after your arrival out there, giving me further information and telling me what I will require and can make all my arrangements while travelling at the expense of the Government.

I would also leave the Major at this time without any one to supply my place and as I came down here in charge of this party [I] do not wish to leave it knowing that it will be of great advantage to me professionally.

I would advise you if possible, to remain until my arrival in Washington or at any rate until cold weather sets in for it will be unpleasant to cross the isthmus at this season of the year.

Let me hear from you soon.

Your affectionate Son

G. Clint. Gardner.

Emory to GCG, Mouth of the Rio Bravo, August 20, 1853

Sir, I leave you in charge of the observatory to carry on the observations for Longitude through four lunations, and to observe for Magnetic Dip and intensity. The verbal instructions are so full and my confidence that you will carry them out fully is such that I do not think it necessary to repeat them in writing.

When you have concluded your work you will take all the instruments in the Depot here, and store them in New York with Messrs. E. & G. W. Blunt.

The Assts. will turn over to you all instruments as they may cease to have use for them. Have them carefully packed in boxes and directed to me. The Zenith Telescope of the Coast Survey, and the Gamboy dip Instrument take to Wurderman in Washington. Take the standard Barometer & the Thermometer, wet and dry bulb to Washington. The chronometers and their boxes to the office in Washington. Also all the books and stationery and drawing instruments. The original of your notes take with you. Send me by mail at the end of each lunation a copy of your observations.

In accordance with the desire of the Commissioner, you will send three wagons in charge of Thompson to him at San Antonio. Send at the same time all the tents and the best mess chest, and such arms as are not required.

I send a transfer list to Gen. Campbell and leave one with you. All the Assts. except Mr. Clark are paid up to Sept. 1st. Mr. Clark is paid up to July 1st. All the men except William, the cook, are paid up to August. William is paid up to July 1st.

The surveyors are ordered to turn over to you all their Instruments and other public property.[77] *All the property not otherwise named or disposed of you will unless otherwise instructed by the Commissioner, advertise and sell at public auction.*

I leave you the Ambulance four mules and the money. When done with them turn them over to the U.S. Commissioner, who has promised to retain them for further use. Write me by every opportunity.

Yours Very Truly,

W. H. Emory.

P.S. The only bill I know of unpaid is that of Conrad Bloom of twenty bushels of oats 40 [cents] = $8. Please pay it.

GCG to his father (telegram), New Orleans, August 22, 1853[78]

I cant join you until January 1st [18]54.

G. Clinton Gardner

Robert B. Campbell to GCG, Ringgold Barracks, August. 22, 1853

G. C. Gardner, Esq.

Sir,

Enclosed herewith please find a check upon Major James Robberts, for one thousand dollars, being a part of the estimate made by you and approved by Majr. Emory of the expences of your party.[79] *I would send a check for the amount you will need hereafter, but fearing you may have some difficulty in cashing them, I have determined to send the money to you hereafter. Please enclose me a receipt at San Antonio.*

Very respectfully,

Yr. Obt. St.

Robert B. Campbell, Com[missioner]

J. Mc. D. Taylor to GCG, New Orleans, August 23, 1853

Capt. G. Clinton Gardner

Dear Sir,

Your favor of 1st Inst. is at hand and in conformity thereto have this day Shipped to Major Emory the Wine & Butter desired, amt. $22.10 to your debit.

I send you also a duplicate of your former Bill amounting to $55.10, making together $77.20 against you which I hope you will find correct.

I am sorry to inform you that not only it is true what you hear and see about the Epidemic in New Orleans, but it is a great deal worse than the papers report it, the Misry and deprivation of hundreds of sick and dying at this moment is truly terrible. The few persons that remain in town have given up business and now occupy themselves visiting and savioring the sick. I presume that there is not any one person in the City who has interested themselves with the Sick that have less than 10 or 12 persons to look after, nurses are not to be had for love or money.

Uncle John is in Attakapers [Attakapas, Louisiana] and I have not heard from him for over a month past.[80] *All well.*

Yours Very truly,

J. Mc. D. Taylor

Sarah Gardner to GCG, Washington, August 25, 1853[81]

My dear Clint:

Father has just received your telegraph stating your inability to accompany him "until the 1st Jan." Father appeared much disappointed, and we are all *sorry you cannot go with him—but there is no evil so great there is not* some *good to come from it. I console myself, that you intend coming to me soon and hope soon to hear from you to that effect. We have a delightful arrangement for you, and for us, if you can only agree to it*—It is, *that you should come home with Col. Emory, or soon after as possible, make us a visit and I will promise you* all sorts *of frolics—and after the fun is all over you can,* like a good boy, *galland [gallant, or escort] your Marian & sister Daisie*[82] *to New Orleans, or Vermillionville [present-day Lafayette, Louisiana], where you can leave them, and as soon after as you choose, you can proceed to* Oregon *and, as Alida says,* "join your Dad" there. *Now* do *come and dont allow us to suffer from* another *disappointment. You are our last hope. We shall be entirely beauless if* you *dont come—Indeed I think it is* too bad *that I have to beg so hard for the folks to attend my wedding.*[83]

Father has written to Frank telling him "tis his *duty to come"; but I dont believe Frank cares enough for* me, *to put himself out of the way to oblige me.*

If you will tell me when you can *come I will arrange matters to oblige you.*

Father will sail certainly, on the 5th, next Monday week, with Charlie & John. We are all going on with him next Tuesday morning. Mother Alida & myself. John will follow in a day or two.

Frank is full of going to Oregon too. He has written urging Father for a situation out there and has serious notions, I believe, of resigning his place in the Army—talks of taking Mathilde with him. Mathilde is now visiting Emma, thinking it will be her last chance if Frank concludes to go to Oregon.[84] *I should not be surprised, before the expiration* *of one year the whole family was out there. Mother, Alida and all. I intend proposing it to Mr. A—*[85] *We will have a new settlement, a colony of our own, as Frank suggests. Father says fortunes have been made, by surveying at $4 a mile in Mo [Missouri]. Father allows $12. in O[regon]. What do you think of that? Mother says she wants you to locate for her, as you have been so lucky in the black Rock property. Mother hopes you have received all the information on* that *subject. You must not fail sending your* "permit"* *to Mother to draw $300 from the B. R. [Black Rock] Company—That is of great importance to me, as I expect it, to assist in my [wedding] arrangements for the fall.*

Write us positively, *when you are coming—Excuse scrawl—I write in great haste. Always direct your letters to Wash. Auntie & Doctor will be here to forward them to me, and we shall not be absent much over a month. With much love from all, I am, most affec. your sister*

Sarah

**Father says, it must be a "power of attorney"*

Rcvd Sept 17th

Sarah Gardner to GCG, Washington City, August 28, 1853[86]

My dear Clint:

Your budget of letters post-marked August 18th Brazos Santiago, *were* punctually *received* this *morning. They gave us much pleasure and we have all concluded you have the correct notion of Fathers going and take the* least selfish *view, of it, of any of the family. Father himself remarked your* letter was extremely sensible *and he does not at all blame you for not resigning your position at present—Father did not know when he accepted the Office that your duties were increased by the charge of an observatory. I imagine he can get along well enough without you until the winter, in fact, I doubt if your duties of surveying will commence much before the spring, though we should all have felt better contented to have you to go with him, as we should place more confidence in your judgment and* care of him.

As you can well imagine I am quite delighted that you intend coming home in December. I consider it a piece of good fortune for me, and truly thank you for the kindness expressed in your letter and the interest I am sure you take in what concerns most, my future welfare and happiness.

As for your response in regard to money matters, I thank you sincerely, *but do not wish you to run the least risk of putting yourself to any inconvenience—You will remember I only requested it as a* loan, *for I am sure Father will return it, and would give me now all that was necessary (which is all I request) if he could possibly afford it—it all counts, your kindness will ever be gratefully remembered by* me *and I am sure by each member of the family.*

It is but due to your generous, disinterested, spirit, to acknowledge, it is fully appreciated by all of us, *now at home—we have felt it from the time of Father's removal from Office until the present time, and you, I hope may be rewarded for it.*

I *consider that Father makes a great sacrifice in going, and have had but one opinion in regard to the matter all along; but do confess I have been somewhat surprised at opinions expressed by some of the other children—They do not appear all to have reflected, as you have done—*

However, we all feel quite encouraged now since it is all decided upon, and have determined to make the best of it and hope for the best. Father is in such fine spirits himself. If he gets safely across the Isthmus, it may be all for the best—He has written to Uncle John offering him business out there. I should not be surprised if I were the only monument of the Gardner race *left here, after a year or two. I wrote to Mr. John J [Almy]—begging* him *not to get the notion too, for I considered the folks at home* Oregon Mad.

I wonder if one friend Ned Kabondle wont want to go too. I expect Father will be glad to have him. I wish you would write him that I shall be much disappointed if he is not here *in December.*

You will I suppose, on your way to Oregon take Mother & Alida with you to visit Emma for a time. You are mistaken about my leaving the "family roof" quite so soon, on the contrary, I am to be left in charge of every thing and Auntie too, until next Spring when Mother returns home with Emma and all her babies.

I would write you if I thought it possible for you to accept, to pay me a visit next winter and see what a fine housekeeper I will make and what nice little dinners and evening parties I will give occasionally. I only wish you could be here.

Father, John & Charlie expect now positively to sail from N. York on Monday the 5th Sept. They have engaged state-rooms in the "Georgia"—Also Frank Wheatley & John Jones, as steerage passengers. Father has promised the two latter some occupation as laborers.[87]

Mother, Alida, & myself will accompany Father as far as N. York. We will all leave here and the Doctor too, I expect, next Wednesday morning, leaving Rosa & Auntie home to keep house during our absence—I suppose we shall visit Little Britain and Rahway and return home in about a month.[88]

Mother received your permit to draw the $300 upon the B. R. [Black Rock] property—If you were here she would go with you to see it and arrange matters there.

I shall be glad for your sake, if you have the opportunity of settling yourself with a nice accomplished little wife any where, but shall be sorry if you live such a distance as Oregon, as I shall not be able to visit you as often as I would like—I think the best plan for you would be to go out there and remain long enough to make a snug little fortune of your own, and then return to the neighborhood of these digings *[diggings] to enjoy it—Get married whenever* you can find *the person exactly suited to your taste; but not before, even though you should have to wait a* long time. *A year of* true *happiness is worth a life time of* uncertain *sort of bliss—so don't hurry the matter—let it take care of itself.*

Mother thinks, since you have decided to keep on, you had better not resign until after *you reach home, for perhaps Father may not require your services* immediately, *and you can stay here a little while doing Office duty—Write to us by every steamer. Your last came in ten days.*

All join me in much love
Ever affectionately yours,
Sarah.

GCG to Emory, Mouth of the Rio Grande, September 1 and 2, 1853[89]

Dear Major

I send by this mail under cover to the Topographical bureau the return of instruments with a receipt attached, and also a duplicate of the Transit observations up to the end of the last lunation, August 24th. The observations for time, which you will find without the numbers for reference, are taken from a small note book and were made while you were observing for latitude on the 9th and 10th of last month.

The *Yatch* [*Yacht*] has been off the bar ever since Monday morning but owing to the want of a pilot did not come in, and since Tuesday the wind has been quite strong from the North so much so that she can't get in and therefore the capt. will not send in the mail. The steamer *Comanche* arrived here on the 26th to repair and I hear from the capt. that Genl. Campbell has gone by land to San Antonio together with his son. I sent the wagons off the day after you left—via Ringgold Barracks thinking they would join the General before his departure, but now I fear they will not. My reason for sending them by Ringgold was because the teamsters said they could not go from here without I would allow them something to pay their expenses back to the Rio Grande, which I had no authority to do, so I compromised the matter by sending them that way so they could get their private horses, which Thompson told me they had at Rio Grande City. They will be but a day or two behind the Genl. and as his son is taking over a herd of sheep he will move slow so I expect the wagons will over take him.

The day after you left us I was compelled to discharge William as he refused to cook for the mess, and then I told James he would have to cook, which he also refused to do, so I paid them both off without allowing them any thing for their passage back to New Orleans as I think they left to take passage in the steamer Louisiana, which was off the bar at this place with troops. I have employed a cook and as Hopkins attends to all our wants in addition to the observatory, I have not employed any other man.

You will find enclosed a letter for you from Mr. Radziminski which I received the other day. I also received one from [J. Hamilton] Prioleau of the 22nd of last month and he says they will be at Brownsville in four weeks. I heard that [Malcolm] Seaton was at Ringgold, but I [have] not heard any thing from [Arthur] Schott.

Permit me to remind you of the Magnetic instructions you told me you would write after getting to Washington. I put up the Mag. Decer [magnetic declinometer] a few days ago but had to bring it in on account of severe blow which took the top off the tent, and the weather has been so bad since I have not put it out.

Sept. 2nd

We have just received the mail from the *Yatch* and I find but one letter for you, which is addressed to this place, so I have enclosed it with the one from Radziminski under cover to the Topographical bureau so as this may go direct to you in case of any accident.

I received a letter from New Orleans from which I learn the fever is much worse than the papers will admit, but I hope you will not meet it. I think by the time we get ready to move it will have passed off or at least I hope so, for I would like very much to make a visit of a day or two to my sisters plantation as I pass thro New Orleans, if you will give me permission so to do. I think I can arrange it so as it will not inconvenience the service without it is necessary that my arrival in Washington should not be delayed, if so I will not permit any thing if possible to deter me.

I fear Genl. Campbell has forgotten my requisition for funds as I have not yet heard from him, however I got some money from Clark to pay those fellows off and should any thing happen that will require more I can get my draft for $500. cashed, but in order to do, so, I will have to pay a discount of 2 per cent which I understand Genl. C— had to do upon his last draft.

We are all in good health and getting on finely. Clark and Winder wish me to present kind regards &c. and with many wishes for your safe arrival home I remain

Very truly yrs.

G. Clinton Gardner

GCG to his sister Alida, Mouth of the Rio Grande, September 2, 1853

Dear Alida,

I have been in anticipation of a letter from you ever since monday when I first got sight of the steamer off the bar at Brazos and road [*sic*] over there for the purpose of getting the mail sooner but was sadly disappointed as the pilots were unwilling to bring her in for some reason or other so they have detained the mail until this morning. I see with my glass that she has gone in so I have sent one of my men over in a hurry and am now waiting to see whether I am doomed to further disappointment at not receiving any thing, and for fear that I will be disgusted should that be the case I will commence this before hand, so you all will be certain of receiving a few lines at any rate.

I suppose ere this reaches you you will have seen the Major who left us more than a week ago for Washington. I gave him a letter for Mother which I am anxious to hear from so I hope some of you will write me as soon as it is received. You must return the Doctor my sincere thanks for some late Newspapers which I received by the last mail from him some were quite late indeed such as Sept. 1852. However they all contained little stories of which some were quite interesting.

I have never yet given you an idea of the house where we live in this delightful watering place which is by far the most comfortable house here with but one exception that is we run the risk of being found afloat some morning. When I first took the house we had about four or five feet of land between the porch and the bank of the river but now it is within a foot of the porch and seems to increase on us every day rapidly so should the U.S. Bdy. Commission be found off the Balize [Belize] in a two story house we will be fortunate in crossing the gulf without taking passage in the favourite steamer *Yacht*. Our house being two stories high we have the benefit of all the air in motion and are not much troubled with the mosquitos which are annoying every one both here and at Brazos. I have never suffered so much as I did at Brazos two nights ago when I staid [*sic*] there all night waiting for the mail. Altho' they have a constant brease there the cister[n]s are so close to the house that the mosquitos which they breed, are blown in the house and live on the inhabitants.

I have just received Father's letter of the 14th august, Postmarked at New Orleans, in which he says he is preparing to leave August 5th which I can't understand. I presume he must mean Sept. 5th. However my letters by the last mail will be recieved by him before he leaves and he will then know my decision on the subject. If I could have supposed for a moment that he would have been in Washington so long after he wrote me that he was to leave on the 5th August I would have been with him by this time and run the risk of the Yellow fever in New Orleans but I had not the remotest idea that I could have joined him. I hope tho' my delay until the first of the coming year will make but little difference and I think now since it can't be helped something like the old woman "every thing is for the best." I hope the Major will arrive in Washington before Father leaves so as he can tell him of the conclusion I came to in case he does not receive my letters. I also received by this mail an answer to my telegraph dated August 14th which seems to be the day after the date of the letter, so there is not much gained by that in the answer yet I hope the message was received by Father in time to give him information of my intentions.

You must still continue to write to me at this place for now it is without doubt that I am to stick it out with the Commission until Dec. at any rate.

I got a letter from Mr. Taylor of New Orleans in which he says the Yellow fever is much worse than the papers will admit and that the business of the City is allmost suspended, so I don't think I shall visit it until November when I hope to pass "Scott free." There is no way for us to go by New York from here more than once a year and that is the reasons the Major went by Pensacola. It was only by accident that he happened to have the opportunity.

Give my love to all and kiss Effie for me.

Your affectionate brother,

G. Clint Gardner

Miss Alida A. Gardner

Washington City.

GCG to his father, Mouth of Rio Grande, September 4, 1853[90]

My dear Father

Your letter of the 3rd July has just been received by what route I can't imagine as it did not come from New Orleans.

I have now received all your letters on the subject and must confess that I regret I did not risk the yellow fever and join you, but now it is, as I stated in my telegraph, out of my power to do so until Jany. next, about which time I will be happy to do so if it is your wish. I would like very much to hear from you and hope you will write as soon as you get out there. Give me a list of what books, if any, that Charlie takes out with him in surveying and engineering that I may not go to the trouble of getting the same if I should come out. And if he has any instruments with him give me a description and list of them for I intend before going out to make my outfit as complete as possible without going in debt.

I presume you will be much occupied on your first arrival if so do have Charlie write me all the news.

My love to all and if you have left before this reaches you I hope Mother will enclose it to you.

Your affectionate son,

G. Clint. Gardner.

Emory to GCG, Pensacola, September 8, 1853

My dear Gardner,

We arrived yesterday after a tedious & stormy passage, and find the whole population either sick or frightened to death.

When I left the mouth of the River by some unaccountable blunder of Bob McWilliams, the washer woman was not paid. I sent twice to get the bill & then came off without receiving it. She washed twice for me—twenty pieces each time. Please pay her forth with and I will refund when we meet.

Take with great care, repeated several times and with different needles, the variation, and send the result to me with your first package of lunations. Give my kindest regards to Clark & [one word missing] & believe me,

Yours truly,

W. Emory

I hope you saw the great comet.[91]

J. Hamilton Prioleau to GCG, Camp 80 on Rio Grande, September 15, 1853

Dear Gardner,

I went down to Brownsville last Saturday evening, and there had the pleasure of receiving your favor of August 21st and also several papers, for which please receive my thanks. I am sorry to learn that you and William [the cook] have fallen out. I certainly expected he would have remained until I returned. As to what he said he told me before I left about not cooking for any mess but mine. I assure you I have no recollection of any such thing. When we were leaving Ringgold, however, I remember his saying he did not wish to cook for you, as James had told him that you were harder to please than all the rest of the mess; but I paid no attention to this, nor would I probably have recalled it to my recollection now, had not your letter, which seems to blame me for not telling you of it, set me to thinking of what he had said. But I regarded it as of so little importance, that I did not think it worth repeating or recollecting, putting it altogether aside, with a great deal more of servent's gossip, as perfect rubbish.

I hear that Major Emory has gone home. Has he left with you or any one else the wages due Wentz since 1st July until he joined Radziminski. By the transfer of him to me, which I now have, there was $25.00 due him and there is now $5.00 more to the 6th August. Radziminski has paid him since then, and will pay him up to the time of his discharge, which will probably be when we reach Brownsville, as we can then get more and better men. He is not worth his salt, and I am very sorry now I consented to bring him at all. But I want to know, how he can obtain his money; I would pay him myself, but I have *vouchers for a larger amount than I expected, and for the payment of which I shall probably have to wait some time.*

"Talking of gingerbread" as a friend of mine used to say, "always puts me in mind of oysters" and talking of Wentz's money, puts me in mind that I owe you some. If you want it before we meet, let me know and I will send it to you by the first safe chance. Moreover, [James] Houston asked me if I had received a letter from you, concerning a draft of mine on Major E[mory]? I have not. The only letters of yours I have got from the Mouth were the one with my clothes, cot &c, and the one which I am now answering. The draft was brought back to me, and paid.

The Grampus *went up the River on Monday last with about 200 troops for Ringgold Barracks and other unknown destinations. I came up to camp on her, as it rained so hard on Sunday, I could not travel on horseback. Some six or eight officers went up on her, but they either did not know their destination, or why they were sent out here, or they would not tell. Artillery is not a usual arm to send on the frontier, so it looks as if something was up.*

I cannot say exactly when we will be in Brownsville with the survey. Ever since I have been up here it has rained almost continually, and has caused much delay. We have had, too a good deal of sickness in the camp among the men. Nor can we form any estimate at all approaching the truth as to the distance we have yet to go; the Steam-boat men, who ought to know, don't agree at all in their estimates, and the Mexicans on the Gila know nothing about it. So my conclusion is, that we will get there when we arrive and not sooner. Houston offered to bet a month ago, we would not get there till the middle of October, and we have had much worse weather since he left us than before. I myself will go down to Brownsville ten days from now, which will be Saturday [a] week, and hope I may hear from you there. I shall send this by the Grampus, *which will be here tomorrow or next day, and then it will get to you by mail, unless she goes directly down.*

Remember me affectionately to Clark & Michler. I mean our Clark, and not your landlord. Give my regards to both the Moreses *and if you see the Capt. at the Brazos, ask if he has heard anything of my claret, which I ordered some time ago from New Orleans. Also say to him I wrote home, just before leaving the mouth for some books to be sent me by Adams' Express to his care. It is time they should be at the Brazos now. As there are some among them which I can not easily get elsewhere, and which I do not wish to lose, I will thank you, if you get hold of the parcel to keep it for me, or send it up a safe hand.*

[James] Houston is at Brownsville, where he has been nearly a month, keeping the Tide Register, but not keeping Mag Evans. The latter & he have quarrelled and he has returned her miniature. So they are like Russia & Turkey in casu belli, *without an actual commencement of hostilities.*[92]

Mag told me a piece of news, which perhaps you may not have heard, and which may or may not interest you but it is that your acquaintance, Mrs. White is in limbo for keeping a brothel "the resort" says the paper "of idle and vicious men and of abandoned women." As she has a very heavy fine to pay, she will probably not be able to get us, in time to receive a visit from us on our return.

Tell Winder [Emory] that there are a great quantity of Antelopes on this river. I have shot several. They differ, however, materially from those that he will find described in natural history, in as much as they are two legged, have black long straight hair, white teeth, rather a yellow belly and no horns, *tho' probably their husbands have.*

Adieu.

Yours as ever,

J. H. Prioleau

GCG to Emory, Mouth of the Rio Grande, September 20, 1853[93]

My dear Major

I received your letter of the 8th by the "*yacht*" and we were all glad to hear of your arrival at Pensacola for we were afraid the storm, which was quite strong at this place, would reach you.

Please find herewith enclosed the variation as taken several times with the only surveyors compasses we have. I tried them several times and Mr. Clark did also without any change.

Your account with the washer woman was presented to me, as I wrote you by the last steamer, and I will settle it. She has meet [met] with a misfortune, the loss of her oldest son, with the yellow fever as the Doctor says he would call it in New Orleans, but here he calls it the bilious fever. I am sorry to say that this was the second case and a great many others have been taken with sym[p]toms, but nothing more. Two of the men of my party have been down but they are now perfectly well. I cannot suppose it to be the fever, or if it is, it must be a very slight form. However I would request that nothing be said about it as I should not like those at home to think we have any signs of it. We all now enjoy good health & of us three Winder [Emory] is the only one who has felt at all unwell.

I have packed all the instruments that will not be required here, keeping the surveying instruments in case of an accident to any of the parties, and am in hopes of being able to ship them on the shr. [schooner] *Minashafer* which sails direct to New York.[94] I have directed them to *you*, to the "care of E. & G. W. Blunt", and will request that he acknowledge the receipt of them to you, as well as myself, and I will send by the next mail after shipping them a complete return.[95]

The Zenith inst. belonging to the Coast survey and boxes &c. belonging to the Chronometers, I intend shipping also. I think this to be the safest way of getting them that far and when I come on I will bring them to Washington. I find a spy glass which is in the return of inst. as belonging to the Coast Survey, that I have retained here to bring to Washington with me.

You told me you were going to tell me what to do about the observatory whether to sell it or leave it in charge of some one here; permit me to remind you of it and also about my instructions for the Magnetic instruments.

The notes of the last lunation I sent by the last steamer together with the inst. receipt and all letters that came out in the mail for you. By this mail I send un[der]cover to the Topogl. bureau all your letters and should the steamer not get off before the end of this lunation I will send the notes. We have had very bad weather so far and have only been able to observe the culminations between the clouds.

The Yacht has been off the bar since sunday morning and they say she may lay out there during the whole week as there is no water, or perhaps they are keeping her in quarantine, as the citizens of Brownsville are rather badly frightened.

Clark and Winder wish me to present their kind regards and with my kindest regards for Mrs. E

Believe yours truly

G. Clinton Gardner

Mathilde Mouton to GCG, Ile Copal,[96] September 21, 1853

My dear Clint,

I suppose you have heard ere this that I am on a visit home. I left Fort Smith on the 25th of July, and reached here two weeks after, having been delayed on the way. Frank did not accompany me, but I expect him here in about a month or six weeks. I found all hands well, and the children much grown. Emma has now her hands full, yesterday week, she gave birth to another son, whom she has named George Clinton in compliment to her dear brother, but the child will go by the name of George for these French people here will never pronounce Clinton *properly. Both mother and child are doing very well.*

Mamma received your last letter a week or two since, she was much pleased to hear that you intend visiting Washington this winter and she says that she hopes and prays that you will let her know when you pass through New Orleans, for she will be very much tempted to join, for she is extremely anxious to visit Washington, particularly now, as Sarah is to be married, she is very desirous of being present at the wedding.[97] Write and tell us when you think of starting for the States and could you not make an effort and come as far as the plantation? If Mamma goes she would then have no trouble. You can come thro' by the stage in thirty-six hours.

I cannot write more at present for I have two other letters to dispatch this morning. All send much love to you and accept the same from

Your affectionate sister

Mathilde

P.S. I forgot to tell you about Odeide's fine boy, Alex. he is now four months old and a very pretty child, the image of his father. C— is here to day—and is very well.

GCG to Emory, Boca del Rio Grande, September 22, 1853[98]

Dear Major

I forgot to tell you in my letter day before yesterday that I received on the 6th of Sept a check from Genl. Campbell for $1000. He writes me that he will send me the balance in money as I will have difficulty in getting checks cashed. He had to pay 5% discount but I got mine cashed at par.

Cloudy weather still prevails no obs. since the 18th

Yours truly

G. Clinton Gardner

Emory to GCG, Washington, September 28, 1853

My dear Gardner,

I arrived home on Saturday, having been detained in Montgomery by an attack of yellow fever taken on the Alabama River.

Your family are all absent, but your aunt, to whom I delivered your letter. She says your father is perfectly satisfied with your arrangement about joining him.

I have no objection to your stopping in Louisiana provided you make suitable arrangements about the instruments and chronometers. Some one should be with them all the time.

In looking over your notes at the mouth of the River, I find no sufficient observations for collimation.[99] *Take a number forthwith on circumpolar stars, by reversing the Instrument, and send them to me at once.*

Genl. Campbell promised me he would send you a check for that money and I desire if you have not received it to write him at once, and I will do the same.

Very truly yours,

W. H. Emory

I have not yet had time to obtain the magnetic notes.

Emory to GCG, Washington, October 10, 1853

My dear Gardner,

I send you all I can obtain in the way of instructions for the intensity apparatus. It is not very satisfactory, but with the Admiralty instructions you may be able to get a satisfactory set of observations.

I can find no good place in N.Y. for the Instruments and Genl. Campbell concurring, I desire they may be brought *to this city.*

Say to the General if he is with you I have sent several communications, and that I have received an order from the Sec[retary] of the In[terior] to make an estimate, and a report of the present state of the work, that I informed the Secty. my report & estimate had already gone to the Commissioner and that he was the person to look to. The Secty. replied that he had been called upon by the Secty. of the Treasury for his estimate and could not wait a moment.

Your people have not returned yet. I have been on my back with the chills and fever. The yellow fever leaves one very sensitive to cold.

Do not omit to take a more perfect set of observations for collimation on slow circumpolar stars and send me the result. I wrote you, that on examination, your notes were deficient in that data.

If a very convenient opportunity presents itself to send the Poney *[pony] to New York or Baltimore, I wish you would do so, first arranging with Genl. Campbell the price I am to pay for him.*

Present my regards to all the Gentlemen & believe me,

Very truly yours,

W. H. Emory

What has become of my bow & arrow for George?[100] *Send it by Winder [Emory].*

GCG to Emory, Mouth of the Rio Grande, October 10, 1853[101]

My dear Major

We have now no regular mail between this and New Orleans, as the *Yacht* was wrecked on her way out from Brazos on the 23rd of last month, and therefore, have to depend on the sailing crafts. I have prepared the duplicates of the notes and also a return of inst. showing the present position of them. The schr. *Mina Shiffer* [schooner *Minna Schiffer*] on which they were shipped, sailed on the 24th of last month and I have requested Messers. Blunt to acknowledge the receipt of them to you as well as myself, as our mails are very uncertain.

Mr. Prioleau has just come down to keep the rise and fall of the river as well as to duplicate the notes & tells me Mr. Schott had arrived at Ringgold Barracks and was on his way down. Mr. Radziminski is below Brownsville and expects to be down here about the 1st of November.

We have had cloudy weather for some time past as the notes will show but we have it now quite clear & have commenced this lunation with fair prospects so I think we can get thro' by the middle of next month, if you think the Sept. lunation is sufficient for one of the four.

I think the most direct way of communication now is by way of San Antonio as there is a weekly mail between that place and Brownsville.

We are all now enjoying good health after a short spell of sickness which each of us had about the last of Sept. Mr. Clark and Winder wish me to present their kind regards and with many good wishes for your health I remain

Yours truly

G. Clinton Gardner

P.S. It is with regret we heard to day of the death of Lt. Whiting at Fort Brown.

Emory to Gardner, Washington, October 25, 1853

My dear Gardner,

I cannot [find] the receipts for the ambulance and four mules and the horse turned over to you. Will you be kind enough to send me receipts by the next mail marked quadruplicate. I do not know certainly that you have ever given me any receipts.

The Instruments I wish brought to this place as I find it impossible to get satisfactory storage at New York. They can come to New York and thence to this place by the Georgetown packet. You or Clark must accompany them to New York and see them on board the Georgetown packet. If necessary Winder can go along to assist.

I am at a loss to know what has become of Michler. He wrote me long since he had finished his work and was in San Antonio. I sent him orders to come to Washington.[102] *It may be he is deterred from coming by the Yellow fever at N[ew] Orleans & Mobile.*

Where is Radziminski now, and Schott I have received no new Culminations or returns from you or any of the party. What is the cause? Pray write by return mail.

Truly yours,

W. H. Emory

GCG to Emory, Mouth of the Rio Grande, October 30, 1853[103]

My dear Major,

I send by this mail the duplicates of the observations up to date. The weather still continues about the same the last three nights are the first clear one for some time.

Major [Richard] Delafield of your corps has asked for the latitude and longitude of the points on the river that have been determined and I referred him to you, being neither *authorized* or capable of furnishing them.

I received a note from Mr. Radziminski this morning, from Palo Alto, in which he says he will be hear about the end of this week. Mr. Schott is somewhere on the river below Mr. R-. I have not heard from Genl. Campbell but we are expecting him at Brownsville daily.

I regret that I have to report to you the loss of the ambulance, which I sent to Brownsville on the 19th to get some flour and forage; and in its return the next evening about dusk the driver, Brothers, informs me the road, near the bank of the river between this place and Camp Belknap,[104] caved in and the ambulance fell into the river together with its contents. The body bolt gave away which cleared it from the mules or perhaps they might have draged it out. There was nothing saved except the front wheels and the back seat. I sent a party up in a boat the next morning to see if anything could be recovered but after drag[g]ing along for some distance from the place of the accident nothing could be found. I should have discharged Brothers immediately on his arrival here but his service being necessary in the attempt to recover it and having since examined the place I am of [the] opinion that it might have been unavoidable so I have concluded to retain him.

I received your letter of the 28th Sept. on the 13th and the one of Oct 10th day before yesterday. I think I will be able to make a good set of Magnetic obs. If I can get the declinometer to work well but I find one or two little things out of order however I am in hopes of fixing it all myself.

The bow and arrow for George was sent down to Brownsville by Palfrey and I will get it the first time I go up and see that it is carried safely to Washington. I could have sent the Poney [pony] to New York by the *Mina Shiffer* [*Minna Schiffer*], the schooner that carried the instruments, at a very small expense if I had known your wishes in time, however I expect some other opportunity will occur.

I regret to hear that you have been sick but I expect the cold weather is too severe after being so long south. With many wishes for your health and kindest regards for Mrs. E. I remain.

Yours truly

G. Clinton Gardner

Emory to GCG, Washington, November 1, 1853[105]

Sir,

In pursuance of a request made by the Superintendent of the Coast Survey to whom the Gambey dip circle belongs I desire you to send, or to take to the [Port] Collector in New Orleans, that instrument and have it left subject to the order of J. E. Hilgard, Esq., Asst. Coast Survey.[106]

If you take the Boundary instruments by the way of New Orleans leave the dip circle there, and take the Collectors receipt. If you do not pass through New Orleans, send it by some vessel and forward me the bill of lading.

I am very Respectfully,

Your Obt. Servant.

W. H. Emory

Astronomer & Surveyor B. C.

E. and G. W. Blunt to Gardner, New York, November 3, 1853

Sir,

Your letter of the 24th Sept. is recd. this day.

The Schooner Minna Schiffer *is now in port & we shall probably receive the Boxes in a day or two.*

Maj. Emory has advised us that he will send a person to attend to their receipt.

We are, Very Respy. Yours &c.

E. & G. W. Blunt

p[er] Saml. A. Cash

Emory to GCG, Duplicate, Washington, November 5, 1853

Sir,

You will please send by the first opportunity that presents itself, the Zenith Telescope for obtaining Latitude (the property of the Coast Survey) to the Collector of the port of New Orleans, to be delivered to J. E. Hilgard, Esq., Asst. Coast Survey, or to his order.

Very Refly. [respectfully]

Yr. Obdt. Servt.

W. H. Emory

Mark the box "up with care" and send letter of advice to Mr. Hilgard Pass Christian Louisiana.

Robert B. Campbell to GCG, Mouth of the Rio Grande, November 9, 1853

Sir,

A board of surveyors having recommended a sale of the property at this place belonging to the U.S. Boundary Commission, and your duties requiring your continuance here until about the 25th inst., you will take charge of the property, and after due notice being given dispose of it to the highest bidder at public sale.

Very respectfully,

Yr. Obt. Srvt.

Robert B. Campbell

Emory to GCG, Washington, November 9, 1853

My dear Gardner,

I have had no more culminations from you except the first set and duplicates.

I wrote you to say you must take a complete set of observations for collimation coron[a?], on circumpolar stars, reversing the instrument after middle wire [?].

Please send with the instruments my bed and books. If I recollect, I asked you to store them in New Orleans. I now request them to be sent here.

I find so many demands for my money, I can't afford to bring the poney [pony] on to Washington. I therefore revoke my former request and beg you will turn him over to Genl. Campbell, to keep until we go back, if we ever go back.

Yrs. truly,

W. H. Emory

Do not break up your observatory without getting the obs for Collimation and a satisfactory set of Moon Culminations.

Minutes of the meeting of the boundary commission, November 10, 1853[107]

At a meeting of the Members of the Boundary Commission, held at the Mouth of the Rio Grande on the evening of the 10th November, 1853, consisting of Genl. Campbell, Commissioner, and Messrs. C. Radziminski, G. C. Gardner, John H. Clark, Jas. H. Prioleau, Edw. Ingraham, Malcolm Seaton, John E. Weyss, T. W. Emory, and Jas. H. Houston, Assistants,

Genl. Campbell was called to the Chair, and Mr. Prioleau appointed Secretary.

The Chairman explained the object of the meeting and appointed Messrs. Radziminski, Prioleau, Ingraham, Seaton, and Emory a Committee to draft and report appropriate resolutions.

Whereupon the Committee retired, and after consultations, reported the following Preamble and Resolutions which were unanimously adopted,

Whereas, it has pleased Almighty God in his wise providence to remove from among us, one, who was endeared to all who knew him, by his many excellent qualities, We the Members of the United States and Mexican Boundary Commission now, for the first time since this melancholy accident, assembled together do resolve,

1st. That we regret and deplore the untimely death of Thos. Walter Jones, late Assistant in the U.S. and M. Bdy. Commission, who was, on the 23d of July last, cut off in the prime of his life and usefulness, and whilst in the discharge of his duty.

2nd. That we will always cherish a tender regard for his memory, and will keep the example of his virtues constantly before us.

3rd. That we deeply sympathise with his family in their affliction, and regret that we have not before had an opportunity of expressing that sympathy.

4th. That a copy hereof, signed by the Chairman and Secretary of this meeting be forwarded to Mr. Jones family, and a copy be also furnished to each member of the Commission.

5th. That the Secretary be instructed to carry into effect the 4th Resolution.

The meeting then adjourned.

(signed) R. B. Campbell, Chairman

J. Hamilton Prioleau, Secretary

Radziminski to GCG, Brownsville, November 15, 1853

I have attended to all your Commissions but Lawrence's bill. I have done the same with Prioleau's.

I send you per Grampus *2 bottles of brandy & some preserves & request you & your mess to accept them from me. No good tobacco. Sent 4 bundles of bad.*

Send Prioleau's blanket, boots, "the handbills" and inside of Prioleau's blanket is Schott's Bird book which I beg you to deliver to him. [Malcolm] Seaton sick, Hurbide &c. the former only slightly so, expects to be up in one or two days. I am so far quite well—have been very busy cannot get a man to carry my letter to Gimeniz [Francisco Jiménez].

I send you also some Mercurial ointment.

You will receive one package & 1 small box, for which please enquire of Mr. Douglass.

Love to all—

Yours,

Radziminski

—Don't either of you come here. It is very sickly. R.

GCG to Emory, Mouth of the Rio Grande, November 18, 1853[108]

My dear Major,

I received your letter of the 25th of October by the last mail and please find herewith the receipts. I am sorry to hear that the notes for the September lunation had not arrived before the date of your letter for they were mailed here about one month previous which I think is ample time for them to go to Washington.

Mr. Schott is here and I told him that you had heard nothing from him at which he expressed great surprise for he says that he has received answers to private letters which were mailed with those to you. Mr. Radziminski wrote to you a few days ago and I suppose told you of Genl. Campbell's arrival here. He was here two or three days and left for San Antonio without stopping at Brownsville on account of the yellow fever, which is still quite bad there. Lt. Dungan I am sorry to say fell victim to the disease and died after a few days sickness.[109] I think the deaths in the Garrison[110] number some twenty five and there is as many more in the hospital.

I read to Genl. Campbell parts of your letter of the 10th of October and he told me that he concurred with you about the instruments. I will on my return go by New York and take those, that I have already shipped to that place, to Washington as directed by you. I am afraid I will not have an opportunity to send the poney [pony] direct to New York or Baltimore from this place, but I think I may be able to get him taken to New Orleans by the steamer *Fashion* free of charge, if so, I will ship him from there to Baltimore. If I were to pay his expenses to New Orleans it would make him rather costly upon his arrival in Washington. The General told me that he would like me if possible to send him to you and if not to sell him at the sale, so if I can arrange his passage to N. O. I will sell him and buy him, which will be the cheapest way of getting him for he will not sell at auction for much.

We have had very fine weather for our observations up to the last two night [s] but I fear the bad weather has now commenced. Last night we had a severe storm from the S.W. rather unusual.

Mr. [W. E.] Greenwell of the Coast Survey has arrived at the Brazos and applied to me for the Lat. & Long. Of this point which I told him I could not give him with accuracy. He says you told him that I would furnish him with the data for his work so I have concluded to give him an approximate result knowing that he is operating with the Bdy. Survey, altho' I am doubtful about doing it as I have received two letters from you since he left Washington in which you say nothing about his coming here. He says it is necessary for him to plot his work here before he leaves and that he must have one point, if only approximate. He also asked me to let him have the small boat and some rope &c. but my orders from you, as well as Genl. Campbell, direct all the property to be sold at public sale and I do not think I have a right to transcend my authority.[III]

I remain with great respect

Yours truly

G. Clinton Gardner

---

1. Bartlett, 1854, 2:512.
2. Emory had known this for some time, learning it first perhaps from Hardcastle. Hardcastle, writing to Emory from Washington on September 10, 1852, provided a full account, including the language of the proviso (box 6, f. 43, EP). Emory's views on the southern route were somewhat ambiguous, but he expressed them without ambivalence in a letter to J. M. Mason on February 16, 1853 (draft, box 6, f. 48, EP). Emory seems to contradict himself in another letter to Mason, this one from Washington on April 21, 1853 (box 6, f. 50, EP).
3. Bartlett (1854, 2:514–16) quotes most of Stuart's letter, dated Washington, October 15, 1852.
4. Bartlett to Emory, Ringgold Barracks, December 21, 1852, box 6, f. 46, EP; Bartlett, 1854, 2:517–18; Werne, 2007, 129–31.
5. Bartlett to Emory, Ringgold Barracks, December 23, 1852, box 6, f. 46, EP; Bartlett, 1854, 2:518–19. Bartlett and Emory both took pains to maintain usable equipment and animals, for, as Emory said, "the boundary commission is suspended, but it is believed only temporarily" (Emory to Lt. Green, Ringgold Barracks, December 24, 1852, box 6, f. 46, EP).
6. We infer this from a draft of a letter from Emory to Bartlett, dated at Corpus Christi, January 1, 1853 (box 6, f. 47, EP).
7. Emory's earliest report might have come from Lieutenant Michler, still in Washington, who had informed Emory on December 20 the bill had passed the Senate and the House the previous Friday, the sixteenth, and had gone to the president (Michler to Emory, Washington City, December 20, 1852, in box 6, f. 46, EP). Bartlett learned of it on January 7 from a newspaper account and wrote to tell Emory (copy, Bartlett to Emory, Decker's Point, Texas, January 7, 1852, box 6, f. 47, EP).
8. Bartlett, 1854, 2:519.
9. Norris, Milligan, and Faulk, 1998, 145.

10. A letter from Emory [to Bartlett], written in Washington, on February 21, 1853, says that Emory directed "Gardner to make for you from my book copies of all the letters addressed to you by me" (box 6, f. 48, EP).
11. GCG sent that letter from New Orleans on May 6, 1853. He had arrived in Havana at the end of April.
12. GCG to his sister Alida, Brazos Santiago, May 17, 1853.
13. Emory's estimate of funds for getting into the field put his own salary at $600 a month and the salaries of GCG and John Henry Clark at $300 a month (Emory to Commissioner Robert Campbell, n.d. [ca. March 1853], box 6, f. 49, EP).
14. Emory to Abert, Mouth of the Rio Bravo, August 1, 1853, box 6, f. 54, EP.
15. GCG to his father, Boca del Rio Grande, Near Brazos Santiago, August 4, 5, and 9, 1853.
16. Emory planned to have GCG stay into November "to continue the observations on the culminations of the moon and the moon culminating stars." With that, "all the field work of the undisputed portion of the Boundary will be completed" (draft, Emory to Abert, Washington, October 5, 1853, box 6, f. 56, EP).
17. Clark's role appears in a draft from Emory to Secretary of the Interior Robert McClelland, who notes that "Gardner & Clark" took observations on "the moon & moon culminating stars running thro four lunations" (Office of the Boundary Survey, March 16, 1854, box 7, f. 61, EP).
18. GCG to Emory, Mouth of the Rio Grande, September 20 and November 18, 1853.
19. Radziminski to Emory, Brownsville, November 28, 1853, box 7, f. 57, EP.
20. Radziminski to GCG, Brownsville, November 15, 1853, box 7, f. 57, EP. Prioleau had been appointed assistant astronomer by Commissioner Campbell (Prioleau to Emory, Charleston, April 22, 1853, box 6, f. 50, EP).
21. The date of GCG's departure is not clear, but Commissioner Campbell told Emory that GCG would remain until "after the 23rd" (Campbell to Emory, Mouth of the Rio Grande, November 10, 1853, box 7, f. 57, EP).
22. This letter may be found in box 6, f. 47, EP.
23. GCG, forgetting that the calendar had moved ahead another year, mistakenly dated this letter 1852.
24. By "get rid of the property," GCG suggests he is selling it. Emory, though, ordered GCG to turn over "the public property of the Body Commission in his possession" to Bartlett, who was making his way overland into Texas (Emory to Bartlett, Corpus Christi, January 1, 1853, box 6, f. 47, EP). This was only a small portion of the commission's property; more was on its way to San Antonio to be auctioned off.
25. James H. Houston was, like GCG, a civilian assistant. He was a nephew of Sam Houston's; Houston had been elected senator from Texas in December 1849 (De Bruhl, 1993, 329).
26. Forbes N. Britton, a West Point graduate, served in the Seventh Infantry from 1834 until July

16, 1850, when he resigned and moved to Corpus Christi. In the Seventh Infantry he knew GCG's brother Frank. In addition to practicing law and speculating in real estate, Britton was cofounder of the Corpus Christi Navigation Company (Frank Wagner, "Forbes N. Britton," *NHBT*, 1:744).

27. Emory had notified Ingraham that the commission had been disbanded and that he should remain at home "and there await further orders" (Emory to Ingraham, New Orleans, January 15, 1853, box 6, f. 47, EP). Before Emory's missive reached Ingraham, however, he had gone back to Brazos Santiago, following Emory's earlier orders of December 25. Still unaware that Emory had ordered him to wait at home for further orders, Ingraham asked Emory for permission "to come on to Washington and be at work rather than spend the winter at home" (Ingraham to Emory, New Orleans January 23, 1854, and telegram, Ingraham to Emory, January 25, New Orleans, both in box 6, f. 47, EP).
28. At the bottom of this letter Sarah has written to her father that they have learned from Uncle Wash of the death by drowning of his oldest son, Charley. "In the midst of our joy at hearing such happy news from Clinton—we can truly sympathize with them. It must be a great blow to poor Rebecca. Mother, however, says, you must certainly visit Rahway [New Jersey] before you return home."
29. We have not been able to identify E. E. This letter is in the Smithsonian Institution.
30. Perhaps Andrew Jackson Smith or Gustavus Woodson Smith (Heitman, 1965, 1:894, 898).
31. Perhaps Robert Seaman Granger (Heitman, 1965, 1:469).
32. The *Yacht*, a small steamer owned by Harris and Morgan, had been put into service in the Gulf of Mexico in 1848 and was, like other Harris and Morgan vessels, much criticized. See, for example, "Gulf Steamers to Matagorda Bay, Harris & Morgan, Things as They Are, Things as They Were," which appeared in the (Victoria, Texas) *Texian Advocate*, on October 10, 1850, p. 2, and rises to the defense of Harris and Morgan.
33. Emory provides a dramatic account of the near wreck of this vessel and his own key role in saving the ship from crashing during the night (1857–59, 1:53).
34. The Harris and Morgan steamship line had a monopoly on shipping in the Gulf ("Legislative," *Texas State Gazette*, January 4, 1854, p. 134).
35. The town of Brownsville, on the Rio Grande across from Matamoros and about seventeen miles up the Rio Grande from its mouth, had grown up around Fort Brown after the U.S.-Mexican War (Alicia A. Garza and Christopher Long, "Brownsville," *NHBT*, 1:776–79).
36. To Ringgold Barracks.
37. William Emory's nephew, Levin Winder Emory, the child of Major Emory's older brother, Thomas Alexander. Levin Winder Emory was born in 1836 and thus was about seventeen, the same age as GCG when he first joined the survey (U.S. Census, Easton, Talbot, Md., roll M432_297, p. 6).

38. Antonio López de Santa Anna, who returned to the presidency of Mexico on April 20, did order troops to the border (Garber, 1923, 72).
39. The Mesilla Valley, from El Paso to Doña Ana, had become a flash point. After Congress repudiated the Bartlett–García Conde compromise, New Mexico governor William Carr Lane traveled to Doña Ana and, on March 13, 1853, proclaimed the Mesilla Valley part of New Mexico. Chihuahua governor Ángel Trías prepared to resist Lane's claim with military force. The Pierce administration replaced Lane as governor and sent Gen. John Garland to replace Edwin Sumner as head of the military department of New Mexico. GCG was wrong. Both men had instructions to avoid conflict. GCG had apparently been influenced by the newspapers he received, many of which spread rumors of an impending war over Mesilla (Rippy, 1921, 728–32; 736–38; Werne, 2007, 157–64).
40. The *(New Orleans) Daily Delta.*
41. With Democrats back in power, Charles Gardner did indeed receive a "fat Office," surveyor general of Oregon.
42. This letter may be found in box 6, f. 52, EP.
43. Although this letter, which is in the G. Clinton Gardner Journals and Correspondence at Yale, appears here in chronological order, GCG received it quite late, on August 10 (GCG to his father, Mouth of Rio Grande, Near Brazos Santiago, August 11, 1853).
44. In an undated note to an unnamed member of his family filed in the DeGolyer collection for no apparent reason between GCG to his father, Near El Paso, December 11, 1851, and GCG to his mother, Frontera, Texas, January 8, 1852, GCG says: "Tell Father to send me by mail one of Col. Lee's formulae, compiled for the use [of] Officers of the Topl Corps, I think he can get one at the Department, and he will much oblige his son." GCG refers to Thomas Jefferson Lee, *A Collection of Tables and Formulae Useful in Surveying, Geodesy, and Practical Astronomy, Including Elements for the Projection of Maps.* Prepared by Order of the Topographical Bureau for the Use of the Corps of Topographical Engineers, by Captain T. J. Lee (Washington, D.C.: Gideon, 1849).
45. GCG's ongoing concern with the Black Rock property.
46. This letter is in box 6, f. 53, EP.
47. John Henry Clark was part of the survey party, but the owner of this house was a different Mr. Clark, as is clear from a letter from J. Hamilton Prioleau to GCG, Camp 80 on Rio Grande, September 15, 1853: "I mean our Clark, and not your landlord."
48. This letter is in the GCGJC.
49. Rittenhouse School in Washington, D.C., which we identified in chap. 1, note 33.
50. Maxwell Woodhull entered the navy in 1832 and became a lieutenant in 1843 (Callahan, 1901).
51. Colonel Gardner interpreted the situation correctly. James Gadsden did obtain territory for the United States south of the Gila, as explained in chap. 9, although not all the land the United States wanted.

52. Frank did not leave the army.
53. References to two of his two brothers.
54. A bowling alley.
55. Building the observatory had required more help from Brownsville. J. H. Prioleau, an assistant surveyor, worked with GCG at the mouth of the river, and GCG sent him back to Brownsville to get "mechanics." On July 9 Prioleau reported that he had engaged one carpenter, Smith (whom GCG had earlier rejected), who traveled back to the observatory with him on the tenth, and one bricklayer, who took the *Grampus* to the mouth of the river the next Monday (July 11) with bricks and lime (Prioleau to Emory, Brownsville, July 9, 1853, box 6, f. 53, EP).
56. Campbell was the new boundary commissioner, who took care of administrative matters and left the work of concluding the survey to Emory (Norris, Milligan, and Faulk, 1998, 135).
57. Sarah's future husband, John J. Almy, a navy lieutenant, had begun his career as a midshipman in 1829 and ended it as a rear admiral. He retired in 1877 and died in 1895 (Callahan, 1901).
58. Maj. Benjamin Brown French, an attorney and fellow resident of Washington City, was appointed commissioner of public buildings, a position he held for the next two years (Hunter, 1853, 38; *New York Times*, August 25, 1853, and June 5, 1855).
59. The daughter was apparently Jessie Mary, described by Arnold as "the only child of Col. Alexander Clinton McLean of Newburgh, New York. She was a great beauty, and had the most distinguished social success in New York, until she went to Europe" (1899, 291). Arnold says this in the context of a story about her tragic later years.
60. The bride is apparently Sarah, who was then planning her wedding. We have not been able to identify Grafton.
61. This letter is in box 6, f. 53, EP.
62. The U.S. surveying schooner *Morris* had arrived at Brazos Santiago to work on the coastal survey. Emory had coordinated this with the superintendent of the coast survey, Professor Alexander D. Bache his father-in-law. Part of its job was to survey a marine league beyond the mouth of the Rio Grande as specified in the Treaty of Guadalupe Hidalgo (Lt. John Wilkinson to Emory, Off Brazos Santiago, July 28, 1853; draft, Emory to Wilkinson, Fort Brown, Texas, July 29, 1853; and draft, Emory to Francisco Jiménez, Fort Brown, July 29, 1853, all in box 6, f. 53, EP; draft, Emory to A. D. Bache, Camp at Mouth of the Rio Grande, August 3, 1853, box 6, f. 54, EP).
63. The German-born and -educated William Würdemann, who moved to Washington to work for the coast survey and became a prominent maker of "mathematical instruments" (Hunter, 1853, 112; Turner, 1991, 97–110).
64. Radziminski, who was with Jones and who "came within a few seconds of time of being drowned myself," reported this to Emory on the evening it occurred, July 23, 1853, then penned a fuller account to Emory the next day. Jones's body was found on the twenty-fifth. The ver-

sion GCG heard corresponds to Radziminski's account (Radziminski to Emory, July 23, July 24, and July 31, 1853, Camp . . . on the Rio Grande, box 6, f. 53, EP). On November 11 a number officers of the commission, led by Commissioner Campbell and with GCG among them, met at the mouth of the Rio Grande and drafted an expression of sympathy to send to the Jones family, which we reproduce later in this chapter (Minutes of the Meeting of the Boundary Commission, November 10, 1853).

65. Emory was at Fort Brown, planning to go to the observatory on Monday, August 1, "to take personal charge of the observations to be made there" (drafts Emory to Lt. John Wilkinson, Fort Brown, Texas, July 29, 1853, box 6, f. 53, EP).
66. This letter is in the GCGJC.
67. GCG had received a note from his father dated July 14. His father's letter of July 5, 1853 (which is reprinted in chronological order in this chapter), spelled out the details but had not yet arrived (GCG to his father, Boca del Rio Grande, Near Brazos Santiago, August 4, 5, and 9, 1853).
68. Emory may have arrived later that day, for he wrote to Abert from the "Mouth of the Rio Bravo," on August 1, 1853 (box 6, f. 53, EP). In another letter, however, Emory indicates that he arrived at the mouth of the Rio Bravo on August 2 (Emory to A. D. Bache, Camp at the Mouth of the Rio Bravo, August 3, 1853, box 6, f. 54, EP). Emory's correspondence indicates that he stayed there through the month of August.
69. This letter is in the GCGJC.
70. This letter is in the GCGJC.
71. The letter of June 30 appears in this chapter; we have not found the letter of July 14.
72. This letter is in the GCGJC.
73. Known to Anglo Americans as Aspinwall but to Hispanics as Colón (the name it retains today), this community had its beginnings in 1850 as northern terminus of a transisthmian railroad.
74. GCG's oldest brother, John, a lawyer.
75. A major epidemic of yellow fever. The first cases had appeared in June. By July it had become an epidemic. By the time it ended in December it had carried away eight thousand to eleven thousand people, or about 10 percent of the population. John Duffy, *Sword of Pestilence: The New Orleans Yellow Fever Epidemic of 1853* (Baton Rouge: Louisiana State University Press, 1966), is the classic account. Henry M. McKiven Jr., "The Political Construction of a Natural Disaster: The Yellow Fever Epidemic of 1853," *Journal of American History*, 94 (December 2007): 734–42, looks at ways that politicians used the epidemic to advance their agendas. We have taken the 10 percent figure from McKiven.
76. This letter is in the GCGJC.
77. Emory did, indeed, leave such instructions (drafts, Emory to Radziminski and Schott, September 1853, Mouth of the Rio Bravo, box 6, f. 55, EP).

78. This letter is in the GCGJC.
79. Neither a James Robberts nor a James Roberts appears in Heitman (1965).
80. With yellow fever raging through New Orleans, Uncle John apparently joined the exodus to the interior of the state.
81. This letter is in the GCGJC.
82. We have not identified Marian and her sister Daisie, but they were apparently part of the Mouton family or friends of the family.
83. Sarah apparently planned her wedding to John Almy for that autumn, but it did not take place until February 2, 1854. It was held at Trinity Church, which the Gardners attended, with Reverend Butler officiating (*[Washington, D.C.] Daily National Era*, February 3, 1854, p. 3). She may have delayed the wedding so that GCG could be present. Then too the death of her uncle, Alexander Clinton McLean, at her parents' home in Washington on January 12, might have prompted her to postpone the wedding.
84. The reference here is to Frank's wife, Mathilde, who was visiting her stepmother, Emma, who was also Frank's sister.
85. John J. Almy, Sarah's future husband.
86. This letter is in the GCGJC.
87. John Jones might have been J. Jones, an African American resident of Washington and a "laborer." Frank Wheatley may have been another Washington resident, Francis Wheatley, whose occupation we do not know (Hunter, 1853, 55, 107).
88. Little Britain, New York, and Rahway, New Jersey, where the family had relatives.
89. This letter may be found in box 6, f. 55, EP.
90. This letter is in the GCGJC.
91. Of four new comets scientists observed in 1853, this appears to be the one designated as C/1853 L1, discovered by Wilhelm Klinkerfues of Gottingen, Germany. Klinkerfues first sighted it in June, and it was apparent to the naked eye throughout August, when the length of its tail increased rapidly (Kronk, 2003, 222–26, 834–35). It was well reported in both the northern and southern hemispheres, although the "C" designation by Kronk indicates that the observations were insufficient to determine whether this was comet in elliptical (likely to return in future) or hyperbolic (unlikely to return again) orbit.
92. Prioleau seems to mean that James Houston and Mag Evans were, like Turkey and Russia, in a state of war but that fighting had not broken out.
93. This letter may be found in box 6, f. 55, EP.
94. The *Minna Schiffer*, belonging to Schiffer and Brothers (*New York Weekly Herald*, October 23, 1853, 352).
95. For the Blunts, see Burstyn (1957), reference courtesy of Charles Hughes.
96. The Mouton plantation, near today's Lafayette, Louisiana.

97. Mamma was apparently Mathilde's mother-in-law.
98. This letter may be found in box 6, f. 55, EP.
99. Emory appears to mean that GCG should adjust the line of sight of his telescope by taking sightings on stars.
100. George Bache Emory was apparently the youngest of Emory's four sons. The 1860 census lists two thirteen-year-old sons, George Bache and William Hemsley. William's name appears before George's in the chronological listing. They might have been twins, but since Emory mentions only George it seems more likely they were a year apart (U.S. Census, 1860, Ward 1, Washington, D.C., roll M653_102, p. 363).
101. This letter may be found in box 6, f. 56, EP.
102. Emory had received a letter from Michler dated August 18, informing him "of the completion of the work," and Emory ordered him to Washington (Emory to Michler, Washington, September 29, 1853, box 6, f. 55, EP).
103. This letter may be found in box 6, f. 56, EP.
104. A puzzling reference. Camp Belknap, established in 1851, was far to the north. GCG may have meant Fort Brown.
105. Emory sent GCG a similar order on November 5, 1853 (box 7, f. 57, EP).
106. A dip needle, or dip circle, used for measuring the angle between Earth's magnetic field and the horizon. Several dip circles, including the French-made Gambey, can be seen at http://physics.kenyon.edu/EarlyApparatus/Electricity/Dip_Needle/Dip_Needle.html.
107. This remained in GCG's family papers and a copy sent to Emory remained with him (box 7, f. 57, EP). From the mouth of the Rio Grande, Prioleau sent a covering letter to Emory on November 11, 1853, a copy of which is also in the Gardner Papers: "In pursuance of the duty assigned to me by the accompanying resolutions, I have the honor to enclose to you for your information, a copy of the proceedings of a meeting of the members of the U.S. & M. Bdy. Commission, held at this place yesterday."
108. This letter may be found in box 7, f. 57, EP.
109. Lt. Hugh E. Dungan died November 11, 1853 (Heitman, 1965, 1:388).
110. Fort Brown.
111. Emory apologized to Greenwell, saying that he had written to GCG, "directing him to give you every thing you required," and explaining that the letter had not reached GCG (Emory to Greenwell, Washington, December 28, 1853, box 7, f. 58, EP).

CHAPTER 9
Resignation and New Adventures
January 25–April 13, 1854

Between 1849 and 1853 the Mexican and U.S. commissions had performed heroically to survey the border under the terms of the Treaty of Guadalupe Hidalgo, despite undependable support from their respective governments. Just as the surveyors completed their work, however, a new treaty made parts of their survey irrelevant. In July 1853, while Clint Gardner was overseeing the construction of an observatory at the mouth of the Rio Grande, the Democratic administration of Franklin Pierce sent James Gadsden to Mexico with instructions to acquire more of northern Mexico.[1] In particular, Pierce wanted to undo the Bartlett–García Conde compromise and regain the Mesilla Valley without fighting a war. Pierce also wanted to acquire an all-weather railway route to the Pacific. In an era of rapid railway expansion, a transcontinental railroad seemed both necessary and inevitable. The land immediately below the Gila River, through Tucson, seemed to hold the best potential for such a route. Moreover, with its logical eastern terminus in the South, it would benefit Southerners, to whom Pierce, a Northerner with Southern sympathies, owed large political debts.

The Treaty of Guadalupe Hidalgo had left the area below the Gila in Mexico, and Gadsden's task was to negotiate a new treaty. By the end of December, Gadsden had succeeded, and the United States and Mexico had a new boundary. The brusque, bullying Gadsden, whose multifaceted career included the presidency of the South Carolina Railroad Company and a keen interest in a transcontinental railroad with South Carolina as its eastern terminus, was prepared to deal quickly with the administration of Antonio López de Santa Anna, which he correctly perceived was weak and likely to be overthrown. That spring Santa Anna had come to the presidency of a country deep in debt and facing war with the United States over the Mesilla Valley. After failing to get help from European nations, he yielded to Gadsden's least offensive offer and sold the Mesilla Valley and the area south of the Gila to the United States. With the United States massing troops on the border, Mexicans had good reason to believe that if Mexico did not sell the railroad route, the United States would take it. Santa Anna later offered that as the reason for the sale, although his acute need for cash to pay the army that kept him in power would have been reason enough.[2]

Mexico and the United States exchanged ratifications of the Gadsden Treaty in Washington on June 30, 1854. The treaty had called for both countries to appoint boundary commissioners, who would meet in El Paso three months after the exchange of ratifications and begin surveying. Once again Maj. William H. Emory won the assignment, but this time he held the positions of commissioner, surveyor, and chief astronomer. He would not have to answer to a civilian political appointee and could work expeditiously. He did. He reached El Paso at the end of November 1854, and by the end of September of the following year he and his Mexican counterpart, once again José Salazar Illaregui, had finished the fieldwork.[3]

Emory worked on the Gadsden line without George Clinton Gardner, who had resigned from the boundary commission that spring. Clint had returned home in December in time for the holidays and stayed on through April. The new year started with two family events, one sad and one happy. First, his uncle Alexander Clinton McLean died at Clint's parents' house on January 12.[4] Less than a month later, on February 2, his sister Sarah's long-planned wedding to John Almy took place—a wedding that may have been postponed until Clint's homecoming (she had originally planned it for autumn, it would appear, then postponed it again with the death of their uncle).[5]

Between those two events, on January 25 Clint sought a leave of absence from the boundary commission for the month of March, to be followed by his resignation at the end of the month. Either he changed his mind or Emory would not accept his resignation. His resignation did not become official until May 1, although Emory put Clint on leave as of April 13, at which point he promptly left for Oregon.[6]

By July he was in Salem, Oregon, where he and his brother Charlie went to work for their father, who was then surveyor general of the Land Office of Washington and Oregon.[7] Clint put his experience to good use. He carried with him a letter of introduction from William Emory to General I. I. Stevens, governor of Washington Territory, that commended Clint for his ability and "for his qualities as a true man and honest friend."[8] Clint fell in love with the country near the mouth of the Tillamook and hoped to make his home there, but by 1856 his father's survey had wound down, and Clint needed work.[9] That year Congress authorized a survey of the boundary between Washington Territory and British Columbia, and either Clint or his father, who had returned to Washington, solicited recommendations for him for the position of surveyor and astronomer. William Emory warmly recommended Clint to President Pierce, identifying him as the son of Colonel Gardner. During the Mexican boundary survey, Emory noted, Clint Gardner had risen from the position of subassistant to become "one of the principal assistants." Emory added: "During the progress of that work many harassing circumstances occurred, & many temptations were offered to induce the assistants to desert their work, but Mr. George C. Gardner remained firm to his duty. To an elevated character as a gentleman, he unites an uncommon sprightliness of mien, and capacity for business, & I have no doubt, if employed on the new Boundary, he will give entire satisfaction."[10]

Meanwhile, Gardner's father, who knew Pierce personally, strenuously lobbied the president for an appointment for his son. The senior Gardner succeeded over the objections of Chief Astronomer John G. Parke, an experienced officer in the Corps of Topographical Engineers who favored West Point–trained candidates over civilians.[11] Clint became the assistant astronomer and surveyor for the Northwest Boundary Survey. Working on this binational line, Clint would have to cooperate with British counterparts, so his personal qualities counted as much as his experience and family connections. Hence, before appointing Clint Gardner, Pierce asked Emory whether Gardner was a "'gentleman' qualified to meet British officers of distinction." Emory assured the president that Gardner was "a very finished gentleman."[12]

In the spring of 1857 Clint shipped out of San Francisco with the survey party, which established a base camp just north of the 49th parallel on Semiahmoo Bay. Whatever reservations Joseph Parke had about working with a civilian assistant quickly disappeared. Twenty-year-old Joseph Smith Harris, who worked under Clint's supervision as an assistant surveyor and admired his experience and ability, reported that Clint quickly proved his worth to his employers:

> Gardner who Mr. Parke and Mr. Campbell both thought was going to be rather in the way the appointment being decidedly against their wishes has turned out a trump.[13] He went all through the Mexican Boundary Survey except the Gadsden boundary and has excellent ideas not only of instruments and observing but also of

> this kind of life. He was invaluable in San Francisco getting up camp equipment & mess furniture. He is exceedingly particular about having nice adjustments and keeping people away from the observing tents &c and I think if our observations here are not reliable it will be the fault of no one in the Astronomical party.
>
> Mr. Parke (this of course and the preceding you will keep to yourself) knew nothing of astronomical work except going out once with Capt. Lee on the Iowa boundary some 8 years ago. He said as much very openly and did not pretend to any more knowledge of the matter than he had.[14]

Gardner did have one liability, however, from Harris's point of view: "He seems to be a clever and efficient officer whose greatest fault in a social point of view is that of telling the longest winded stories imaginable of his previous campaigns [on the U.S.-Mexico survey]. They are interesting enough but one gets tired of hearing them when given in doses of 1 or 2 hours to be repeated every evening."[15] Clint, who once described Marine Chandler as a "bag of gas," seemed to have become a gasbag himself.[16]

Plagued by difficult terrain, sometimes-hostile Indians, frigid high altitudes, and summertime mosquitoes, the Northwest Boundary Survey wound down its difficult fieldwork in the spring of 1861. Clint and Charlie returned to Washington to help prepare the survey's written report.[17] Clint retained his position as assistant astronomer and surveyor and oversaw the office work for the next eight years, joined for a time by Charlie, who had found a position as a draftsman on the survey in 1859.[18] During those years Clint believed that he performed the duties of a chief astronomer without the commensurate salary, and, although he lodged a number of complaints, he received no satisfaction.

In February 1869, although the work was not yet completed (the final report was not submitted to Congress until 1890!), Clint resigned from the Northwest Boundary Survey and abruptly switched careers.[19] At thirty-eight he took a position as assistant engineer for the Philadelphia and Erie Railroad, in Erie, Pennsylvania, and began a meteoric ascent in railroad administration.[20] Within four years, in 1873, he was promoted to general superintendent of the Pennsylvania Railroad Division. He occupied the position with distinction, and, when he resigned in 1879, it was to take charge of another railroad in Massachusetts. That would be the pattern for the rest of his career. In 1881–84 he served as the general manager of the Mexican National Construction Company, the Mexican National Railroad Company, and the Texas-Mexican Railroad Company. In

1884–88 he presided over the Buffalo, New York and Philadelphia Railroad. In 1892 to 1896 he managed the Ohio River Railroad, and in 1896 he oversaw the work of an American railroad company in Peru.[21]

George Clinton Gardner died at his home at 416 North Beach Street, Richmond Hill, on Staten Island, on August 12, 1904, just short of his seventy-third birthday. His relatives transported his body from New York to Washington over the Pennsylvania Railroad, and he was buried with an Episcopal service in the Congressional Cemetery on August 17.[22] His wife and five of his six children survived him. In 1863 he had married a woman eight years younger than he, Mary Frances (Fanny) Brodhead, the sister of Delia Brodhead, who was married to his friend Joseph Smith Harris. The couple had five children: Julia Brodhead (b. July 13, 1866) and Alida (b. October 8, 1868), both born in the District of Columbia, and four children born in Pennsylvania: Delia Brodhead (b. February 23, 1870), Charles Kitchel (b. March 28, 1872; died April 16, 1872), George Clinton (b. March 5, 1873), and Mary Frances (b. April 14, 1875).[23] George Clinton Gardner's obituary in the *New York Times* identified his surviving children as "Mrs. George H. Brodhead, Miss Alida Gardner, Mrs. Edward B. Finch, G. Clinton Gardner, and Miss Mary Gardner."

It is possible that Gardner had at least one more child, unbeknown to the *New York Times* but perhaps known to his family. On May 15, 1860, at an Indian settlement called Matsqui on the banks of the Fraser River in British Columbia, an Indian woman name Selaamia gave birth to a son. The child received the tribal name of Quotaseltill, along with the name of his surveyor father, George Clinton Gardner. As the story goes, the boy never met his father, who had apparently left the area before the birth of his child but arranged for him and his mother to receive supplies regularly at Bellingham and left instructions for his education. When Quotaseltill reached the age of four, his father's agents at Bellingham talked about sending the boy away to school. Rather than lose the child, Selaamia told them he had died and stopped coming to Bellingham for supplies. If the story is true, and if George Clinton Gardner had continued to receive news about his son, he probably stopped sending support at that point. Years later Selaamia sent Quotaseltill to a mission school, where the priest christened him Charles Alphonsus Gardner. Thereafter people in the white world knew him as Charlie Gardner. He went on, in the words of one writer who knew him, to become "one of the really great river boat captains of Canada."[24]

Four letters from the spring of 1854 follow. Two concern Clint's resignation, and one is a report from the survey schooner. The most interesting by far is a newsy letter from Edward Ingraham, who had shared Clint's adventures from the start. Ingraham's letter suggests that the two had become good friends and that Ingraham had also come to know Clint's family and had corresponded with Clint's sister Sarah, with whom he was apparently smitten.

Clint had wanted a military commission but instead became a surveyor and a railroad executive; Ingraham entered the army and came to an untimely end. The eve of the Civil War found him stationed in the Indian Territory of Oklahoma at Fort Washita, where he was a member of the First U.S. Cavalry and post adjutant. In the spring of 1861 Ingraham went over to the Confederate side. Along with other deserters he made his way to Indianola, Texas, where he became part of Company A, Confederate States Cavalry. First Lt. Edward Ingraham, in the judgment of one historian, became "one of the most active of the Confederate recruiters at Indianola." He was the first of Company A to die in the Civil War, killed at Farmington, Mississippi, in May 1862.[25]

GCG to Emory, Washington, D.C., January 25, 1854[26]

Dear Sir

My private affairs making it necessary for me to leave Washington shortly I have most respectfully to request, through you, a leave of absence from the 28th of Febry 1854 to the end of March. At the expiration of which leave I desire to offer my resignation as assistant on the U.S. & Mex. Bdy. Survey.

I have the honor to be
With great respect
Your Obt. Svt.
G. C. Gardner
Ast. U.S.B.C.

Edward Ingraham to GCG, Ashwood,[27] ***February 3, 4, and 6, 1854***

Dear Clint

I rec.d your two favors of Jan. 20 & 22nd also Sarah[s] last *favor for now I suppose she will have to quit writing to single gentlemen. I am sorry for it but can't help it. But I must beg you to present my compliments and congratulations and hope her all manner* of happiness which is possible for human beings to be blessed with. I would have come on before but I wrote to the Major [Emory] and asked him what to do. [Nathaniel] Michler gave me no orders and I did not want to go on and have to return again—but now I wish I had done so. I am sorry to hear such bad news and will not attempt to offer any consolation for I am a very poor hand at talking. I can think, but very often can't express myself. Give my love to Mother. I wish I could raise the "wind." I would go to Oregon with you. Altho' I don't know what I could do after I got there—I am but an indifferent surveyor & in fact have no head for it but if it comes to throwing my heels in the air I am "thar." Have you seen any thing of my friend [F. W.] Tansill. And tell Phillips to write & let me know whether he recd. his trunk or not. I would like to know. This is a pretty dull place only the young people have got into the marrying mania 5 weddings & accompaniments last week but one & next week a grand party to be given to one of the "brides." I have three promised me if I will get married but can't fool me into any such arrangement altho' $75,000 is a great inducement. I prefer ranging around and when I get tired I will perhaps come into the noose. I hope your Mother is not as uneasy to see you married as mine is me. I wish I was rich, then I could do as I pleased. It is 12 o'clock P.M. which is late for the country. I will close now and write more tomorrow. E.[dward] I.[ngraham]

4th. I have just read your letters over again & to tell the truth I don't know what I am going to do. I don't want to go away from here yet awhile & I want to be with you in Washington. I don't much care whether I am in the Bdy. Com. or not yet I don't want to lose the pay. I think Lieut. Michler has acted very meanly towards me anyhow, and I do not want to serve with him any more. I am very much surprised that I don't receive something from the Major. I wish you would find your way down here before you go to Oregon if I come out there some day you must not be surprised for I don't know when or where I will bring up. If I should not see you again you must remember me to all my Cali[fornia] friends & acquaintances, and if you stop in San Francisco you will find Judge Sutherland and his lady living there. You must not fail to write me when you get there for I will always be happy to hear from you, in fact I am uneasy unless I get some kind of letter from you. When you reach Oregon and get settled you must let me know what your outfit consists of and what is best for that kind of Climate—it may be useful to me. Up to what date have you drawn your pay & have you got your[s] due us for Jan., Feb. & March 1853—I am waiting for an answer from the Major—if it costs me my place I can't help it.[28] If I do leave the Commission, I shall be on in the summer, and collect (if possible) what is due me.

6th. I go down to the Gulf [Grand Gulf] tomorrow with Father who goes to New Orleans next week there will be nobody home but Charlotte & myself. She has kicked the "Swede" so say nothing about it to any one. I was astonished to receive your sister Sarah's notice of marriage, altho' I did have my suspicions last winter. I did not think he was favorably recd. by her, however there is no finding out such things unless they choose to tell it, which was not likely to be the case in this instance for I know she would have been plagued nearly to death how I would liked to have known it for certain last winter—what fun we would have had. Whew! she may be very thankful I did not come on in the early part of the winter, tho' I expect I would have caught a pair of boxed ears if I had mentioned such a thing.

You must remember me to all the friends of my acquaintance who enquire after me. If I find a letter from any one in the Gulf tomorrow, I will write more.

Love to Mother, Aunty, Sarah & all at home.

I remain, Yours truly

Edward Ingraham.

P.S. February 8, 1854. I have just recd. yours dated 24. I shall come on as soon as I can leave.

Yours, E. Ingraham

W. E. Greenwell to GCG in Washington, D.C., Surveying Schooner Guthrie, Rio Grande, February 28, 1854

My dear Sir,

Your letter of Jany. 19th reached me this morning. By the same mail too I received a letter from Major Emory giving me the Latitude and Longitude of his Station at the Mouth of the River. His letter is dated Dec. 28, just two months in getting from Washington here. Quick work, for this age of progressiveness.[29]

I will do as you request in reference to the key. Capt. Van Fleet is in Washington at this time, but will return probably before I leave here. Should he not, however, I will leave it with Capt. Moses for him.

I expect to get through here about the middle of April. My work thus far comes out well, but we have had a hundred difficulties to contend with, and the worse is not yet over.

Ben Moses hasn't improved the slightest since you left.

Mr. West has gotten so fat can hardly walk. The rest remain in status quo.

I expect to reach Washington about the middle of May. Hope I shall find you there.

Yours very truly,

W. E. Greenwell

Love to Seaton, Radziminski & Prioleau.

Emory to GCG, Washington, April 13, 1854[30]

Sir

Your resignation to take effect May 1st 1854 and application for leave of absence for that time, has been approved by me & forwarded to the Commissioner.

In accepting your resignation, I have the pleasant duty of thanking you for your able and faithful service on the work, and to wish you success in your new undertaking.

I remain with great regard yr. friend & servant

W. H. Emory

---

1. Gadsden was appointed minister to Mexico in May and received his written instructions on July 15 (Garber, 1921, 82–84).
2. Werne (2007, 175–92) provides a crisp overview. Two older sources also remain valuable: Rippy (1921) and Garber (1923).
3. Werne, 2007, 193–218.
4. *(New York) Weekly Herald*, January 21, 1854, p. 24.
5. *(Washington, D.C.) Daily National Era*, February 3, 1854, p. 3. See chap. 8, note 83.
6. GCG to Emory, Washington, D.C., January 25, 1854; Emory to GCG, Washington, April 13, 1854.
7. GCG left Washington for New York in April 1854 to begin his journey to Oregon (M. Seaton to my Dear Harrison, Washington, April 13, 1854, GCGJC). GCG received his letter of appointment as deputy surveyor from his father at Oregon City, July 7, 1854, GCGP.
8. GCG quotes this in an undated and untitled summary of his career, written in his own hand in the mid- to late 1860s (Career of George Clinton Gardner, in GCGJC).
9. GCG to his sister Sarah, Fort Yamhill, Indian Reserve, October 5, 1856, and other letters from 1856 in GCGJC.
10. Emory to Pierce, Washington, August 13, 1856. GCG also received an endorsement from Joseph Lane, Oregon's first territorial governor (1849–51), then in Washington representing Oregon in Congress (Lane to Franklin Pierce, August 9, 1856, GCGJC).

11. Lane to Pierce, August 9, 1856, refers to GCG as the "son of our mutual friend, Col. G[ardner]." For the lobbying, see, for example, Charles K. Gardner to GCG, October 19 and November 4, 1856, and March 3, 1857, in GCGJC.
12. Charles K. Gardner to GCG, Washington, March 18, 1857, recounting a conversation with Emory. GCG's letter of appointment, made by Pierce on February 28, was forwarded to him by the Department of State on March 2, 1857, and addressed to him in Salem, where GCG received it in early April. Both documents are in the GCGJC.
13. Parke and Alexander Campbell, the American boundary commissioner, made no secret of their opposition to GCG. Harris knew of their opposition on first meeting GCG in San Francisco in May and reported that it began to soften when they met him: "He will turn out they now think much better than their fears" (Harris to his brother Stephen, San Francisco, May 18, 1857, in Harris and Anderson, 2000, 6). Harris appreciated GCG's skill: "He I judge is a good observer from what little I have seen of the way in which he handles a transit" (Harris to his brother Stephen [himself a surveyor], San Francisco, June 11, 1857, Harris Papers, box 1, f. 22; and Harris and Anderson, 2000, 9).
14. Harris to his brother Stephen, Camp Simiahmoo [Semiahmoo], July 31, 1857, in box 1, f. 23, Harris Papers, and in Harris and Anderson (2000, 19).
15. Harris to his mother, June 2, 1857, box 1, f. 22, Harris Papers, and in Harris and Anderson (2000, 8).
16. GCG to his father, Frontera, Texas, February 6, 1852.
17. GCG's lengthy and vivid manuscript report on the fieldwork, 1857–61, is in GCGJC. Harris and Parke remained for the summer to complete the fieldwork.
18. Harris and Anderson, 2000, 135 and 4. Charles K. Gardner had tried to get positions for both sons, but Commissioner Campbell was reluctant initially to hire two members of the same family. Charles Thomas Gardner enlisted in the Union army in 1862 and was honorably discharged in 1865. The U.S. Census for 1870 shows him at age forty, living in Washington, D.C., working as a treasury clerk, and living with his mother, his brother John, his sister Alida, and his wife, Anna (U.S. Census, Ward 5, Washington, D.C., roll M593_125, p. 75). By 1880 he and Anna had three daughters, Helen, Gertrude, and Sarah. He continued to work as a clerk in the Treasury Department (U.S. Census, Washington, D.C., roll T9_124, p. 81). Charles died on December 25, 1895 (Heitman, 1965, 1:445).
19. GCG, letter of resignation, to Archibald Campbell, Office of the Northwest Boundary Survey, February 1, 1869, GCGJC.
20. The inspiration for this move might have come from Joseph Smith Harris, who had worked under GCG on the 49th parallel and who became GCG's brother-in-law when Harris and GCG married sisters. Sometime in 1869 Harris sent a ringing endorsement of GCG to R. H. Sayre, in Bethlehem, Pennsylvania, as the survey was winding down. Harris recommended GCG for a position on the Union Pacific (letter in GCGJC).

21. Wilson, 1900, 41–47. GCG left behind ample documentation of his later life. Collections of his papers from the years following his work on the U.S.-Mexico boundary survey are at the Bancroft Library, University of California, Berkeley (focusing largely on his work on the Northwest Boundary Survey); at the Beinecke Rare Book and Manuscript Library, Yale University, which has two Gardner collections, one consisting of nineteen items, most treating the Northwest Boundary Survey, Indian claims, and land that GCG purchased in Buffalo (GCGJC), and another comprising 385 pages of journals and letters from the Northwest Boundary Survey, 1853–59 (GCGP); the William L. Clements Library, University of Michigan (150 items, treating his dredging of the Erie Canal in 1869–70, and his work in Mexico and Peru); and in the Pennsylvania Railroad Company's archives, Hagley Museum and Library (which has GCG's letterbook from his years as superintendent).
22. *New York Times*, August 15, 1904, p. 7 (this obituary drew from Wilson [1900] and *Washington Post*, August 18, 1904).
23. Harris and Anderson, 2000, xxvin31. *The Genealogy of the Vanderpool Family, Part II* (1912), pp. 520, 526, available on Ancestry.com, provides birthdates for the children and the birth date for Fanny Brodhead—November 9, 1839. The Vanderpool family genealogy spells Brodhead as Broadhead. Contemporary sources, however, use Brodhead (U.S. Census, 1880, Supervisor's Dist. # 60, Enumeration Dist. #267, Greenfield, Franklin County, Massachusetts, June 4, 1880, Roll T9_533, p. 8). Harris's wife, Delia, visited the Gardners' home in Washington, D.C., in late 1865 or early 1866 when Fanny was ill. While there she received a series of letters from her husband imploring her to return to their home in Pottsville. Those letters that remain (and he may have penned more) are Harris to D.B.H., c/o G. Clinton Gardner, 566 New Jersey, "south of Capitol," Pottsville, January 7, 9 (2 letters), 10, and 11, 1866, box 3, f. 55, Harris Papers.
24. Charlie Gardner retired in 1930. Edna Brandon Hanson interviewed him and wrote his life story in a series of articles, "The Strange Life Story of Captain Charlie Gardner," which ran in a series in the *Vancouver Sun:* November 29, and December 6, 13, 20, and 27, 1941. We are grateful to Cathy Craft for calling this story to our attention. GCG had close relations with Indians during his work on the survey, employing them as porters and traveling through villages. Joseph Smith Harris, for example, describes returning to Camp Semiahmoo with GCG from a strenuous hike. As night fell, they were still four miles from the camp, "and Gardner proposed that we should stop and get something to eat at an Indian settlement we were passing" (Harris to his mother, February 12, 1858, in Harris and Anderson, 2000, 70). The great-great-grandson of GCG, Steven Halliday, reported that GCG's illegitimate son has remained in "surviving family memories" and that "certainly it is family lore" (Halliday to Weber, December 10, 2008).
25. Weinert, 1962, 100; post orders and letters, Fort Washita, 1859–67, Yale Collection of Western Americana, Beinecke Rare Book and Manuscript Library. The Ingrahams lost all three of their sons in the Civil War, and their plantation, run by slave labor, was raided by federal troops

and their circumstances greatly reduced (Cleaves, 1960, 355–56; U.S. Census, 1860, Port Gibson, Claiborne, Miss., roll M653_580, p. 0 [*sic*]; ; U.S. Census, 1870, New Orleans Ward 5, Orleans, La., roll M593_521, p. 193).

26. This letter may be found in box 7, f. 59, EP.
27. A small community on Mississippi's southern border, about ninety miles down the Mississippi Valley from Ingraham's home at Grand Gulf, Mississippi. Ingraham and his father were apparently staying in Ashwood as they made their way home to Grand Gulf.
28. Ingraham had written to Emory the previous September to inquire about pay due him from January 1 to April 9, 1853 (Ingraham to Emory, San Antonio, September 23, 1853, box 6, f. 55, EP).
29. Greenwell acknowledged receipt of Emory's letter in a response dated February 28, 1854, and written aboard "Surveying Schr. *Guthrie*, Rio Grande" (box 7, f. 60, EP).
30. This letter may be found in box 7, f. 62, EP.

Appendix
A Chronological List of the Correspondence

1. To New Orleans and Panama, February 13–June 22, 1849

Emory to GCG
Letter of Appointment,
Washington, February 13, 1849

GCG to his mother, Mrs. C. K. Gardner
Setting out.
Washington, Pennsylvania, February 16, 1849

GCG to his father, Col. Charles K. Gardner
To Cincinnati, then down the Mississippi to New Orleans.
Evansville, Indiana, February 26, 1849

GCG to his mother
Down the river to New Orleans.
Steamer *Glencoe*, March 1, 1849

GCG to his father
Arrival at New Orleans. Financial stringency of Colonels Weller, Emory, and Gray.
New Orleans, March 6, 1849

GCG to John McCalla Jr.
In New Orleans without fashionable clothing.
New Orleans, March 16, 1849

GCG to his brother Charles
Affairs at New Orleans. Cholera epidemic. Plans to set out for Chagres with Gray.
New Orleans, March 18 and 19, 1849

GCG to his father
Departure of Colonel Weller. Failure to provide funds. Gray's kindness.
New Orleans, March 19, 1849

GCG to his father and mother
Of Chagres. Incidents of the voyage. Major McKinstry.
Steamer *Falcon*, March 27, 1849

GCG to his father
Money matters.
n.p. [Panama], n.d. [April 1849]

GCG to his sister Sarah
Journey across the Isthmus. Railroad survey in progress. Colonel Weller's neglect. Emory's arrangements.
Panama, April 23, 1849

Emory to Col. C. K. Gardner
Testifying to the good character, intelligence and deportment of GCG, the recipient's son.
Panama, April 27, 1849

GCG to his sister Alida
Description of Panama. Arrival of the California steamer. Wild scenes of excitement. Skyrocketing prices. Mails looted and mail agent murdered. Pretty señoritas and hopeful flirtations.
Panama, May 8, 1849

GCG to his father, mother, John, Jacob, Sarah, Charlie, and Aunty
Further details of scenes and events at Panama. Weller leaves. Preparation for voyage to San Diego.
Panama, May 13, 1849

GCG to his father
Sending money home. Lieutenant Whipple borrows $25. Departure of Emory and Gray for California. Likeable Whipple takes command. Theft of clothes. Needs a Spanish dictionary. Unsavory rumors regarding Colonel Weller.
Panama, ca. May 20, 1849

GCG to his mother
Sending a box of Panama trinkets.
Panama, May 21, 1849

GCG to his sister Sarah
Arrival of General Smith; headquarters at Tobago. *Crescent City* comes to port. Fearful toll of cholera. Craven conduct of the priests. Returning gold hunters stricken. Wholesale gambling. Thirteen members of survey still remain. Their positions and duties.
Panama, June 12, 13, and 14, 1849

GCG to his brother Charles
Announcing departure for San Diego. Hart dies of cholera. Pays fare of stranded emigrant.
Panama, June 22, 1849

2. Initial Points—Camp Riley to the Gila, July 23–October 31, 1849

GCG to his sister Sarah
Harrowing trip. Cholera and storm. Arrival in San Diego. Move to Camp Riley. Forty-niners.
Camp Riley, California, July 23, 1849

GCG to his uncle John McLean
Fruitful reflections of a youthful mind in Panama. Acknowledgment of receipt of the Treaty of Guadalupe Hidalgo.
Camp Riley, California, July 23, 1849

GCG to his mother
Writing on his birthday. Sharp practice on Weller's part. Selling government supplies to Survey members. Shortages. Emory's upright course. Difficulties in Gray's camp. News from the goldfields. Meteorological details.
Camp Riley, California, August 18, 1849

GCG to his brother John
Distress rife among the gold seekers. Shipwrecked emigrants. Advice for the journey. Best advice: stay home.
Camp Riley, California, September 15, 1849

GCG to his father
Frémont rumored to be the next commissioner. Emory threatens to leave. Underground politics. Off to the mountains.
n.p., September 15 and 20, 1849

GCG to his father
Establishment of boundary's initial point. Journey to the mountains and among the Indians. Frémont on the way overland. Weller's continued duplicity—"loves women, cards and liquor too much." Melancholy affair on the Gila. The Collier overland expedition. Frayed nerves. Weller and Gray quarrel—Weller chokes Gray and Gray shoots Weller in the thigh.
Camp Riley, California, October 15, 1849

GCG to his father
More on the Collier disaster, out of food and money ($100,000 spent). Captain Thorn and two Indians drowned crossing the Gila. Lieutenants Whipple and Couts out of supplies. Women and children facing starvation.
San Diego, October 28 and 31, 1849

GCG to his sister Sarah
With Hardcastle to Tecate, Mount Los Pinos, and Mount Wicammon. Flashing. Wretched Indians.
Camp Riley, October 30, 1849

GCG to his mother
Thanking her for gift.
n.p., October 31, 1849

3. "Glorious State of Uncertainty," November 15, 1849–July 2, 1850

GCG to his father
Appointed clerk to Captain Hayden. Improvement in finances. Frémont's plans.
San Diego, November 15, 1849

GCG to his father
Frémont and Weller in political war. Emory promises Gardner eventual berth in the coast survey. Scarcity of mail.
San Diego, November 30, 1849

GCG to his sister Sarah
Christmas celebration. Reading family letters. Dreaming over family daguerreotypes. Church services and pretty señoritas. New Year's dances, parties, and balls.
San Diego, January 4, 1850

GCG to John M. McCalla
A wedding. Beautiful young ladies. His room. Brandy.
San Diego, February 3, 1849 [1850]

GCG to his father
Forwarding draft by Captain Hardcastle for $500.
San Diego, February 9, 1850

GCG to his sister Sarah and to his father
Departure of Weller and Whipple. Granted three-month leave. Weller to retain post. Picnic with the Bandinis. Dancing the polka with Ysidora till the dawn. Emory to act as commissioner in Weller's absence. Prevalence of gambling.
San Diego, February 19 and 23, 1850

GCG to his mother
Large estates of the local grandees. Fertility of the land. Americans plan a "Territory of Southern California." Mormon claims to the region. Ferry on the Gila. Rival company. Murders and outlawry. Arrival of General Smith. Fred Emory makes $20,000 speculating. Gardner quarrels with Emory and is ready to leave.
San Diego, April 6, 9, and 20, 1850

GCG to his father
Asking to have *Weekly Intelligencer* sent to Captain Hardcastle.
n.p., n.d.

GCG to his brother Charles
Ordering artists' materials. Duties and life in San Diego. Expenses, parties.
San Diego, April 20, 1850

GCG to his father
Completion of boundary line to the Gila. Hardcastle and Gardner to place the monuments. Major Fitzgerald takes command of dragoons. General Smith planning to take over the missions. Personnel on hand. Reconciles with Emory.
n.p., April 20 and May 3, 1850

GCG to his father
Sending $90. Captain Hayden leaving for the East. Asking that Hayden be received socially by the family.
San Diego, May 14, 1850

GCG to his mother and father
Urging father to push boundary line bill through Congress. No pay in months. From comfort to a buttonhole existence. Emory, Parry, and Gardner now messing together. Report on the work done on the boundary line.
San Diego, June 30 and July 2, 1850

GCG to his uncle John McLean
Asking his uncle to use his influence in Congress for a commission in the newly organized California Dragoons.
San Diego, July 2, 1850

4. Finishing the California Line, July 23, 1850–June 1, 1851

GCG to his uncle Clinton McLean
Asking about the $50,000 boundary appropriation. Listing clothing needed. Severe straits.
San Diego, July 23, 1850

GCG to his father
In charge of a party. Sketching topography. Hopes Uncle Clint will replenish his wardrobe.
Camp Odeide near U.S. & Mexican Bdy., August 1, 1850

GCG to his mother and father
Bartlett has taken the $50,000 congressional appropriation. Emory borrows $1,000.
San Diego, October 2, 1850

GCG to his uncle Clinton McLean
For the third time, cannot meet him in San Diego.
Camp at Vallecito on route to Gila and Colorado rivers, November 8, 1850

GCG to Lt. E. Murray, Dr. H. Hewit, and Lt. Thomas Sweeny
Asking them to attest to the propriety of his conduct in his quarrel with Captain Hardcastle.
San Diego, November 18, 1850

Dr. Hewit et al. to GCG
Advising of their favorable view of his character. Hardcastle will not dismiss him.
Vallecito, California, November 21, 1850

GCG to Captain Hardcastle
Demanding written discharge and reasons for Hardcastle's actions.
San Diego, November 18, 1850

Capt. E. L. F. Hardcastle to GCG
Claiming disobedience of order, but withholding discharge.
Vallecito, California, November 20, 1850

GCG to his father
His version of the fracas with Hardcastle.
San Diego, November 30, 1850

GCG to his father
Cautioning against supporting Colonel Weller in the imbroglio with Horace Greeley. Weller's affair with a prostitute at Panama.
n.p. (San Diego), December 2, 1850

GCG to his sister Sarah
Appointed to survey and lay out the new town of San Diego. History of the town.
San Diego, December 2, 1850

GCG to his father
Regarding preemption laws. Plans to locate some land.
San Diego, January 2 [1851]

GCG to his mother
Completion of the Gila notes. Local surveying work. Marriage of Dolores Bandini, a grand affair.
San Diego, January 17, 1851

GCG to his mother and father
Arrival of Uncle Clint on the *California*. General Smith gone to the Gila with Ingraham. Dr. Parry leaving for Washington. Gardner left in charge at San Diego.
San Diego, February 18, 1851

GCG to his father
Remits $500 by Dr. Parry. Asks to be transferred to Lieutenant Whipple's party. Arrival of the monuments. Future post on the coast survey arranged for Gardner by Emory.
San Diego, February 28, 1851

GCG to his father
Colonel Weller's mistakes in computing pay roll of boundary survey.
San Diego, March 23, 1851

Emory to Col. Charles K. Gardner
Advising that Captain Hardcastle will be obliged to retain GCG in California for some months—"This is not unfavorable to your son's interest—quite the contrary."
Washington, March 15, 1851

GCG to his sister Alida
The trip from San Diego to the Rio Colorado and Gila. Murder at the ferry. Emigration disasters. Placing monuments on boundary line. Fort Defiance.
Camp at the Little Lagoon, April 28, 1851

GCG to his sister Alida
Dr. Parry's visit to the family.
Camp near Initial Point of U.S. & Mexican Boundary, June 1, 1851

5. Home for a Day and Off to El Paso del Norte, September 7–October 20, 1851

GCG to his mother
Arrived in New York. In search of Uncle John.
New York, September 7, 1851

GCG to his father
At Cousin Eliza's. In poor health.
Little Britain, September 12, [1851]

C. K. Gardner to GCG
Hardcastle looking for Clint Gardner.
Washington, September 15, 1851

GCG to his mother
Arrives at Montgomery, Alabama, en route to El Paso.
Montgomery, September 24, 1851

GCG to his father (telegram)
Emory gone. Ordered to San Antonio. No funds. What to do?
New Orleans, September 29, 1851

Col. Charles K. Gardner to GCG
Wiring $100 via S. F. Butterworth.
Washington, October 1, 1851

GCG to his father.
Received telegram. Insulting attitude shown toward the boundary commission at New Orleans. Funds low and credit poor.
New Orleans, October 1, 1851

GCG to his sister Alida
Journal of the trip through Alabama by stage coach.
New Orleans, October 3, 1851

GCG to Samuel Butterworth
Has missed connecting with Emory. Acknowledging communication. Sailing for Indianola and thence to San Antonio.
New Orleans, October 3, 1851

GCG to his mother
March from San Antonio. Preparations for crossing the Plains. Establishment of a mail route to San Antonio.
Camp near Ft. Inge, Texas, October 20, 1851

6. Frontera, Texas, a "Derned Mean" Country, November 27, 1851–May 16, 1852

GCG to his sister Sarah
Announcing arrival at El Paso after a forty-one-day march. Conference with Colonel Graham. Bartlett drawn on next year's appropriation and can't be located. Report has it that he has gone north to write a book. Several murders committed, allegedly by discharged members of the commission.
El Paso, November 27 and December 11, 1851

GCG to his father
Appointed Emory's clerk. Financial matters.
Near El Paso, December 11, 1851

GCG to his mother
Severe weather. Health good. Family should write. Needs boots. Emory to search for Bartlett and may go with him.
Frontera, near El Paso, December 19, 1851

GCG to his mother
Appointed clerk of the chief astronomer and head of the scientific corps. Suspicious of Bartlett's movements. Holy days, cavorting Mexicans, and bullfights. Dinner for the Mexican commission.
Frontera, Texas, January 8, 1852

GCG to his father
In bitter attack on Commissioner Bartlett and the *Intelligencer,* a newspaper that supports the commissioner, and fighting Emory and the California contingent. Requesting Uncle John to "say something in reply." Asserts Gray has claimed undue credit. Cites gross errors in the survey. States he was not within ninety miles of the Gila. His map a copy of that drawn by Whipple and Foster. Charges Gray with issuing a "false publication." Pleads to be allowed to expose the incompetence and duplicity of the Bartlett-Gray faction.
Frontera, Texas, January 11, 1852

GCG to his father
Use influence to get him a commission in the army. Encloses letters to William Gwinn and Joseph Lane.
Frontera, Near El Paso, January 27, 1852

GCG to Honorable William M. Gwinn
Soliciting commission in the newly formed western regiment.
Frontera, Texas, January 27, 1852

GCG to General [Joseph] Lane
Soliciting a commission in the Oregon Regiment.
Frontera, Texas, January 27, 1852

GCG to his father
Needs to add to his wardrobe. Wants news. Emory en route to the Gila. Gardner in charge of observatory.
Frontera, Texas, January 27, 1852

GCG to Major Emory
Reporting on his fulfillment of his duties.
Frontera near El Paso, February 1, 1852

GCG to his father
A barren country. Murderous and thieving Apaches. Magoffin offers $20 for an Apache scalp. Gardner's solution to the Apache problem. Use influence to get a commission in the army.
Frontera, Texas, February 6, 1852

GCG to his father
Indians daily committing depredations upon the surrounding settlements. Daring murders perpetrated within gunshot of the post. Attacking the copper mines, dragoons killed, horses carried off, boy taken captive.
Frontera, Texas, February 16, 1852

GCG to his sister Sarah
Eager for news from home. Mail intercepted by Indians.
Frontera, Texas, February 16, 1852

GCG to his brother Charles
Avoids camp followers. Seeks women from old El Paso society. Not as beautiful as girls in San Diego but accomplished.
Frontera, Texas, March 18, 1852

GCG to his father
Waste and inefficiency of the boundary survey.
n.p., March 1852

GCG to his sister Alida
Society—or lack of it—in El Paso. Visits with members of the commission and with ladies without beauty, education, or anything else that constitutes a lady by his reckoning.
Frontera, Texas, March 21, 1852

GCG to his father
Sending $250. Shameful usury; a public disgrace; Washington taking no notice of the thieving rascals. Commissioner Bartlett's influence. Commission now $100,000 in debt. Magoffin adopts Gardner as his son. Indian outrages multiplying.
Frontera, Texas, March 27, 1852

GCG to his sister Sarah
Pete Brady survived fight with Indians. Hardships of camp life. A delicious dinner intervenes. Ordered on a reconnaissance to parts unexplored.
Frontera, Texas, April 9, 1852

GCG to his father
Gray and Bartlett in San Diego. Hopes Congress will not give Bartlett all the funds; Emory needs them. Emory should be in charge of commission.
Frontera, Texas, April 9, 1852

GCG to his family
Acknowledging receipt of letters.
Frontera, Texas, April 10, 1852

GCG to his sister Alida
Alone with servant and horse. Emory, Ingraham, and Dr. Parry at Fort Filmore. Heads reconnaissance to Lake Guzmán, a barren and desolate region, ninety miles without water. Rain saved mules. Indians about. Geography of the region.
Frontera, Texas, May 8, 1852

GCG to his mother
Black Rock property. Preparing to move down the Rio Grande. His wardrobe.
Frontera, Texas, May 11 and 16, 1852

GCG to his father
Colonel Weller elected senator; use influence with him. General conspiracy against Emory, with inside account of the animosities. Teamsters armed. Scandalous and false orders. Shameful charges.
Frontera, Texas, May 16 and 18, 1852

7. "God Knows What Will Become of the Survey": Down the Rio Grande to Ringgold Barracks, May 30–December 11, 1852

GCG to his sister Sarah
Packing instruments. Needed as Emory's clerk. Chandler returned with report of visit to Gardner's family.
San Elizario, Texas, May 30, June 2, and June 5, 1852

GCG to his father
Wardrobe, finances, and commission in the army. Emory's troubles with army officers.
San Elizario, Texas, June 5 and 7, 1852

GCG to his sister Alida
A hasty note written to report impending departure.
San Elizario, June 10, 1852

GCG to his sister Sarah
Setting up observatory. Work proceeding nicely. Looks forward to finishing.
Camp [?]eminst the Cañon on the Rio Grande, June 18 and 20, 1852

GCG to his father
The journey from San Elizario to the Canyon of the Rio Grande. Completion of the survey to Presidio del Norte. Astronomical points determined at Fort Fillmore, Frontera, San Elizario, Rio Grande Camp, and Eagle Pass—work completed without money. Emory deserves credit.
Camp at the Canon of the Rio Grande, June 20, 1852

GCG to his sister Alida
A flood. Good tasting water. Not Indians, yet.
Camp on the Ciballo River [Cibolo Creek], July 6, 1852

GCG to his father
Arrival opposite the presidio, luxuriating in watermelons and credit.
Camp opposite Presidio del Norte, July 10, 1852

GCG to his sister Sarah
Expected Indian outbreak. Emory puts GCG in charge of a survey party. Family matters. Remind father to get him a commission in the army.
Camp Sumner, opposite Presidio del Norte, July 23, 1852

GCG to his sister Alida
Unfinished letter. In charge of surveying party and ordered to proceed down the river.
Camp on the Rio Grande above Presidio del Norte, July 29, 1852

GCG to Emory
Took charge of surveying party. Begins work.
Surveying Camp Opposite Presidio del Norte, August 1, 1852

GCG to his sister Alida
Needed by Emory; relieved of command of survey party. Two dances for the surveyors, arranged by Colonel Justiniani. Delicious melons.
Camp, August 9, 1852

GCG to his father
Thanks his father for boots and shoes.
n.p., August 10, 1852

GCG to his mother
Expected move eastward. Attends two dances. Emory fears for his assistant's youthful heart.
Camp Opposite Presidio del Norte, Mexico, August 13, 1852

GCG to his sister Sarah
En route to Eagle Pass. Indian fight at Devil's River. His wardrobe. Pierce will win.
Camp on Las Moras Creek, September 21, 1852

GCG to his sister Alida
Report on trip from Presidio del Norte over unexplored country to Leona Spring and eastward. In command of the advance guards. Indian tracks in abundance. Making twenty miles a day.
n.p. (Las Moras Creek), September 21, 1852

GCG to his father
Personal finances and investment. Use influence with senators to get per diem for officers.
n.p., n.d. (Las Moras Creek, September 21, 1852)

GCG to his father
Reporting arrival on the twenty-third. Personal finances.
Camp near Fort Duncan, Texas, September 27, 1852

GCG to his sister Alida
Comfortable camp. Report on the march from Presidio del Norte. No military escort. Comanche signs daily. Indians attack mail coach. Regrets office work.
Camp near Eagle Pass, Texas, September 30, 1852

GCG to his father
Personal finances.
Camp near Fort Duncan, Texas, October 2, 1852

GCG to his sister Sarah
Attends "rural" dance. Girls not attractive but dance well. Will stay here for some time. Family and friends.
Camp, October 3, 1852

GCG to his father
In command of a party. Desolate situation. Bartlett thought to be on way. Lies published in the *Union*. Clash with the Mexican commission.
Camp near Fort Duncan, Texas, October 8 and 10, 1852

GCG to his father
Emory's drafts protested by the department. Personal finances. Weekly mail delivery from San Antonio.
n.p., n.d. (ca. October 15, 1852)

GCG to his sister Alida
Family matters. Departure of Michler and Ingraham for Washington. Dances—Emory disapproves. Bartlett expected.
Camp near Fort Duncan, Texas, October 16, 1852

GCG to his aunt
Asking for a letter and news of home and family.
Camp near Fort Duncan, Texas, October 16, 1852

GCG to his sister Sarah
Dull place. Hopes U.S. will invade Cuba and he will get a commission in the army. Family matters. Bartlett expected.
Fort Duncan, Eagle Pass, Texas, October 22, 1852

GCG to his sister Alida
No money, no credit, no Bartlett.
Fort Duncan, Texas, October 31, 1852

GCG to his father
Work up in the air. Will remain with Emory, in charge of instruments. Hopes not return to Washington without employment.
Camp near Fort Duncan on the road to Laredo, November 12, 1852

GCG to his sister Alida,
Will not be home. No letters from Frank. Fine quarters. Commission in the army. Wardrobe.
Ringgold Barracks, Texas, November 28, 1852

GCG to his father.
Election of Pierce. Revolution in Chihuahua. Cuban situation. No money. Bartlett "on a pleasure trip."
Ringgold Barracks, Texas, November 28, 1852

John H. Ingle to GCG
Local news and gossip from a Washington friend.
Alexandria, Virginia, December 3, 1852

GCG to his sister Sarah
Army personnel. Agreeable society. Camp life in general.
Ringgold Barracks, Texas, December 11, 1852

8. Back to the Rio Grande, Then Home Again, January 2–November 18, 1853

Emory to GCG
Ordered to turn public property over to Bartlett's designee.
Corpus Christi, Texas, January 2, 1853

GCG to his father
Disposing of the property and heading home.
Corpus Christi, Texas, January 2, 1853

GCG to his father
Advising of his temporary transfer to Washington. En route.
New Orleans, January 14, 1853

James Green to GCG
Boundary finances.
New York, January 18, 1853

E. E. to GCG at Washington
Ingraham's horse. Indian depredations and murders.
Fort Inge, Texas, February 16, 1853

GCG to Francis M. Holley
Repayment of loan.
Washington City, March 9, 1853

GCG to his father
Arrival in New York and headed to New Orleans. Gifts for the family and servants.
New York, April 22, 1853

E. A. Palfrey to GCG
Filibuster uprising. Arrest of the ringleaders—"our sweet village is now quite respectable." Pleased GCG is returning.
Ringgold Barracks, Texas, April 23, 1853

GCG to his sister Sarah
Voyage to New Orleans by way of Havana, with account of visit there.
New Orleans, May 6, 1853

GCG to his sister Alida
Voyage to Brazos in "a small, dirty and dangerous tub of a sea boat." Narrow escape at the bar. Emory's movements. Pleasantries and family matters.
Brazos Santiago, May 17, 1853

GCG to his sister Sarah.
Rumors that President Pierce has ordered the occupation of the disputed valley near El Paso. Santa Anna collecting troops in all directions.
Ringgold Barracks, Texas, June 18, 1853

GCG to his father
Inquiring as to possibility of war with Cuba. The projected railroad to the Pacific.
n.p., June 16, 1853

GCG to Emory
Leaving to build observatory. A house for Emory.
Brownsville, Texas, June 29 and 30, 1853

Col. Charles K. Gardner to GCG
Asking his son to join him in Oregon.
General Land Office, Washington, June 30, 1853

GCG to his father
Real estate matters. Trip to the Brazos.
Mouth of Rio Grande, July 3, 1853

GCG to Emory
Arrived at the mouth of the Rio Grande. Situating the observatory.
Mouth of the Rio Grande, July 3, 1853

Col. Charles K. Gardner to GCG
Advantages of moving to Oregon to work for him. Needs him immediately.
Capitol Hill, Washington, July 5, 1853

GCG to his sister Alida
Description of his quarters. Erection of the observatory. Emory gone to Ringgold Barracks.
Boca de Rio Grande, July 18, 1853

GCG to his sister Alida
Political affairs in Washington. Quarters in a ten-pin alley. Our rights in the matter of the initial point of this boundary. Unfriendly attitude of the Mexican commissioners.
Boca de Rio Grande, July 19, 1853

GCG to Emory
Observatory built. Coast survey arrived.
Boca de Rio Grande, July 27, 1853

GCG to his sister Sarah
Death of Thomas Jones by drowning. Gloom. Up all night observing.
Boca de Rio Grande, July 30 and August 1, 1853

GCG to his father
Needs to stay to finish his work. Asks for more information about Oregon.
Mouth of Rio Grande, Near Brazos Santiago, August 1, 1853

J. Pioleau to GCG
Receipt for goods delivered to Rancho Blanco.
Brownsville, August 2, 1853

GCG to his father
Reasons not to join his father immediately. Emory agrees.
Boca del Rio Grande, Near Brazos Santiago, August 4, 5, and 9, 1853

GCG to his father
Plans for obtaining surveying equipment and joining his father.
Advice for crossing Panama.
n.p., August 11, 1853

Col. Charles K. Gardner to GCG
Planning to leave for Oregon with son Charlie. Hopes Clint is on the way to join him.
Capitol Hill, Washington, August 13, 1853

GCG to his father
Reiterates reasons for remaining at the observatory.
n.p., August 1853

Emory to GCG
Leaves Gardner in charge of the observatory for next four months. Instructions for disposing of equipment and shipping instruments. Confident in Gardner's ability.
Mouth of the Rio Bravo, August 20, 1853

GCG to his father (telegram)
Cannot go to Oregon until January.
New Orleans, August 22, 1853

Robert Campbell to GCG
Sends Gardner $1,000 for expenses.
Ringgold Barracks, August 22, 1853

J. Mc. D. Taylor to GCG
Gardner's and Emory's accounts. Yellow fever worse than the newspapers say.
New Orleans, August 23, 1853

Sarah Gardner to GCG
Father approves of Clint's plan; fortunes to be made in Oregon. Clint should come home for a visit.
Washington, August 25, 1853

Sarah Gardner to GCG
Whole family approves of Clint's plan; father heading to Oregon. Whole "Gardner race" may go, too!
Washington City, August 28, 1853

GCG to Emory
Return of instruments. Fired the cook.
Mouth of the Rio Grande, September 1 and 2, 1853

GCG to his sister Alida
Return of Emory to Washington. Account of life at the Rio Grande quarters.
Mouth of the Rio Grande, September 2, 1853

GCG to his father
Plans to join his father in January.
Mouth of Rio Grande, September 4, 1853

Emory to GCG
Arrival at Pensacola, yellow fever. Pay washerwoman. Take care in making observations. Hope that he "saw the great comet."
Pensacola, September 8, 1853

J. Hamilton Prioleau to GCG
Discussing various matters of the boundary survey workers. Movement of troops to Ringgold Barracks. Heavy rains cause survey delay. Sickness prevalent. Houston at Brownsville "keeping the Tide Register—but not keeping Mag Evans. They have quarreled." Mrs. White in limbo for keeping a brothel.
Camp 80 on Rio Grande, September 15, 1853

GCG to Emory
Shipping instruments. Bilious fever in Brownsville; his men sick.
Mouth of the Rio Grande, September 20, 1853

Mathilde Gardner to GCG
Family matters.
Ile Copal, September 21, 1853

GCG to Emory
Gets check from Campbell. Clouds halt observations.
Boca del Rio Grande, September 22, 1853

Emory to GCG
Announcing his arrival in Washington. Contracted yellow fever on the Alabama River. Magnetic notes, observations. General Campbell to send more money.
Washington, September 28, 1853

Emory to GCG
Relating to the preparation of a report and estimate of the present state of the Survey. Still ill with fever. Necessity of a perfect set of observations.
Washington, October 10, 1853

GCG to Emory
No mail service. Report on activities.
Mouth of the Rio Grande, October 10, 1853

Emory to GCG
Listing instruments to be brought north.
Washington, October 25, 1853

GCG to Emory
Sending observations; loss of the ambulance. Pony and bow and arrow.
Mouth of the Rio Grande, October 30, 1853

Emory to GCG
Send the dip circle to New Orleans.
Washington, November 1, 1853

E. and G. Blunt to Gardner
Receipt of instruments.
New York, November 3, 1853

Emory to GCG (duplicate)
Ship Zenith telescope to New Orleans.
Washington, November 5, 1853

Robert B. Campbell to GCG
Ordering Gardner to sell the boundary survey's property at the Rio Grande station.
Mouth of the Rio Grande, November 9, 1853

Emory to GCG

Requesting various observations, and to keep certain other things "till we go back—if we ever go back."

Washington, November 9, 1853

Minutes of the meeting of the boundary commission

General Campbell presiding; Gardner in attendance. Lamenting the death of Thomas Walter Jones.

November 10, 1853

Radziminski to GCG

Advising on various shipments.

Brownsville, November 15, 1853

GCG to Emory

Yellow fever in Brownsville. Shipping instruments and the pony. Coast survey needs his help.

Mouth of the Rio Grande, November 18, 1853

9. Resignation and New Adventures, January 25–April 13, 1854

GCG to Emory

Asks for leave through end of February, then will resign.

Washington, D.C., January 25, 1854

Edward Ingraham to GCG

Replying to Gardner's news that he is going to Oregon. Regrets he cannot join him. Sarah's marriage.

Ashwood, February 3, 4, and 6, 1854

W. E. Greenwell to GCG

Reporting on survey duties. Schooner Guthrie.

Surveying Schooner Guthrie, *Rio Grande, February 28, 1854*

Emory to GCG

Gardner's resignation to take effect May 1.

Washington, April 13, 1854

Sources Cited

Almada, Francisco R. 1955. *Resumen de historia del estado de Chihuahua*. México: Libros Mexicanos.

Arnold, Howard Payson. 1899. *Historic Side-Lights*. New York: Harper.

Arrington, Leonard J. 1966. *Great Basin Kingdom: An Economic History of the Latter-day Saints, 1830–1900*. Lincoln: University of Nebraska Press.

Audubon, John Woodhouse. 1852. *Illustrated Notes of an Expedition through Mexico and California*. New York: J. W. Audubon.

Austerman, Wayne R. 1985. *Sharps Rifles and Spanish Mules: The San Antonio–El Paso Mail, 1851–1881*. College Station: Texas A&M University Press.

Bancroft, Hubert Howe. 1884–1990. *History of California*. 7 vols. San Francisco: History Co.

Bartlett, John Russell. 1854. *Personal Narrative of Explorations and Incidents in Texas, New Mexico, California, Sonora, and Chihuahua, Connected with the United States and Mexican Boundary Commission, during the years 1850, '51, '52, and '53*. 2 vols. London: George Routledge.

———. 2006. *Autobiography of John Russell Bartlett, 1805–1886*. Ed. Jerry E. Mueller. Providence, R.I.: John Carter Brown Library.

Berthold, Victor M. 1937. *The Pioneer Steamer California, 1848–1948*. Boston: Houghton Mifflin.

Brown, Lenard E. 1969. *Survey of the United States–Mexico Boundary, 1849–1855: Background Study*. Washington, D.C.: National Park Service.

Bryan, Jimmy L. 2006. "The American Elsewhere: Adventurism and Manliness in the Age of Expansion, 1815–1848." Ph.D. diss., Southern Methodist University.

Burstyn, Harold L. 1957. *At the Sign of the Quadrant: An Account of the Contributions to American Hydrography Made by Edmund March Blunt and his Sons*. Mystic, Conn.: Marine Historical Association.

Callahan, Edward W. 1901. *List of Officers of the Navy of the United States and the Marine Corps from 1775 to 1900*. New York: L. R. Hamersly. Revised and digitized as "Officers of the Continental and U.S. Navy and Marine Corps, 1775–1900," Naval Historical Center Home Page, www.history.navy.mil/books/callahan/reg-usn-a.htm.

Chaffin, Tom. 2002. *Pathfinder: John Charles Frémont and the Course of American Empire*. New York: Hill and Wang.

Clark, A. Howard. 1902. *A National Register of the Society Sons of the American Revolution*. New York: Louis H. Cornish.

Cleaves, Freeman. 1960. *Meade of Gettysburg*. Norman: University of Oklahoma Press.

Cleland, Robert Glass. 1941. *The Cattle on a Thousand Hills: Southern California, 1850–1870*. San Marino, Calif.: Huntington Library.

Conrad, Glenn R., ed. 1988. *A Dictionary of Louisiana Biography*. 2 vols. New Orleans: Louisiana Historical Association and the Center for Louisiana Studies, University of Southwestern Louisiana.

Couts, Cave Johnson. 1932. *From San Diego to the Colorado in 1849: The Journal and Maps of Cave J. Couts*. Ed. William McPherson. Los Angeles: Arthur M. Ellis.

Davies, Charles. [1853] 1866. *Elements of Surveying and Navigation with Descriptions of the Instruments and the Necessary Tables*. Rev. ed. New York: A. S. Barnes.

Davis, William Heath. 1967. *Seventy-five Years in California. Recollections and Remarks by One Who Visited These Shores in 1831, and Again in 1833, and Except When Absent on Business was a Resident from 1838 until the End of a Long Life in 1909*. Ed. Harold A. Small. San Francisco: J. Howell.

Dawson, Joseph G. III, ed. 1990. *The Louisiana Governors from Iberville to Edwards*. Baton Rouge: Louisiana State University Press.

De Bruhl, Marshall. 1993. *Sword of San Jacinto: Life of Sam Houston*. New York: Random House.

DeBuys, William. 1999. *Salt Dreams: Land and Water in Low-down California*. Albuquerque: University of New Mexico Press.

Derby, George Horatio. 1969. *Derby's Report on Opening the Colorado, 1850–1851. From the Original Report of George Horatio Derby*. Ed. Odie B. Faulk. Albuquerque: University of New Mexico Press.

Diccionario Porrúa: Historia, biografía y geografía de México. 1971. 3rd ed., 2 vols. Mexico: Editorial Porrúa.

Ellison, William Henry. 1950. *A Self-Governing Dominion, California, 1849–1860*. Berkeley: University of California Press.

Emory, William H. 1851. "Notes on the Survey of the Boundary Line between Mexico and the United States." Pamphlet. Cincinnati: Morgan & Overend.

———. 1857–59. *Report on the United States and Mexican Boundary Survey Made Under the Direction of the Secretary of the Interior*. 2 vols. Washington, D.C.: Cornelius Wendell.

Etter, Patricia A. 1998. *To California on the Southern Route, 1849: A History and Annotated Bibliography*. Norman, Okla.: A. H. Clark.

Faulk, Odie B. 1967. *Too Far North, Too Far South*. Los Angeles: Westernlore.

Fireman, Janet, ed. 2004. "R. P. Effinger's Excellent Adventure: The Unknown Letters of a Young Ohio Lawyer, 1849–1850." *California History* 82:2–75.

Forbes, Jack D. 1965. *Warriors of the Colorado: The Yumas of the Quechan Nation and Their Neighbors*. Norman: University of Oklahoma Press.

Foreman, Grant. 1937. *The Adventures of James Collier, First Collector of the Port of San Francisco*. Chicago: Black Cat.

Frazer, Robert W. 1965. *Forts of the West: Military Forts and Presidios and Posts Commonly Called Forts West of the Mississippi River to 1898*. Norman: University of Oklahoma Press.

Garber, Paul Neff. 1923. *The Gadsden Treaty*. Philadelphia: Press of the University of Pennsylvania.

Goetzmann, William H. [1959] 1991. *Army Exploration in the American West, 1803–1863*. Austin: Texas State Historical Association.

Goode, James M. 1979. *Capital Losses: A Cultural History of Washington's Destroyed Buildings*. Washington, D.C: Smithsonian Institution Press.

Green, Constance McLaughlin. 1967. *The Secret City: A History of Race Relations in the Nation's Capital*. Princeton, N.J: Princeton University Press.

Greenberg, Amy S. 2005. *Manifest Manhood and the Antebellum American Empire*. New York: Cambridge University Press.

Harlow, Neal. 1982. *California Conquered: War and Peace on the Pacific, 1846–1850*. Berkeley: University of California Press.

Harris, Joseph, and Samuel Anderson. 2000. *Letters from the 49th Parallel, 1857–1873: Selected Correspondence of Joseph Harris and Samuel Anderson*. Ed. C. Ian Jackson. Toronto: Champlain Society.

Harrison, Donald H. 2005. *Louis Rose: San Diego's First Jewish Settler and Entrepreneur*. San Diego: Sunbelt.

Hartmann, Clinton P. 1993. "Charles Wright: Botanizer of the Boundary, Part II: At the Edge of the Power Struggle in the U.S. Boundary Commission." *Password* 37:171–87.

Haskins, Charles Warren, ed. 1975. *Libera Martina Spinazze's Index to the Argonauts of California*. New Orleans: Polyanthos.

Heitman, Francis Bernard. [1903] 1965. *Historical Register and Dictionary of the United States Army from Its Organization, September 29, 1789, to March 2, 1903*. 2 vols. Urbana: University of Illinois Press.

Hendricks, Rick, and W. H. Timmons. 1998. *San Elizario: Spanish Presidio to Texas County Seat*. El Paso: Texas Western Press.

Hewitt, Harry P. 1990. "The Mexican Boundary Survey Team: Pedro García Conde in California." *Western Historical Quarterly* 21:171–96.

———. 1991. "The Mexican Commission and Its Survey of the Rio Grande Boundary, 1850–1854." *Southwestern Historical Quarterly* 94:555–80.

Hine, Robert V. 1968. *Bartlett's West: Drawing the Mexican Boundary*. New Haven, Conn.: Yale University Press for the Amon Carter Museum, Fort Worth, Texas.

Horgan, Paul. 1954. *Great River: The Rio Grande in North American History*. 2 vols. New York: Rinehart.

Howe, Daniel Walker. 2007. *What Hath God Wrought: The Transformation of America, 1815–1848*. New York: Oxford University Press.

Hughes, Charles W. 2007. "'La Mojonera' and the Making of California's U.S.-Mexico Boundary Line, 1849–1851." *Journal of San Diego History* 53:126–47.

Hunter, Alfred. 1853. *The Washington and Georgetown Directory. Strangers' Guide-Book for Washington*. Washington, D.C.: Kirkwood & McGill.

Hurtado, Albert L. 1988. *Indian Survival on the California Frontier*. New Haven, Conn.: Yale University Press.

Kajencki, Francis C. 1988. "Charles Radziminksi and the United States-Mexican Boundary Survey." *New Mexico Historical Review* 63, no. 3:210–40.

Kemble, John Haskell. [1943] 1972. *The Panama Route, 1848–1869*. New York: Da Capo.

Kerrigan, William T. 1998. "Race, Expansion, and Slavery in Eagle Pass, Texas, 1852." *Southwestern Historical Quarterly* 101:275–302.

Kingsbury, John M. 1996. *Trading in Santa Fe: John M. Kingsbury's Correspondence with James Josiah Webb, 1853–1861*. Ed. Jane Lenz Elder and David J. Weber. Dallas: SMU Press.

Kronk, Gary W. 2003. *Cometography: A Catalogue of Comets. Vol. 2, 1800–1899*. Cambridge: Cambridge University Press.

Langley, Lester D. 1968. *The Cuban Policy of the United States: A Brief History*. New York: John Wiley.,

Lesley, Lewis B. 1930. "The International Boundary Survey from San Diego to the Gila River, 1849–1850." *California Historical Society Quarterly* 9 (March): 1–15.

Marshall, Don B. 1978. *California Shipwrecks: Footsteps in the Sea*. Seattle: Superior Publishing.

May, Robert E. 2002. *Manifest Destiny's Underworld: Filibustering in Antebellum America*. Chapel Hill: University of North Carolina Press.

Moody, Richard. 1958. *The Astor Place Riot*. Bloomington: Indiana University Press.

Morgenthaler, Jefferson. 2007. *La Junta de los Rios: The Life, Death and Resurrection of an Ancient Desert Community in the Big Bend Region of Texas*. Boerne, Texas: Mockingbird.

Norris, L. David, James C. Milligan, and Odie B. Faulk. 1998. *William H. Emory: Soldier-Scientist*. Tucson: University of Arizona Press.

Padilla Corona, Antonio. 2004–2005. "The Rancho Tía Juana (Tijuana) Grant." *Journal of San Diego History* 50 (Winter–Spring): 30–41.

Phillips, George Harwood. 1975. *Chiefs and Challengers: Indian Resistance and Cooperation in Southern California*. Berkeley: University of California Press.

Pitt, Leonard. 1966. *The Decline of the Californios: A Social History of the Spanish-Speaking Californians, 1846–1890*. Berkeley: University of California Press.

Pomfret, John Edwin. 1954. *California Gold Rush Voyages, 1848–1849: Three Original Narratives*. 2 vols. San Marino, Calif.: Huntington Library.

Poore, Ben Perley. 1886. *Perley's Reminiscences of Sixty Years in the National Metropolis*. 2 vols. Philadelphia: Hubbard Brothers.

Pourade, Richard F. 1965. *The History of San Diego, vol. 3: The Silver Dons*. San Diego: Union-Tribune.

Powell, H. M. T. 1931. *The Santa Fé Trail to California, 1849–1852: The Journal and Drawings of H. M. T. Powell*. Ed. Douglas S. Watson. San Francisco: Book Club of California.

Rebert, Paula. 2001. *La Gran Línea: Mapping the United States-Mexico Boundary, 1849–1857*. Austin: University of Texas Press.

Rippy, J. Fred. 1921. "The Boundary of New Mexico and the Gadsden Treaty." *Hispanic American Historical Review* 4:715–42.

Rittenhouse, Jack. 1965. *Disturnell's Treaty Map: The Map That Was Part of the Guadalupe Hidalgo Treaty on Southwestern Boundaries, 1848*. Santa Fe: Stagecoach.

Rosenberg, Charles E. 1962. *The Cholera Years: The United States in 1832, 1849, and 1866*. Chicago: University of Chicago Press.

Scharf, Thomas L. 1973. "Amiel Weeks Whipple and the Boundary Survey in Southern California." *Journal of San Diego History*, no. 19:18–31.

———. 1976. "Pages from the Diary of Cave Johnson Couts: San Diego in the Spring and Summer of 1849." *Journal of San Diego History* 22 (Spring): 9–19.

Smith, Michael L. 1987. *Pacific Visions: California Scientists and the Environment, 1850–1915*. New Haven: Yale University Press.

Starr, Raymond, ed. 1983. "Emigrants and Indians: Selections from C. J. Cout's Military Correspondence, 1849." *Journal of San Diego History* 29:165–84.

Stewart, George Rippey. 1937. *John Phoenix, Esq., the Veritable Squibob: A Life of Captain George H. Derby, U.S.A.* New York: Henry Holt.

Stuart-Wortley, Emmeline. 1851. *Travels in the United States . . . during 1849 and 1850*. New York: Harper.

Sweeny, Thomas W. 1956. *Journal of Lt. Thomas W. Sweeny: 1849–1853*. Ed. Arthur Woodward. Los Angeles: Westernlore.

Taft, Robert. 1953. *Artists and Illustrators of the Old West, 1850–1900*. New York: Scribner's.

Taylor, Bayard. [1850] 1949. *Eldorado, or, Adventures in the Path of Empire*. New York: Alfred A. Knopf.

Taylor, Morris F. 1971. *First Mail West: Stagecoach Lines on the Santa Fe Trail*. Albuquerque: University of New Mexico Press.

Timmons, W. H. 1990. *El Paso: A Borderlands History*. El Paso: University of Texas at El Paso.

Timmons, W. H. 1999. *James Wiley Magoffin: Don Santiago, El Paso Pioneer*. El Paso: Texas Western Press.

Trafzer, Clifford E. 1989. *California's Indians and the Gold Rush*. Sacramento: Sierra Oaks.

Tupper, Martin Farquhar. 1851. *Tupper's Complete Poetical Works*. Boston: Phillips, Sampson.

Turner, Steven. 1991. "William Würdemann: First Mechanican of the U.S. Coast Survey." *Rittenhouse* 5:97–110.

Tyler, Ronnie C., ed. 1968. "Exploring the Rio Grande: Lt. Duff C. Green's Report of 1852." *Arizona and the West* 10:43–60.

———. 1975. *The Big Bend: A History of the Last Texas Frontier*. Washington, D.C.: National Park Service.

Waite, Edward, comp. 1850. *The Washington Directory and Congressional and Executive Register for 1850*. Washington, D.C.: C. Alexander.

Ware, Joseph E. [1849] 1972. *The Emigrants' Guide to California*. Ed. John Walton Caughey. New York: Da Capo.

Warner, Ezra J. 1959. *Generals in Gray: Lives of the Confederate Commanders*. Baton Rouge: Louisiana State University Press.

Weber, William A. 1997. *King of Colorado Botany: Charles Christopher Parry, 1823–1890*. Niwot: University Press of Colorado.

Weinert, Richard P. 1962. "The Confederate Regular Army." *Military Affairs* 26:97–107.

Weisenburger, Francis P. 1937. *The Life of John McLean, a Politician on the United States Supreme Court.* Columbus: Ohio State University Press.

Wheat, Carl I. 1957–1963. *Mapping the Transmississippi West.* 5 vols. in 6. San Francisco: Institute of Historical Cartography.

Werne, Joseph Richard. 1987. "Partisan Politics and the Mexican Boundary Survey, 1848–1853." *Southwestern Historical Quarterly* 90:329–46.

———. 2007. *The Imaginary Line: A History of the United States and Mexican Boundary Survey, 1848–1857.* Fort Worth: Texas Christian University Press.

Whipple, A. W. 1961. *The Whipple Report: Journal of an Expedition from San Diego, California, to the Rio Colorado, from September 11 to Dec. 11, 1849.* Ed. E. I. Edwards. Los Angeles: Westernlore.

Wilson, William Bender. 1900. "George Clinton Gardner, Eighth General Superintendent, March, 1873–April 1, 1879." In *General Superintendents of the Pennsylvania Railroad Division, Pennsylvania Railroad Co.* 41–47. Philadelphia: Kensington.

Wiltsee, Ernest A. 1938. *Gold Rush Steamers of the Pacific.* San Francisco: Grabhorn.

Woodward, Arthur. 1955. *Feud on the Colorado.* Los Angeles: Westernlore.

Zevely, Douglass. 1903. "Old Residences and Family History in the City Hall Neighborhood." *Records of the Columbia Historical Society* 6:104–22.

Index

The initials GCG indicate George Clinton Gardner.

The author of several prize-winning books on the American Southwest, DAVID J. WEBER directs the William P. Clements Center for Southwest Studies at Southern Methodist University in Dallas. He is a Fellow of the American Academy of Arts and Sciences and both the Mexican and Spanish governments have given him the highest award they bestow on foreigners.

Hillsman Jackson

JANE LENZ ELDER is a theological reference librarian at Bridwell Library, Perkins School of Theology, SMU. She worked as a researcher and exhibits curator at SMU's DeGolyer Library, publishing a number of articles and catalogs, then became the founding associate director of the William P. Clements Center for Southwest Studies at SMU, where she served until 2001.

Missy Collins

Number Six in the DeGolyer Library Series has been designed by
Tom Dawson and printed in an edition of 500 copies by Thomson-Shore.
The body text is set in Adobe Caslon Regular and chapter titles
are set in Bernhart Modern Italic.